SYSTEMATIC BOTANY MONOGRAPHS

VOLUME 43

AF784047

Revision of the Mexican Species of Cyperus (Cyperaceae)

Gordon C. Tucker

THE AMERICAN SOCIETY OF PLANT TAXONOMISTS
22 August 1994

SYSTEMATIC BOTANY MONOGRAPHS
ISSN 0373-8211

ISBN 0-912861-43-6

Printed in the United States of America

Publication of volume 43 of Systematic Botany Monographs was partly funded by Walter L. Meagher, whose generous support is gratefully acknowledged.

REVISION OF THE MEXICAN SPECIES OF CYPERUS (CYPERACEAE)

Gordon C. Tucker
Biological Survey
New York State Museum
Albany, New York 12230

ABSTRACT. A revision, based on herbarium study and field work, of the 89 species of *Cyperus* (Cyperaceae) occurring in Mexico is presented. Discussions concerning morphology, infrageneric classification, and habitats and distribution are included. Five subgenera are represented in Mexico; 14 species are endemic (*C. arsenei, C. aschenbornianus, C. calderoniae, C. dioicus, C. duripes, C. matudae, C. michoacanensis, C. nayaritensis, C. perennis, C. pycnostachyus, C. semiochraceus, C. sordidus, C. tempeae, C. wilburii*). Two new species, *C. hypopitys* and *C. tempeae*, are described, and one new combination, *C. megalanthus*, is proposed. Several nomenclatural corrections are made: *C. sordidus* is the oldest name for *C. howellii*; *C. sphaerolepis* displaces *C. rusbyi*; *C. flavicomus* is the correct name for the species commonly called *C. albomarginatus*; and *C. aggregatus* is the correct name for *C. cayennensis* [sometimes also called *C. flavus* (Vahl) Boeck. or *C. huarmensis*]. Extensive intergradation among the variants of several wide-spread species showed recognition of certain segregates untenable; *C.apiculatus* is here included in *C. manimae*, and *C. eggersii* and *C. macrocephalus* in *C. odoratus*. The apparent hybridization of *C. microbrunneus* and *C. uncinulatus* in southern Mexico and northern Central America is discussed. The following species are reported for the first time from Mexico: *C. erythrorhizos*, *C. lundellii*, *C. microbrunneus*, *C. panamensis*, *C. retrorsus*, and *C. uncinulatus*.

INTRODUCTION

The pantropic and temperate genus *Cyperus* comprises about 600 species. It is the second largest genus in the Cyperaceae; only *Carex* L., with about 1200 species, is larger. *Cyperus* is readily recognized by the distichous arrangement of the scales on the rachilla of the spikelets. Nearly all other genera of this family represented in the New World have spirally arranged scales.

Six subgenera have been recognized on the basis of characters of the spikelets, achenes, and vegetative anatomy. Some workers have treated these subgenera as distinct genera, whereas most recent American workers have given *Cyperus* a broad circumscription. Subgenera *Cyperus, Diclidium, Juncellus, Pycnostachys*, and *Pycreus* are represented in Mexico. *Kyllinga* Rottb., sometimes considered a subgenus of *Cyperus*, is here excluded and considered a separate genus distinguished by its 2-scaled, 1-flowered spikelets (Tucker 1987). Three species of *Kyllinga* occur in Mexico (Tucker 1984b).

Examination of herbarium specimens and field observations form the basis of the descriptions presented in this revision of *Cyperus* in Mexico. Some 9000 herbarium specimens from Mexico and ca. 5000 from Central America, South America, the West Indies, the United States, and the Old World were examined. Field work in Mexico during the summers of 1983 and 1984 resulted in some 700 collections of *Cyperus*.

Eighty-nine species of *Cyperus* are here recognized in Mexico. Of these, 26 are pantropic, and 6 species are both pantropic and temperate; numerous others are widespread in the neotropics. Fourteen species are endemic to Mexico (*C. arsenei, C. aschenbornianus, C. calderoniae, C. dioicus, C. duripes, C. matudae, C. michoacanensis, C. nayaritensis, C. perennis, C. pycnostachyus, C. semiochraceus, C. sordidus, C. tempeae, C. wilburii*).

FLORISTIC ACCOUNTS OF CYPERUS IN MEXICO

Several studies have contributed to our knowledge of the Mexican species of *Cyperus*. The most important of these is Kükenthal's worldwide monograph of the genus in *Das Pflanzenreich* (1935–36), in which descriptions are complete but keys to the sections and species are brief and synoptical. Although Kükenthal had relatively few specimens available for study, his monumental treatise is the primary reference for all later studies of the genus.

In the nineteenth century, several authors published papers pertaining to the genus *Cyperus* in Mexico. J. Presl and K. Presl (1828) described five new species [*C. canus*, *C. sordidus*, *C. regiomontanus*, *C. tenerrimus*, and *Mariscus haenkei* (=*C. dentoniae*)] based on collections made by Haenke in 1790–91 in western Mexico. In 1831, Junghuhn enumerated the Cyperaceae collected by Schiede and Deppe in 1828–29 in eastern Mexico, and described two new species, *C. ciliatus* and *C. thyrsiflorus* (both accepted here). The most important work was Liebmann's *Mexicos halvgraes* (1850), an account of the sedges of Mexico. Liebmann's monograph was based on his extensive collections in the states of Veracruz, Puebla, and Oaxaca from 1841–43. Liebmann's Mexican collections of *Cyperus* have been restudied and an account of the current status of the 40 species named by him has been prepared (Tucker 1986). Four of Liebmann's species are currently recognized: *C. fugax*, *C. dipsaceus*, *C. sanguineoater*, and *C. piceus* [as *C. flavescens* var. *piceus*].

Several regional treatments of the Mexican species of *Cyperus* have appeared in the twentieth century. O'Neill (1940), in his account of the Cyperaceae of the Yucatan Peninsula, recorded and described 14 species of *Cyperus* from the states of Campeche, Quintana Roo, and Yucatán. The present revision records 19 species from these three states. Matuda (1958), in his account of the Cyperaceae of the Valle de México, treated 16 species, whereas the present revision recognizes 26 species. Shreve and Wiggins (1964) recorded 12 species that occur in the Sonoran Desert south of the U.S. boundary; 18 species are recorded here from that area. Wiggins (1981) in his *Flora of Baja California* provided keys to 17 species (two of which, *C. tenuis* and *C. flavicomus*, could not be confirmed in this study). This revision records six additional species from the peninsula. (Five of these new records, *C. amabilis, C. arsenei, C. dipsaceus, C. manimae*, and *C. mutisii*, were collected in the 1890's by T. S. Brandegee, but had never been correctly identified). Tucker and McVaugh (1993), in the *Flora Novo-Galiciana*, recognized 53 species of *Cyperus* in western Mexico; their treatment essentially coincides with the revision presented here.

The most recent survey of the genus in Mexico appeared in 1946, when Brother Benedict Ayers published his dissertation, "The Genus *Cyperus* in Mexico." Benedict Ayers worked at the Catholic University of America in Washington, D.C., under the supervision of Father Hugh T. O'Neill, the latter also a student of this genus. Benedict Ayers provided keys, some descriptions, and brief synonymies for 74 species. His study was based on only about 1000 collections. His keys to the species work fairly well, but Benedict Ayers provided descriptions for only about half the species and referred to recently published descriptions (by O'Neill or his students) for the other species.

MORPHOLOGY

Underground parts. Roots are uniformly fibrous; the largest are at most 1 mm thick. Many species are rhizomatous. In many of these, the rhizome internodes are short (less than 1 cm), and the culms are clustered or approximate, e.g., *C. luzulae* and *C. involucratus*. About 20 species have longer rhizomes, with internodes as long as 10 cm, e.g., *C. articulatus*. Rhizomatous species occur in all the subgenera. Two species, *C. esculentus* and *C. rotundus*, produce tuberiferous stolons. These are the only Mexican species with this characteristic.

Culms. Most species are caespitose, producing several to many clustered culms. The stoloniferous species produce solitary culms connected by subterranean stolons. The rhizomatous species produce a single culm at each node, e.g., *C. ischnos* and *C. manimae*, or several culms clustered at the summit of an erect or oblique rhizome, e.g., *C. entrerianus* or *C. ligularis*. Culms range in height from 5 mm in *C. pennellii* to ca. 3 m in *C. prolixus* and *C. giganteus*.

In transverse cross section, the culms may be trigonous (three-sided with acute to slightly obtuse angles), as in *C. digitatus*, triquetrous (three-sided with winged angles), as in *C. nayaritensis*, or round, as in *C. luzulae*. The shape of the culm is sometimes useful in distinguishing related species. For example, *C. nayaritensis* differs from *C. tenerrimus* and *C. michoacanensis* by its triquetrous culm; the two other species have terete culms.

In many species, the culms are scabrous or scabrellate due to the presence of crystalline prickles. These prickles point upward or outward. The only exception is *C. surinamensis*, in which the prickles point downward. Generally, the presence or absence of prickles is constant within a species.

Leaves. Expanded leaf blades occur in at least some individuals of most species. In three Mexican species, however, leaf blades are always absent: *C. involucratus*, *C. canus,* and *C. giganteus*. In several other species, one finds plants with and without leaf blades, sometimes in the same population, e.g., in *C. bipartitus* and *C. haspan*. In length, leaf blades range from about 1 cm to about 130 cm, and in width from about 0.5 mm to 15 mm. The margins, nerves, and keel usually bear crystalline prickles. Five species have conspicuous cross veins between the main veins; this is a distinctive characteristic of *C. virens*, *C. articulatus*, *C. prolixus*, *C. ligularis*, and *C. elegans*.

The shape of the leaf in transverse cross section shows some variation. Terminology in this treatment follows that of Metcalfe (1977). The following shapes occur: flat (e.g., *C. michoacanensis*); v-shaped (e.g., *C. entrerianus*); flanged v-shaped (e.g., *C. rotundus*); inversely w-shaped (e.g., *C. laxus*); and involute (e.g.,

C. elegans). Within a species, usually only one leaf shape is found. Closely related pairs of species usually have the same leaf shape. Leaf shape is easily noted in the field. In herbarium specimens, it is necessary to rehydrate leaf fragments in boiling water or 3% Contrad solution to observe this feature.

Inflorescence. The inflorescence of *Cyperus* is a system of approximate lateral branches (rays), each borne in the axil of a leaflike bract. There may be as few as 1 or 2 bracts (*C. arsenei* and *C. articulatus*) or as many as 22 (*C. involucratus*). The bracts usually resemble the leaves in size, shape, nerves, and margins; however, the bracts are usually sheathless, or they may have a short collar-like sheath up to 3 mm long (e.g., *C. ochraceus*). A species may be characterized by bracts that are erect (e.g., *C. articulatus*), all horizontal (*C. involucratus*), or ascendent at angles from 30–60° (–75°). In some species, the longest bract is erect and appears as a continuation of the culm, whereas the remaining bracts are horizontal to ascendent (e.g., *C. breedlovei* or *C. laevigatus*). The orientation of the bract is useful in distinguishing certain pairs of species, e.g., *C. sphaerolepis* and *C. fendlerianus*.

The inflorescence has 2–22 primary rays (peduncles) (Fig. 2B). The longest ray of an inflorescence ranges from as little as 1 cm in *C. amabilis* to about 24 cm in *C. hermaphroditus*. In nearly all species with rays, the inflorescence consists of several spikes borne on elongate rays and also of 1–3 sessile spikes. In about ten species, e.g., *C. seslerioides* (subg. *Pycnostachys*) (Fig. 1D) and *C. svensonii* (subg. *Cyperus*) (Fig. 2D), the primary rays are so much reduced as to appear to be absent. Thus, the inflorescence in such species is a single dense head of seemingly sessile spikes or spikelets. In about two-thirds of the species, the rays are terminated by an elongate rachis bearing the spikelets. In subg. *Pycnostachys*, however, the rachis is reduced to less than 1 mm in length, so that the clusters of spikelets are digitate (Fig. 1C) or glomerulate rather than spicate (Fig. 2C). In many species, the longer rays are branched, forming secondary and sometimes tertiary rays. Examples of species with tertiary rays are *C. surinamensis* and *C. virens*. *Cyperus matudae* sometimes produces quaternary rays (Fig. 2C).

In several species that normally have both primary and secondary rays, individual plants can be found in which the rays have failed to elongate. Thus, the plant has a compact sessile inflorescence rather than an open, diffuse one. Rayless individuals occur in *C. luzulae*, *C. odoratus*, and *C. hermaphroditus*, and collections of these variants have been described as "new" species.

Spikelets. The distichous arrangement of scales on the spikelets is a diagnostic characteristic of *Cyperus*. Other New World genera of the family (with the exception of *Kyllinga* of tribe Cypereae, and *Schoenus* L. and *Dulichium* Pers. of tribe Schoeneae Dumort.; see Tucker, 1987) have spirally arranged scales. The axis of the spikelet, the rachilla, represents a terminal branch of the inflorescence. The rachilla has two sterile scales at its base. The lower of these is a rachis bract subtending the spikelet, while the upper is the prophyll of the spikelet. The rachis bract and prophyll are generally similar in shape, but often the prophyll is broader and shorter than the rachis bract. The remaining fertile scales number from 1 to about 80, the particular range depending on the species.

The denuded rachilla may be straight, flexuous, or geniculate. The rachilla usually remains attached to the plant, but in about half the species of subg.

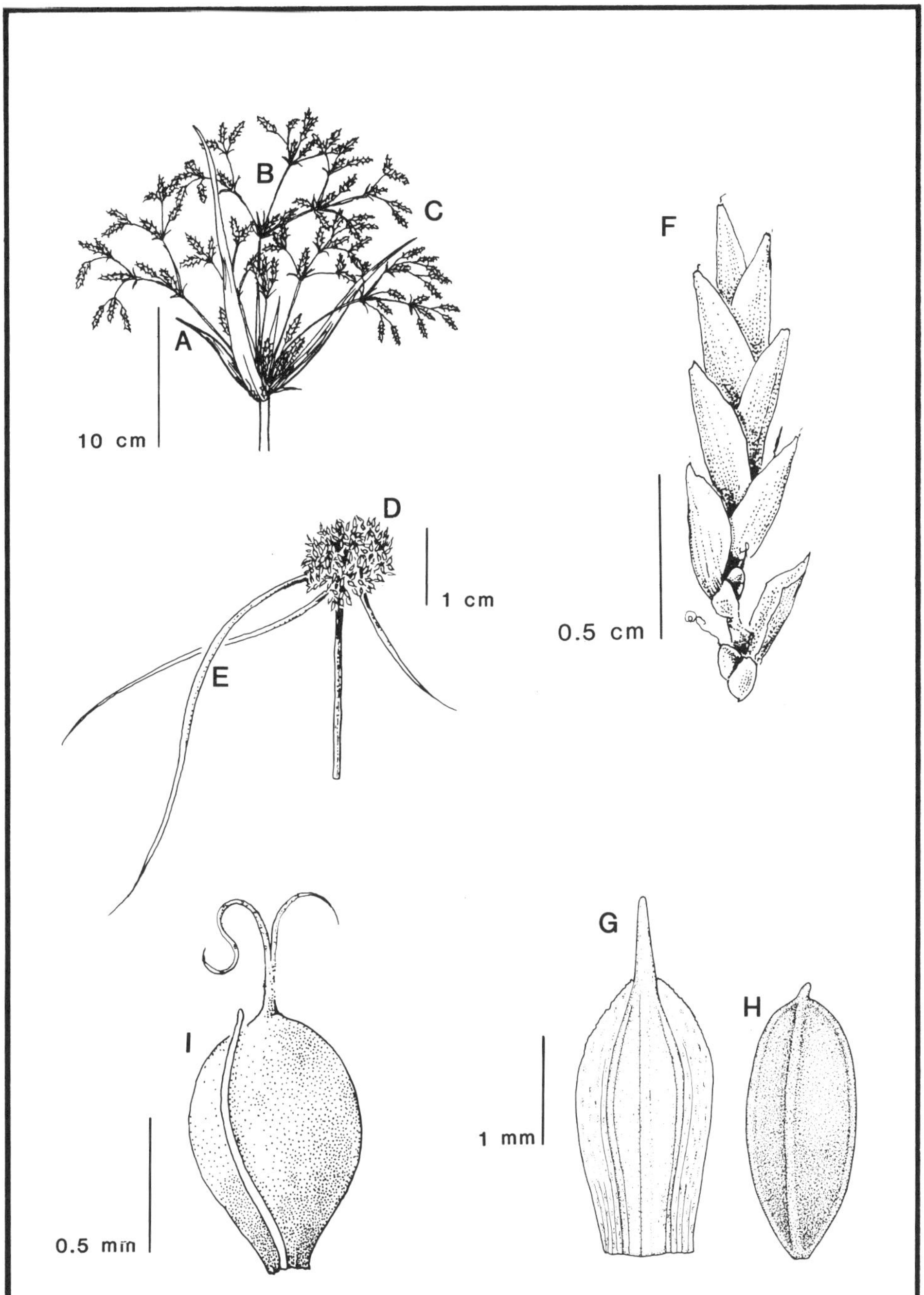

FIG. 1. Representative diagnostic illustrations of Mexican species of *Cyperus*. A–C, *C. matudae* (subg. *Pycnostachys*), example of an open, highly branched inflorescence. A. Inflorescence bract. B. Secondary ray. C. Two spikelets terminating quaternary ray. D, E, *C. seslerioides* (subg. *Pycnostachys*). D. Capitate inflorescence. E. Inflorescence bracts, somewhat reflexed. F, *C. odoratus* (subg. *Diclidium*), spikelet; note thickened rachilla clasping achene, visible at base of two lowest scales. G, H, *C. matudae* (subg. *Pycnostachys*). G. Cuspidate scale, abaxial view. H. Trigonous achene, abaxial view. I, *C. laevigatus* (subg. *Juncellus*), abaxial view of achene with persistent filament, style, and stigmas. Drawn by Charlotte M. Taylor (A, B, C, G, H) and Patricia M. Kernan (D, E, F, I).

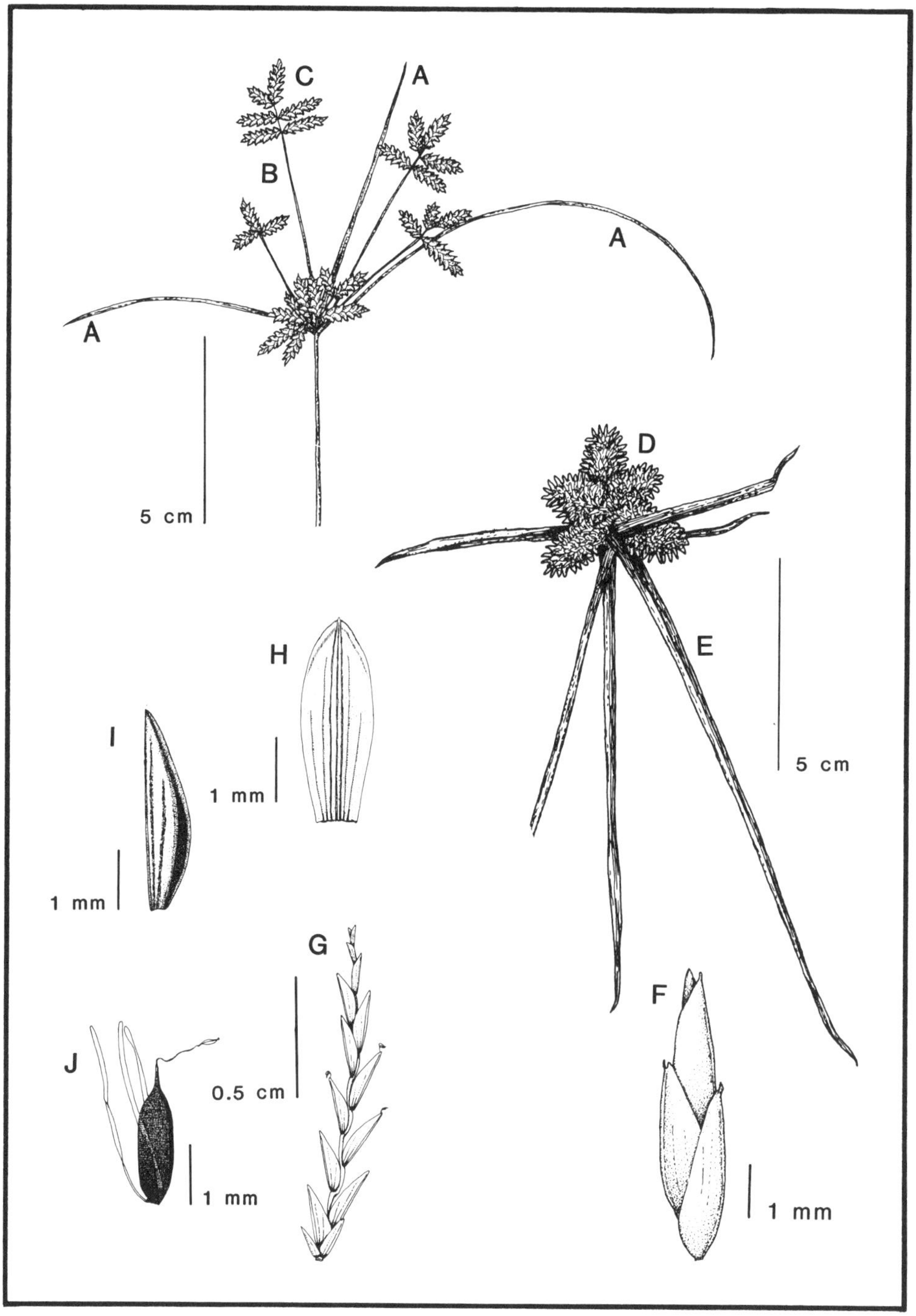

FIG. 2. Representative diagnostic illustrations of Mexican species of *Cyperus* subg. *Cyperus*. A–C, *C. wilburii*; several large spikelets (C) borne on short rachis terminating primary ray (B); inflorescence subtended by three inflorescence bracts (A). D–F, *C. svensonii*; several dense subcylindrical spikes (D) sessile at the summit of the culm and subtended by inflorescence bracts (E); deciduous spikelet with three persistent scale (F). G–J, *C. camphoratus*; single deciduous spikelet with spreading scales (G); scale, detached from rachilla, abaxial view (H), lateral view (I); achene (J) with three persistent filaments and dried style, removed from spikelet and scale. Drawn by Charlotte M. Taylor; layout by Patricia M. Kernan.

Cyperus, e.g., *C. camphoratus* (Fig. 2G) and *C. svensonii* (Fig. 2F), it is articulated at its base and deciduous at maturity. In many species of subg. *Cyperus*, the internodes of the rachilla bear hyaline wings (up to 0.6 mm wide) that clasp the achene. In subg. *Diclidium*, which also has clasping wings, the rachilla is articulated at the base of each scale, and the mature spikelet breaks up into segments composed of a rachilla internode, scale, and achene clasped by thickened rachilla wings (Fig. 1F). The articulation of the scales and rachilla are important features in the subgeneric classification.

The shape of the spikelets shows considerable variation, from ovate to elliptic or linear. In cross section, the shape varies from flatly elliptic to broadly elliptic or quadrate to subcylindric. Mature spikelets range in length from ca. 2 to ca. 40 mm, and in width from 0.7 to 7.0 mm.

Scales. Bracts on the spikelets that subtend the individual flowers are termed scales [called "glumes" by some cyperologists]. The floral scales may be closely appressed, e.g., *C. svensonii* (Fig. 2F), or spreading, e.g., *C. camphoratus* (Fig. 2G). The shape of the scale varies considerably within the genus, but is generally constant within any particular species and thus an important character in the delimitation of species. It varies from lanceolate to narrowly elliptic, oblong, or ovate. The dimensions of the scales range from (0.6–) 1.5 to 2.5 (–4.7) mm long, and from (0.6–) 1.0 to 1.8 (–2.8) mm wide.

Nerves number from (2–) 3 up to 13; the most frequent number is 5. In those species with (2–) 3–5 nerves, the nerves are confined to the medial part of the scale, and the lateral parts are nerveless. In species with more than 5 nerves, 3–5 nerves are medial, the remaining nerves lateral. Colors of the scales include white, stramineous, hyaline, light green, light to dark brown, reddish brown or red, castaneous or purple. In a few species the scales have a conspicuous hyaline border, e.g., *C. flavicomus* and *C. mutisii*.

The apex of the scale is usually broadly rounded, but in several species is acute, acuminate, or muticous. In a few species the apex is emarginate, e.g., *C. amabilis* (Fig. 23A). In many species, the scale has a mucronulate or mucronate apex. In about 10 species, the apex is cuspidate (awned), e.g., *C. matudae* (Fig. 1G). The cusp ranges from 0.3 to 1.2 mm long, and may be straight, curved outward, or even recurved.

The shape in cross section falls into three classes. Most species have conduplicate scales. Bicarinate scales have two conspicuous nerves with a groove between them. Such scales characterize species 1–11 of subg. *Pycnostachys*, including the Luzulae group as treated by Denton (1978), and most New World species of subg. *Pycreus*. A small number of species have semicircular scales, e.g., *C. odoratus* and *C. panamensis*.

Flowers. The flower consists of a gynoecium and hypogynous stamens borne in the axil of a scale. As is typical in many genera of the Cyperaceae, a perianth is absent. The number of stamens per flower is usually three, but in subgenera *Pycreus* and *Pycnostachys* many species bear only 2 or even 1 stamen. The filaments are ribbonlike, from (0.3–) 1.0 to 2.0 (–4.5) mm long. Generally, in a given species the length of the filament approximately equals the length of the scale. The anthers are elliptic, oblong, or linear in outline and range from 0.1 to 2.5 mm long. The length of the anthers is a neglected character often useful in distinguishing closely related species. In some species, the connective is prolonged beyond the apex of the anther cells. The size and shape of the apex of the connective is sometimes an important taxonomic character; for example, it is useful in distinguishing *C. giganteus* from *C. digitatus*.

Within each subgenus, the gynoecium is either bicarpellate or tricarpellate. The style of a tricarpellate ovary always bears 3 stigmas, that of a bicarpellate ovary bears 2 stigmas. Tricarpellate gynoecia mature into trigonous achenes, and bicarpellate gynoecia develop into lenticular or biconvex achenes. Styles range from (0.1–) 0.8 to 1.8 (–4.8) mm long. In a few species, e.g., *C. laxus*, *C. iria,* and *C. fugax*, the style is so greatly reduced (ca. 0.1 mm long) that the stigmas appear sessile at the apex of the ovary or achene. The stigmas range in length from as little as 0.2 mm in *C. iria* to 3.0–6.0 mm in *C. articulatus*.

Achenes. The features of the achenes are very important in the delimitation of subgenera and in the recognition of species. Achenes are trigonous or lenticular in cross section within a subgenus. The shape varies from ovoid to obovoid to oblong or ellipsoid. The angles may be rounded (e.g., *C. laevigatus*, Fig. 1I) to rather acute (e.g., *C. matudae*, Fig. 1H). The faces may be flat or slightly concave to slightly convex. The surface is most often papillose or punctate, but in some species is smooth, granular, or reticulate. Achenes are usually brown or reddish brown; a few species have black achenes, and *C. simplex* has white achenes. In size, the achenes range from 0.6 to 2.8 mm long and from 0.3 to 1.6 mm wide. The apex may be acute, obtuse, apiculate, or entire. The base may be sessile, cuneate, or conspicuously stipitate.

INFRAGENERIC CLASSIFICATION

Cyperus is here considered to consist of six subgenera: subg. *Pycnostachys* C. B. Clarke (subg. *Protocyperus* Lye), subg. *Pycreus* (P. Beauv.) A. Gray, subg. *Juncellus* (Griseb.) C. B. Clarke, subg. *Cyperus*, subg. *Diclidium* (Desv.) Britton [subg. *Torulinium* (Desv.) Kükenthal], and subg. *Fimbricyperus* Lye. The last is monotypic and restricted to East Africa (Lye 1981). The remaining five subgenera are all represented in Mexico and Central America. The largest of these is subg. *Cyperus*, which comprises about 350 of the 600 species in the genus, and 46 of the 87 Mexican species. Second largest is subg. *Pycnostachys*, with about 120 species worldwide and 29 species in Mexico. Subgenus *Pycreus* also has about 100 species worldwide, but only 9 of these occur in Mexico. Subgenus *Diclidium* has about 8 species worldwide; only one of these, the pantropic *C. odoratus*, occurs in Mexico. Additional species in the New World are *C. hayesii* (C. B. Clarke) Standley of the Pacific Coast of Central America (Tucker 1983), and *C. filiformis* Sw. and *C. correllii* (T. Koyama) G. C. Tucker of the West Indies. Subgenus *Juncellus* comprises only six species; of these, only *C. laevigatus* occurs in Mexico.

Some workers have treated *Kyllinga* as a subgenus of *Cyperus*; however, *Kyllinga* differs from *Cyperus* in its 2-scaled, 1-flowered spikelets. The dense, sessile spikes of *Kyllinga* give the plants a different aspect from most species of *Cyperus* with their open, branched inflorescences. The majority of recent specialists of *Cyperus* and the Cyperaceae have recognized *Kyllinga* at the generic level. Some proponents of *Kyllinga* as a distinct genus include: Koyama (1978) in *The Flora of the Lesser Antilles*; Raynal (1973), who at different times worked on the Cyperaceae of Africa, Madagascar, and New Caledonia; Vorster (1978), who studied the southern African species and *Cyperus*; Lye (1981), who has revised the East African sedges and rushes; and Tucker (1984) in his treatment of the Mexican and Central American species. Three species of *Kyllinga* occur in Mexico.

The subgenera of *Cyperus* are distinguished by the structure of the spikelet and the gynoecium, and by the anatomy of the leaves and culms. Subgenera *Pycreus* and *Juncellus* differ from the remaining subgenera in having the derived condition of lenticular achenes and bifid styles, whereas the other three subgenera have primitively trigonous achenes and trifid styles (Raynal 1973). This division into bicarpellate and tricarpellate subgenera is a common feature in treatments of several cyperaceous genera, e.g., *Carex* and *Bulbostylis* C. B. Clarke.

Subgenus *Diclidium* differs from both subgenera *Cyperus* and *Pycnostachys* in that the rachilla of the spikelet is articulate at the base of each scale, whereas in subgenera *Cyperus* and *Pycnostachys* the rachilla is continuous or, in some species of subg. *Cyperus* (subg. *Mariscus*, fide Kükenthal, 1935–36), articulate only at the base of the spikelet. Thus, the mature spikelet of subg. *Diclidium* breaks up into 1-fruited segments consisting of an internode of the rachilla, a scale, and an achene.

The subgenera *Diclidium*, *Juncellus*, and *Pycreus* are easily distinguished from each other and from the remainder of the genus. Yet, the subgeneric classification of the remaining species of the genus has long been a matter of debate. Traditionally, the species here recognized as constituting the subgenera *Pycnostachys* and *Cyperus* (Lye 1981) have been grouped with different circumscriptions as the subgenera *Mariscus* and *Cyperus*. Clarke (1884, 1908) and Kükenthal (1935–36) defined subg. *Cyperus* as differing from subg. *Mariscus* in that the rachilla of the spikelet of subg. *Cyperus* remains firmly attached to the rachis while the scales are deciduous, falling from the spikelet as the achenes mature, whereas in subg. *Mariscus* the scales remain firmly attached to the rachilla even after the spikelet has fallen from the rachis. O'Neill (1942) listed some 20 species which have characters of both subgenera. There are 13 Mexican species (*C. ischnos* through *C. manimae* and *C. camphoratus*) in which both the scales and the rachillas are deciduous. Kükenthal placed these species in his subg. *Mariscus*, but clearly they are transitional between subg. *Cyperus* and subg. *Mariscus*. Also, as O'Neill observed, *C. rotundus* and *C. esculentus*, both of which have been placed in subg. *Cyperus*, have persistent scales, a characteristic attributed to subg. *Mariscus* by Kükenthal. I agree with O'Neill's conclusion that there is no single character that consistently separates subg. *Mariscus* from subg. *Cyperus*. O'Neill (1942, p. 47) stated: "It is ill-advised to maintain *Mariscus* as a genus when it is ill-defined even as a subgenus." Yet, Koyama (1978) recognized *Mariscus* as a genus.

Rikli (1895) surveyed the anatomy of the leaves and culms of the genus *Cyperus*. He proposed dividing it into two genera, which he named *Eucyperus* and *Chlorocyperus*. Lerman and Raynal (1972) examined the distribution of C_4 photosynthesis in the Cyperaceae, and reported that the genus *Cyperus* contains both C_3 and C_4 species. These physiological differences were correlated with the division that Rikli had based on anatomical evidence. In view of this evidence, Lye (1981) proposed subg. *Protocyperus* (a superfluous name for subg. *Pycnostachys*) to contain those species of *Cyperus* with non-kranz anatomy and C_3 metabolism. Lye (1981, 1992) concurred with O'Neill that *Mariscus* should not be maintained even at the subgeneric level, but should be included in *Cyperus* subg. *Cyperus*. The recognition within the genus *Cyperus* of the subgenera *Pycnostachys* and *Cyperus*, rather than subgenera *Mariscus* and *Cyperus* (as delineated by Kükenthal, 1935–36), reflects current knowledge of the phylogeny of the genus, as outlined by Raynal (1973) and Goetghebeur (1989).

Van der Veken (1965) made an extensive investigation of the embryology of the subfamily Cyperoideae, which includes *Cyperus* and *Scirpus* L. In his study, 162 of the approximately 600 species of the genus *Cyperus* were examined. He reported that the embryos of *Cyperus* were ellipsoidal, with a basal coleoptile and lateral radicle. The morphology of the embryos was quite uniform throughout the genus. In the closely related genus *Scirpus*, however, six different embryo types were noted. Their distribution coincided with subgeneric lines recognized on other morphological characters. This indicated that *Scirpus* might be subdivided into several genera, as was done by Koyama (1978) and Oteng-Yeboah (1976). In conclusion, Van der Veken stated (1965, p. 346): "On peut voir dans l'uniformité des embryons de *Cyperus s.l.* un nouvel argument, quoique pas absolu, pour une conception large du genre." The work of Van der Veken thus further supports the recognition of *Cyperus* in the broad sense.

Harborne, Williams, and Wilson (1982) surveyed the distribution of flavonoids in South American, African, and Australian species of *Cyperus*. They examined a total of about 150 species and reported that each subgenus had a distinctive profile with regard to this class of compounds. Subgenus *Pycnostachys* is characterized by flavonols, absent in the remaining subgenera, which have flavones instead. Aurones, which give a yellowish hue to the inflorescences, are present in the subgenera *Cyperus* (including subg. *Mariscus*) and *Diclidium*, but are lacking in subgenera *Pycreus* and *Pycnostachys*. It was the opinion of these investigators that these differences support the recognition of *Pycnostachys* as a subgenus distinct from *Cyperus*. They also concluded that the flavonoid data did not indicate that any subgenus was sufficiently distinct from the others to merit generic status. Thus, these authors also favored a broad concept of the genus.

The genus *Cyperus* is here accepted in the broad sense. The classification and circumscription of the subgenera takes into account the morphological and chemical data acquired since Kükenthal's work, as well as O'Neill's comments on the status of subg. *Mariscus*. Thus, subg. *Pycnostachys* is here defined by its digitate spikelets. Subgenera *Pycreus* and *Juncellus* can always be distinguished by their lenticular achenes, whereas subg. *Diclidium* can be recognized consistently by the peculiar disarticulation of the rachilla at the base of each scale. Subgenus *Cyperus* includes the ill-defined subg. *Mariscus*.

HABITATS AND DISTRIBUTION

Most species of *Cyperus* occur in disturbed habitats. Many of them grow in roadside ditches and pastures along main highways. A small percentage (perhaps

20%) regularly occur in forested habitats. Some species typically occurring in lowland tropical forest are *C. laxus*, *C. simplex*, *C. tenuis*, and *C. thyrsiflorus*; however, these species are usually more abundant in clearings, and along roadsides, stream banks, and trails in these forests. About twice as many species are typical of the montane pine forests; they are *C. manimae*, *C. spectabilis*, *C. hypopitys*, *C. fendlerianus*, *C. ischnos*, *C. dipsaceus*, *C. sanguineoater*, and *C. pallidicolor*. All but the last species are a closely related group distinguished by their spreading deciduous scales, deciduous rachillas, and rhizomatous habit.

Many of the species of open habitats occur in permanently or temporarily wet habitats. *Cyperus articulatus* and *C. giganteus* grow as emergent aquatics when they occur in marshes. Numerous other species are encountered in and around shallow water holes (tinajas), stock ponds, and roadside ditches. About half of the Mexican species have been collected in such situations.

Sixteen species are annuals: *C. acuminatus*, *C. surinamensis*, *C. difformis*, *C. humilis*, *C. uncinulatus*, *C. amabilis*, *C. cuspidatus*, *C. flavicomus*, *C. fugax*, *C. bipartitus*, *C. squarrosus*, *C. iria*, *C. compressus*, *C. wilburii*, *C. erythrorhizos*, and *C. oxylepis*. *Cyperus odoratus* and *C. ciliatus* appear to grow both as annuals or perennials. Thus, there are annual species in all subgenera except subg. *Juncellus*. The annual species are frequently found at sandy scraped roadsides, edges of pools in pastures, and on sand bars along streams. *Cyperus iria* and *C. compressus* are weedy annuals and can be found in such highly disturbed habitats as cracks in sidewalks, parking areas, edges of pavement, and ditches.

Eight species, all C_4 species, occur in maritime habitats. *Cyperus polystachyos*, *C. oxylepis*, *C. elegans*, *C. ligularis* and *C. odoratus* grow in brackish marshes, at edges of creeks in tidal marshes with *Avicennia*, and in swales between dunes. *Cyperus sordidus* occurs on rocky bluffs and arroyos bordering the Pacific, *C. duripes* in brackish swales and on hillsides in the Revillagigedo Islands, and *C. planifolius* on coral sand beaches on the Yucatan Peninsula.

About half the Mexican species of *Cyperus* are montane, growing at elevations of 1000 m or more. Five species occur at elevations exceeding 3000 m (*C. seslerioides*, *C. niger*, *C. pennellii*, *C. fendlerianus*, and *C. hermaphroditus*). Thus three of the subgenera are represented at these elevations. Tieszen et al. (1979), in a study of East African grasses, showed that the C_4 species were predominantly lowland, whereas the C_3 species were predominantly montane. They showed that there was an abrupt transition between grasslands dominated by grasses of these two photosynthetic types at about 2800 m. Thus, one might predict that the alpine species of Mesoamerican *Cyperus* would be predominantly C_3, but in fact only one, *C. seslerioides*, is C_3—the remainder are C_4. Theories about the advantages of C_4 photosynthesis postulate that C_4 plants have greater photosynthetic efficiencies at higher temperatures.

It is unclear why the majority of the Mexican species of *Cyperus* that grow above 3000 m are C_4 and not C_3 species. Tieszen et al. (1979) hypothesized that the few species of C_4 grasses that occur at 3000 to 4000 m survive at these elevations because they grow only in sheltered microhabitats. This could partly explain the presence of C_4 species of *Cyperus* above 3000 m in Mexico. Of the six species that do occur above 3000 m, all are less than 20 cm tall (at least, all collections from these higher elevations are). In fact, *C. pennellii*, the shortest of the Mexican species, is apparently never more than 10 cm tall.

TAXONOMY

The sequence of presentation of species within the subgenera generally follows the arrangement of Kükenthal (1935–36) with more recent modifications as discussed by Tucker (1987). The reliability of morphological criteria alone to determine phylogeny within *Cyperus* has not yet been tested by comparison with other data sets over a large sample of the genus. Thus, the keys to species, although possibly indicating areas of affinity, do not purport to be natural classifications.

The height of the culm is measured from the base of the culm to the insertion of the lowest inflorescence bract. The width of leaves, bracts, spikelets, and scales is measured at the widest point. The length of a ray, whether primary, secondary, or tertiary, is measured from the base of the ray to the point of attachment of the lowest spikelet. The rachis is that portion of the axis of the ray that bears spikelets. The width of the rachilla does not include the wings, if any. Measurements of spikelets were made on dried material only. When moistened, the scale and its apex often straighten out, bending outward from the rachilla, thus increasing the width of the spikelet. The width of scales was measured at the widest point of a detached scale pressed against a flat surface. The width may be estimated as twice the distance from the margin to the keel of an intact, attached scale. The length of a scale includes the length of the mucronate or cuspidate tip, if any, unless otherwise stated. The length of the scale cusp is determined from measurement of distal scales of mature spikelets. The basal scales generally have shorter cusps than the distal scales. The width of achenes is measured in the abaxial view for species with trigonous achenes. This is especially important in subg. *Cyperus* where the achenes are often slightly asymmetrical. The width of lenticular achenes is measured from the lateral view, which shows the widest dimension of the achene.

Cyperus L., Sp. pl. 1: 44. 1753; Gen. pl. 27. 1754.—Lectotype: *C. esculentus* L. (see Tucker 1987; Goetghebeur 1989; Lye 1991).

Tufted or rhizomatous herbaceous plants, perennial or less often annual, smooth throughout (papillose-surfaced in *C. sordidus* and *C. ligularis*, viscid in *C. oxylepis* and *C. elegans*). Roots fibrous, light to dark brown (bright red in *C. erythrorhizos*). Culms triquetrous, trigonous, or terete, bearing leaves near the base only; bulbous-thickened basally in a few species. Leaves with elongate blades (absent in *C. involucratus*, *C. canus*, and *C. giganteus*; often absent in *C. articulatus*); cross veins (septa) present only in a few species (*C. virens*, *C. prolixus*, *C. ligularis*). Inflorescence bracts leaflike, unequal in length (equal in *C. involucratus* and *C. canus*), closely spaced, spirally arranged at the apex of the culm, forming a conspicuous involucre. Rays (peduncles) much shorter than culm (except in *C. simplex*), unequal in length, produced singly from the axils of the inflorescence bracts; secondary (and tertiary) rays present in many species. Spikes ovoid to cylindric, flattened when pressed (in subg. *Pycnostachys*, the spikelets grouped in digitate clusters or glomerulate heads); rachis (3–) 10–50 mm long, usually visible between spiklelets when pressed, smooth (scabridulous in *C. ciliatus*, *C. dentoniae*, and *C. tempeae*). Spikelets spreading (ascendent in *C. iria*, *C. prolixus*, and *C. imbricatus*), cylindric to compressed, usually appearing smooth-margined owing to the closely appressed scales; rachilla wingless or winged. Scales distichous, deciduous or persistent, muticous, mucronulate, or cuspidate, laterally nerveless or several-nerved (infrequently speckled), with smooth (infrequently scabrellate or ciliate) midveins, conduplicate, bicarinate, or semicircular in cross section; scales usually closely appressed (or spreading), and the apex overlapping the base of the next scale; terminal scale sterile, conduplicate (rarely involute). Perianth absent. Flowers perfect (imperfect only in the dioecious *C. canus* and *C. dioicus*), borne in the axils of the distichous scales. Stamens 3 (1 or 2 in some species), hypogynous; filaments free, ribbonlike or slender; anthers ovoid or ellipsoid to linear, the connective sometimes prolonged as an acute appendage up to 0.2 mm long (subulate and 0.2–0.5 mm long in *C. giganteus* and *C. duripes*). Gynoecium bicarpellate or tricarpellate; style smooth, slender; stigmas 2 or 3, always equal in number to the carpels, slender, capillary. Fruit an achene, trigonous or lenticular, the style base not persistent.

Key to the Subgenera of Cyperus in Mexico

1. Stigmas 2; achenes biconvex or lenticular.
    2. Achenes dorsiventrally flattened, borne with a face toward the rachilla. III. *Cyperus* subg. *Juncellus* (39. *C. laevigatus*).
    2. Achenes laterally flattened, borne with an edge toward the rachilla. II. *Cyperus* subg. *Pycreus*.
1. Stigmas 3; achenes trigonous.
    3. Spikelets borne in digitate clusters (rarely singly), or in umbellate or glomerulate heads. I. *Cyperus* subg. *Pycnostachys*.
    3. Spikelets borne in spikes on a rachis (sometimes concealed by bases of spikelets in pressed specimens, evident upon dissection).
        4. Rachilla continuous, or articulate only at the base. IV. *Cyperus* subg. *Cyperus*.
        4. Rachilla articulate at the base of each scale, the mature spikelet disarticulating into segments, each consisting of a scale, internode, and rachilla wings. V. *Cyperus* subg. *Diclidium* (89. *C. odoratus*).

Key to the Species of Cyperus Subgenus Pycnostachys in Mexico

1. Scales conspicuously mucronate to cuspidate, the apex (0.3–) 0.4–1.2 mm long.
    2. Plants (18–) 30–60 (–75) cm tall; scales laterally 2–4-nerved; achenes 1.2–2.0 mm long, puncticulate.
        3. Scales ovate-orbiculate; style 0.1–0.2 mm long; achenes obovoid to ovoid; widespread in lowlands. 12. *C. laxus*.
        3. Scales oblong-elliptic or oblong-lanceolate; style 0.2–0.8 mm long; achenes ellipsoid; southern Chiapas.
            4. Secondary and tertiary rays present; achenes 1.8–2.0 mm long. 13. *C. matudae*.
            4. Secondary rays sometimes present, tertiary rays absent; achenes 1.2–1.4 mm long. 14. *C. breedlovei*.
    2. Plants 0.5–18 cm tall; scales laterally nerveless (or weakly 1-nerved); achenes 0.4–0.9 mm long, papillose.
        5. Achenes globose-ovoid to obovoid; anthers 0.4–0.5 (–0.6) mm long.
            6. Plants perennial; cusp of scale strongly recurved to uncinate; stamens 1, 2, or 3; Baja California and Sonora. 22. *C. perennis*.
            6. Plants annual; cusp of scale straight or slightly excurved; stamen 1; Pacific Coast of Chiapas and Oaxaca and southward. 23. *C. uncinulatus*.
        5. Achenes oblong-obovoid; anthers 0.1–0.3 mm long.
            7. Cusp of scale (0.3–) 0.4–0.5 mm long; body of scale 1.6–2.0 mm long; achenes 0.8–0.9 mm long. 28. *C. amabilis*.
            7. Cusp of scale 0.6–1.2 mm long; body of scale (0.6–) 0.9–1.2 mm long; achenes (0.5–) 0.6 mm long. 29. *C. cuspidatus*.
1. Scales muticous or mucronulate, the apex less than 0.4 mm long.

8. Inflorescence of several to many, loose to dense, umbellate or glomerulate heads or digitate clusters of spikelets, one or more of which are borne on conspicuous rays up to 20 (–30) cm long.
9. Spikelets in digitate clusters (rarely solitary).
10. Longest ray equalling or exceeding the culm in length; scales ovate-lanceolate. 15. *C. simplex*.
10. Longest ray much shorter than the culm; scales ovate to ovate-deltate.
11. Leaf blades and/or inflorescence bracts with conspicuous cross veins. 7. *C. virens*.
11. Leaf blades and/or inflorescence bracts without conspicuous cross veins.
12. Inflorescence bracts (4–) 10–22, approximately equal in length.
13. Flowers perfect; scales laterally nerveless; inflorescence bracts approximately horizontal. 1. *C. involucratus*.
13. Flowers imperfect; scales 5–7-nerved laterally; inflorescence bracts ascendent at 30–60°. 2. *C. canus*.
12. Inflorescence bracts 2–5, decidedly unequal in length.
14. Flowers imperfect; Baja California. 3. *C. dioicus*.
14. Flowers perfect; mainland of Mexico.
15. Achenes puncticulate; stigmas 2.3–2.5 mm long; anthers ca. 1.0 mm long. 14. *C. breedlovei*.
15. Achenes granular to papillose; stigmas 0.5–1.0 (–1.3) mm long; anthers 0.3–0.5 mm long.
16. Achenes ellipsoid, dorsiventrally flattened, ca. 1.0 mm long; scales 1.4–1.7 mm wide; leaf blades present, 15–60 (–90) cm long. 16. *C. lundellii*.
16. Achenes globose to obovoid or broadly ellipsoid, not dorsiventrally flattened, 0.5–0.6 (–0.7) mm long; scales 0.8–1.1 mm wide; leaf blades usually absent (if present, at most 30 cm long). 17. *C. haspan*.
9. Spikelets in dense spherical, hemispherical, pyramidal, or umbellate heads.
17. Culms triquetrous, scabrous or scabrellate on the angles only.
18. Leaf blades and inflorescence bracts with conspicuous cross veins; scales laterally pale reddish brown or greenish; anthers 0.8–1.4 mm long. 7. *C. virens*.
18. Leaf blades and cross veins without conspicuous cross veins; scales laterally white or yellowish white; anthers 0.4 mm long. 25. *C. nayaritensis*.
17. Culms trigonous or roundly trigonous, smooth throughout, or scabrellate on angles and surface, especially apically.
19. Scales 0.6–0.8 mm long; achenes 0.6–0.8 mm long, 0.3–0.4 mm wide; anthers ca. 0.1 mm long. 18. *C. difformis*.
19. Scales 1.0–3.0 mm long; achenes 0.7–1.5 mm long, 0.2–0.6 mm wide; anthers 0.5–1.2 mm long.
20. Inflorescence bracts 3–4 (–6), the longest erect or nearly so, appearing as a continuation of the culm; achenes stipitate.
21. Scales laterally yellowish, yellowish green or light reddish brown, the apex (straight to) excurved; achenes tapering to a slender beak; anthers ca. 0.5 mm long. 4. *C. acuminatus*.
21. Scales laterally reddish, smooth, the apex straight; achenes acute to acuminate; anthers 0.6–1.0 mm long. 5. *C. reflexus*.
20. Inflorescence bracts (3–) 6–12 (–15), all approximately horizontal to ascendent at 30–60° (–75°); achenes sessile or substipitate.
22. Culms scabrellate with retrorse prickles. 6. *C. surinamensis*.
22. Culms smooth.
23. Heads of spikelets ordinarily proliferous; spikelets ovate-orbiculate; scales 2.0–2.5 mm long, oblong-spatulate; anther 0.7–1.0 mm long. 11. *C. megalanthus*.
23. Heads of spikelets rarely proliferous; spikelets oblong to linear-oblong; scales 1.0–2.0 mm long, oblong-lanceolate, oblong-ovate, or broadly ovate; anther 0.5–1.2 mm long.

24. Achenes broadly ellipsoid; rachilla with successive scale scars 0.6–0.8 mm apart; anther 0.8–1.2 mm long. 8. *C. ochraceus.*
24. Achenes narrowly ellipsoid; rachilla with successive scale scars 0.4–0.5 mm apart; anther 0.5–0.7 mm long.
25. Heads loosely globose; secondary rays present, 5–15 (–23) mm long; rachilla dark red at anthesis; scales laterally off-white to dull grayish green or light golden-brown; inflorescence bracts and primary rays ascendent at 45–60° (–75°). 9. *C. entrerianus.*
25. Heads densely oblong to pyramidal; secondary rays absent; rachilla green at anthesis; scales laterally whitish to reddish white; inflorescence bracts and primary rays approximately horizontal to ascendent at 30°. 10. *C. luzulae.*
8. Inflorescence without rays, a single capitate (or irregularly lobate) head or small digitate cluster.
26. Flowers imperfect; Baja California. 3. *C. dioicus.*
26. Flowers perfect; widespread in Mexico.
27. Achenes with apical beak and stipitate base. 4. *C. acuminatus.*
27. Achenes beakless, bases sessile to substipitate.
28. Longest inflorescence bract ± vertical (erect).
29. Anther 0.6–1.0 mm long. 5. *C. reflexus.*
29. Anther(s) 0.1–0.3 mm long.
30. Spikelets 30–120; scales laterally greenish brown to purplish brown. 18. *C. difformis.*
30. Spikelets 2–6 (–12); scales laterally light brown, bronze colored, or whitish. 21. *C. arsenei.*
28. Longest inflorescence bract ascendent, horizontal, or reflexed.
31. Achenes ovoid, obovoid, broadly obovoid, or broadly ellipsoid.
32. Stamens 3.
33. Achenes granular to verrucose; leaf blades often absent. 17. *C. haspan.*
33. Achenes finely reticulate, or papillose; leaf blades always present.
34. Scales 1.6–3.0 (–3.4) mm long, ovate to ovate-lanceolate; achenes not dorsiventrally flattened; anthers (0.4–) 0.6–0.8 (–1.1) mm long. 20. *C. seslerioides.*
34. Scales 1.3–1.6 mm long, oblong-ovate; achenes dorsiventrally flattened; anthers 0.25–0.5 mm long. 27. *C. michoacanensis.*
32. Stamen 1.
35. Annual; scales laterally hyaline to greenish; achenes ovoid-deltoid. 19. *C. humilis.*
35. Perennial; scales laterally reddish brown to brown; achenes ellipsoid. 24. *C. microbrunneus.*
31. Achenes linear-oblong to narrowly ellipsoid.
36. Culms triquetrous, the angles densely scabrellate especially distally; stamens 2; achenes concavo-convex. 25. *C. nayaritensis.*
36. Culms trigonous to round, smooth or very sparsely scabrellate; stamen 1; achenes roundly trigonous or trigonous.
37. Culms 1–2 (–2.5) mm thick; leaves 1.5–5.0 mm wide; achenes 0.9–1.1 mm long. 10. *C. luzulae.*
37. Culms 0.3–1.1 mm thick; leaves 1–2 mm wide; achenes 1.3–1.5 (–1.6) mm long. 26. *C. tenerrimus.*

Key to the Species of Cyperus Subgenus Pycreus in Mexico

1. Achenes oblong to oblong-obovoid, the apex truncate, apiculate.
2. Anthers 0.1–0.2 mm long; style 0.1–0.3 (–0.4) mm long; stigmas 0.3–0.4 mm long. 31. *C. fugax.*
2. Anthers (0.4–) 0.6–0.8 mm long; style 0.6–1.0 (–1.4) mm long; stigmas (1.0–) 1.4–2.0 mm long. 32. *C. polystachyos.*

1. Achenes ovoid, obovoid, or ellipsoid, the apex acute to rounded, apiculate or entire.
   3. Spikelets appearing serrate-margined owing to the spreading scales; scales with conspicuous hyaline margins; achenes lenticular, the faces flattened-depressed. 30. *C. flavicomus.*
   3. Spikelets appearing smooth-margined owing to the closely appressed scales; scales without conspicuous hyaline margins; achenes biconvex or lenticular, but faces not depressed.
      4. Anthers 1.0–2.4 mm long.
         5. Spikelets 3.0–5.0 mm wide; anthers 2.0–2.4 mm long. 33. *C. unioloides.*
         5. Spikelets 2.0–3.2 mm wide; anthers 1.0–1.2 mm long. 34. *C. aschenbornianus.*
      4. Anthers 0.4–0.8 mm long.
         6. Achene surface with isodiametric cells forming transverse undulations.
            7. Scales laterally brown, black, or castaneous.
               8. Perennial with slender rhizome; spikelets (3–) 5–25 (–60); scales 1.5–2.1 mm long, laterally castaneous, blackish or light brown; anthers 0.6–0.8 mm long; achenes brown. 35. *C. niger.*
               8. Annual with fibrous roots; spikelets 3–5 (–8); scales 1.9–2.7 mm long, laterally brown; anthers 0.4–0.5 mm long; achenes black. 36. *C. bipartitus.*
            7. Scales laterally yellow or yellowish brown. 37. *C. lanceolatus.*
         6. Achenes surface with rectangular, longitudinally elongate cells forming transverse undulations. 38. *C. flavescens* var. *piceus.*

Key to the Species of Cyperus Subgenus Cyperus in Mexico

1. Distal scales of spikelets with cuspidate apex (0.3–) 0.4–1.2 (–1.6) mm long.
   2. Scales persistent; spikelets subcylindric to subterete; rachilla wings 0.3–0.7 mm wide.
      3. Scales 3–7 (–14), terminal sterile scale conduplicate; style 0.5–1.0 mm long.
         4. Scales ovate, not laterally speckled, the cusp of the distal scales 0.7–1.4 mm long; successive scales 4.5–5.6 mm apart; Pacific coastal lowlands. 85. *C. panamensis.*
         4. Scales oblong-lanceolate, laterally densely red-speckled, the cusp of distal scales 0.4–0.6 mm long; successive scales (2.2–) 2.6–3.4 mm apart; Caribbean lowlands. 86. *C. lentiginosus.*
      3. Scales 1–3 (–5), terminal sterile scale involute, frequently uncinate; style 0.4–0.5 mm long. 87. *C. retroflexus.*
   2. Scales deciduous; spikelets quadrangular to compressed; rachilla wingless (except *C. dipsaceus* and *C. sanguineoater*, with wings 0.2–0.4 mm wide).
      5. Spikes (1–) 3–10 per ray, narrowly cylindrical; spikelets ascendent; achenes dorsiventrally compressed. 52. *C. imbricatus.*
      5. Spikes 1 per ray (or rays absent), ovoid to oblong; spikelets spreading; achenes not dorsiventrally compressed.
         6. Caespitose annuals; rhizomes absent.
            7. Scales (0.5–) 0.7–0.8 (–1.0) mm wide; stamen 1; achenes (0.2–) 0.3–0.4 (–0.5) mm wide. 40. *C. squarrosus.*
            7. Scales 1.5–3.6 mm wide; stamens 3; achenes (0.5–) 0.9–1.4 mm wide.
               8. Achenes ellipsoid, with beak 0.5–1.2 mm long; leaves involute in proximal portion of culm; living plants viscid. 66. *C. oxylepis.*
               8. Achenes obovoid, without beak; leaves flat to v-shaped throughout culm; living plants not viscid.
                  9. Spikelets 2.0–3.0 (–3.6) mm wide; scales with cusp (0.3–) 0.5–0.7 (–0.8) mm long; achenes substipitate to sessile, (1.0–) 1.2–1.4 (–1.5) mm long. 47. *C. compressus.*
                  9. Spikelets (3.5–) 5.0–6.5 (–7.0) mm wide; scales with cusp (1.1–) 1.3–1.5 (–1.6) mm long; achenes stipitate, 1.7–2.2 mm long. 48. *C. wilburii.*
         6. Perennials, culms single, rhizomes up to 15 cm long.
            10. Longest inflorescence bract erect or nearly so (other bracts erect to ascendent).
               11. Spikes densely oblong-cylindrical; achenes 1.4–1.5 (–1.6) mm long. 56. *C. dipsaceus.*
               11. Spikes loosely ovoid; achenes 2.0–2.4 mm long. 57. *C. schweinitzii.*

10. Longest inflorescence bract ascendent, approximately horizontal or reflexed (other bracts similar).
    12. Scales ovate-orbiculate, 2.4–2.8 (–3.4) mm wide. 59. *C. fendlerianus.*
    12. Scales ovate-lanceolate, oblong-ovate, or oblong-elliptic, (1.0–) 1.4–2.2 (–2.5) mm wide.
        13. Rays present or absent; scales oblong-elliptic, laterally light brown to reddish brown; anthers 0.4–0.6 mm long; style (0.2–) 0.4–0.8 mm long. 65c. *C. manimae* var. *divergens.*
        13. Rays absent; scales ovate-lanceolate to oblong-ovate, laterally light to deep red; anthers 0.6–1.3 (–1.4) mm long; style 0.8–2.0 mm long. 55. *C. sanguineoater.*

1. Distal scales of spikelets entire or emarginate, with a mucronulate apex at most 0.3 mm long.
   14. Spikelets compressed their entire length, i.e., 2 or more times wider than thick (or compressed proximally, becoming subterete distally in *C. distans*).
       15. Spikelets strongly ascendent to appressed, their apices pointing toward the apex of the spike.
           16. Scales obovate to nearly orbiculate, emarginate; style less than 0.1 mm long; secondary rays absent; culms (5–) 20–60 cm tall. 41. *C. iria.*
           16. Scales oblong-ovate, acute to obtuse but not emarginate; style 0.8–1.3 mm long; secondary rays 1–10 cm long; culms 0.5–3.0 m tall. 42. *C. prolixus.*
       15. Spikelets spreading to slightly ascendent.
           17. Rachilla with hyaline, whitish or stramineous wings 0.2–0.9 mm wide.
               18. Culms terete (occasionally becoming trigonous apically), with transverse septa conspicuous when dried; inflorescence bracts 2 (–4), erect; leaf blades absent (occasionally present). 44. *C. articulatus.*
               18. Culms trigonous (or triquetrous), without septa; inflorescence bracts (2–) 3–7, approximately horizontal to ascendent at 45° to reflexed; leaf blades present.
                   19. Scales persistent; rachilla persistent; stolons up to 15 cm long, bearing tubers.
                       20. Scales laterally yellowish brown to dark brown; base of culm soft; stolons spongy, flexible when dried. 45. *C. esculentus.*
                       20. Scales laterally purple to reddish brown; base of culm indurate; stolons indurate, wiry, springy when dried. 46. *C. rotundus.*
                   19. Scales deciduous; rachillas deciduous (rarely persistent); rhizomes up to 10 cm long, without tubers.
                       21. Scales lanceolate (ovate-lanceolate), laterally whitish (faintly greenish white at anthesis). 54. *C. ischnos.*
                       21. Scales elliptic or ovate-lanceolate to suborbiculate, laterally stramineous, bronze-colored, brown, reddish brown, or reddish, .
                           22. Culms scabrous or ciliate-scabrous on the angles distally.
                               23. Scales laterally stramineous to bronze-colored, midvein ciliate-scabrellate; achenes ovoid to very broadly ellipsoid. 60. *C. ciliatus.*
                               23. Scales laterally reddish, the midvein smooth; achenes ellipsoid. 61. *C. hypopitys.*
                           22. Culms smooth.
                               24. Scales laterally reddish brown to brown; rachilla persistent; achenes 0.75–1.0 (–1.1) mm wide, coarsely punctate. 68. *C. planifolius.*
                               24. Scales laterally brown to stramineous; rachilla deciduous; achenes 0.6–0.7 mm wide, finely papillose. 81. *C. camphoratus.*
           17. Rachilla wingless (or with wings at most 0.2 mm wide).
               25. Spikelets 1.2–1.5 (–1.6) mm wide; apex of a scale barely reaching or as much as 0.4 mm distant from the base of the next scale. 43. *C. distans.*
               25. Spikelets (1.5–) 2.0–3.0 (–4.0) mm wide; apex of a scale equalling or surpassing the base of the next scale.

26. Scales broadly ovate, oblong-ovate, ovate, or suborbiculate (about as wide as long).
    27. Longest inflorescence bract ± erect; scales strongly laterally nerved. 58. *C. sphaerolepis.*
    27. Longest inflorescence bract ± horizontal; scales weakly laterally nerved.
        28. Plants (12–) 30–80 (–90) cm tall; spikelets (2.4–) 2.8–4.0 mm wide; successive scale scars (2.0–) 2.4–3.0 mm apart; scales 1.6–2.6 mm long, (1.6–) 1.8–2.4 (–2.7) mm wide; anthers 0.4–0.6 mm long; achenes 1.7–2.2 mm long, 0.8–1.1 mm wide, dark brown to black. 62. *C. spectabilis.*
        28. Plants (1–) 5–20 (–35) cm tall; spikelets 2.0–2.5 mm wide; successive scale scars 1.4–1.8 mm apart; scales 1.7–2.0 mm long, 1.6–2.0 mm wide; anthers ca. 0.3 mm long; achenes 1.3–1.5 mm long, 0.8–1.0 mm wide, light yellowish brown. 63. *C. calderoniae.*
26. Scales ovate, ovate-lanceolate, broadly lanceolate, lanceolate, or oblong-elliptic (1.5–2 times longer than wide).
    29. Culms single, borne 1–15 cm apart on horizontal rhizomes; rachilla ± deciduous as the spikelet matures; montane species at (700–) 1100–3230 m elevation
        30. Scales 1.0–1.6 mm wide, laterally white (and often reddish lineolate), greenish white at anthesis; rays usually absent, or, if present, no more than 2.5 cm long. 54. *C. ischnos.*
        30. Scales (1.2–) 1.4–2.0 (–2.5) mm wide, laterally light brown, reddish brown or deep red; rays sometimes present, up to 16 cm long.
            31. Culms 0.5–3.0 (–10) cm tall, bulbous-thickened basally, smooth; spikes loosely ovoid; spikelets strongly compressed, nearly flat; style 0.8–1.3 mm long; achene base sessile. 64. *C. pennellii.*
            31. Culms (4–) 30–70 (–95) cm tall, not appreciably thickened basally, densely scabrous or hirtellate (sometimes smooth); spikes densely ovoid to oblong-ovoid; spikelets compressed-quadrangular; style (0.2–) 0.4–0.8 mm long; achene base substipitate to stipitate. 65. *C. manimae.*
    29. Culms caespitose, rhizomes erect to oblique or absent; rachilla persistent after the scales fall; lowland and coastal species, mostly at less than 200 m elevation.
        32. Leaves involute; scales 2.8–3.6 mm long, laterally pale greenish white, stramineous or reddish brown, medially scabrous at apex; achenes (1.4–) 1.8–2.0 mm long. 67. *C. elegans.*
        32. Leaves flat to broadly v-shaped; scales 2.2–3.0 mm long, laterally reddish to grayish brown, medially smooth at apex; achenes 1.2–1.4 (–1.6) mm long. 69. *C. sordidus.*

14. Spikelets quadrate or subterete, i.e., 1 to 1.5 times wider than thick.
    33. Scales deciduous; rachillas persistent; rachilla wings detaching laterally but remaining firmly attached basally even after the achenes fall; spikelets with (6–) 12–20 (–30) scales.
        34. Culms 2.0–3.5 m tall; leaves bladeless; secondary rays 5–15 cm long; anther connective subulate, 0.2–0.5 mm long. 53. *C. giganteus.*
        34. Culms (5–) 50–150 cm tall; leaves with blades; secondary rays absent or at most 5 (–7.5) cm long; anther connective blunt, at most 0.1 mm long.
            35. Annuals, ordinarily less than 25 cm tall; roots bright red; culms 1.0–2.5 mm thick; anthers 0.2–0.3 mm long; rare in Mexico, known only from Baja California near the U.S. border. 49. *C. erythrorhizos.*
            35. Perennials, 50–150 cm tall; roots brown; culms 2–15 mm thick; anthers 0.4–1.5 mm long; frequent, widespread in Mexico.

36. Anthers 0.4–0.5 mm long; stigmas ca. 0.4 mm long; successive scales 0.9–1.2 mm apart. 50. *C. digitatus*.
36. Anthers 1.1–1.3 (–1.5) mm long; stigmas 0.8–1.4 mm long; successive scales ca. 0.8 mm apart. 51. *C. semiochraceus*.
33. Scales persistent; rachillas deciduous; rachilla wings not detaching laterally; spikelets with (1–) 2–5 scales.
37. Culms hispidulous distally; rays and rachises hispidulous.
38. Spikelets oblong-ellipsoid to oblong-lanceolate, (0.8–) 0.9–1.2 (–1.3) mm wide; scales 1.6–2.0 mm wide; achenes ellipsoid, (0.7–) 0.8–0.9 mm wide; rachis not visible when pressed; Pacific Coast. 72. *C. dentoniae*.
38. Spikelets oblong-lanceolate to subulate, 0.4–0.5 mm wide; scales 0.8–1.1 mm wide; achenes narrowly ellipsoid, ca. 0.4 mm wide; rachis visible when pressed; northeastern Mexico. 73. *C. tempeae*.
37. Culms smooth (or occasionally with a few prickles on the angles only); rays and rachises glabrous.
39. Culms, rays, leaves, and bracts papillose; leaves harshly scabrous on the margins and mid-veins. 80. *C. ligularis*.
39. Culms, rays, leaves, and bracts not papillose; leaves scabrellate or smooth on the margins and mid-veins.
40. Rays present (or at least some spikes borne on rays).
41. Spikes loosely ovoid to cylindrical, the rachis visible between the spikelets when pressed.
42. Spikelets linear (to narrowly ellipsoid), 4–10 times longer than wide, with (2–) 3–8 (–20) scales.
43. Scales laterally greenish white, the apex overlapping the lower 1/4–1/3 of the next scale. 82. *C. tenuis*.
43. Scales laterally light brown, tawny, reddish brown, deep brown, golden-brown (or pale greenish white), the apex barely reaching the base of the next scale.
44. Achenes narrowly oblong, the stipitate base 0.1–0.2 mm long; stigmas (1.5–) 2.0–3.0 (–4.0) mm long; Caribbean slope and lowlands. 83. *C. thyrsiflorus*.
44. Achenes ellipsoid, sessile or nearly so; stigmas 1.0–1.5 mm long; Pacific slope and mountains. 84. *C. hermaphroditus*.
42. Spikelets ellipsoid to subulate, 2–4 times longer than wide, with 1–2 (–4) scales.
45. Bracts horizontal to reflexed downward parallel to the culm; scales laterally golden-brown to dull grayish white; spikes 2–3 times longer than wide, ± flattened when pressed. 77. *C. pallidicolor*.
45. Bracts ascendent at ca. 45°; scales laterally brownish to reddish brown; spikes 2–10 times longer than wide, not much flattened when pressed. 79. *C. mutisii*.
41. Spikes densely cylindrical (or oblong-ovoid), the rachis not visible between the spikelets when pressed.
46. Anther connective prolonged into a linear subulate apex 0.1–0.5 mm long.
47. Leaves 3–5 (–7); anther connective apex 0.1–0.15 mm long; mainland of Mexico. 70. *C. pycnostachyus*.
47. Leaves 10–25 (–60); anther connectives apex 0.3–0.5 mm long; Revillagigedo Islands. 71. *C. duripes*.
46. Anthers connective not prolonged.
48. Spikes cylindric; scales 2.4–3.4 mm long, laterally golden-brown; widespread in lowlands. 76. *C. aggregatus*.
48. Spikes oblong-ovoid; scales 1.8–2.5 mm long, laterally greenish white; rare, northeastern Mexico. 88. *C. retrorsus*.

40. Rays absent, spikes all sessile.
  49. Spikes flattened in pressing, the rachis visible between the bases of the spikelets. (Return to 42, above).
  49. Spikes not much flattened in pressing; rachis not visible between bases of spikelets.
    50. Longest bract erect; spikes 3–7 mm long. 78. *C. swartzii.*
    50. Longest bract slightly ascendent, horizontal, or reflexed; spikes 10–30 mm long.
      51. Spikelets subulate-lanceolate; achenes narrowly ellipsoid. 75. *C. regiomontanus.*
      51. Spikelets ellipsoid to oblong-ellipsoid; achenes ellipsoid to oblong-ellipsoid.
        52. Scales laterally whitish to whitish green; stigmas 0.8–1.4 mm long; achenes 0.6–0.7 mm wide; uplands of Chiapas and southward. 74. *C. svensonii.*
        52. Scales laterally stramineous to golden-brown; stigmas 1.4–2.1 mm long; achenes 0.8–1.0 mm wide; widespread at low (to middle) elevations. 76. *C. aggregatus.*

I. Cyperus subgenus **Pycnostachys** C. B. Clarke in Hooker f., Fl. Brit. India 6: 597. 1893.—Lectotype, designated by Tucker, 1987: *Cyperus diffusus* Vahl.
Cyperus subgenus *Protocyperus* Lye, Nord. J. Bot. 1: 57. 1981.—Type: *Cyperus difformis* L.

Culms and leaves with non-kranz (eucyperoid) anatomy. Spikelets borne in digitate or glomerulate clusters (rarely borne singly at the ends of the rays) or, in some species, the rays absent and the inflorescences dense, capitate, sessile. Rachilla persistent, wings absent. Scales bicarinate or conduplicate, deciduous at maturity. Carpels 3, stigmas 3. Achenes trigonous (rarely subterete), often slightly dorsiventrally compressed.

Subgenus *Pycnostachys*, found in pantropical and warm- (to cool-) temperate regions, comprises about 130 species. The greatest diversity is in the moist lowland tropical forests of Central and South America, Africa, and Malesia. There are 29 species in Mexico, of which 6 are endemic.

1. Cyperus involucratus Rottbøll, Descr. pl. rar. 22. 1772. *Cyperus flabelliformis* Rottbøll, Descr. icon. pl. rar. 42. 1773, nom. superfl. *Cyperus alternifolius* subsp. *flabelliformis* (Rottbøll) Kükenthal, Pflanzenreich 4(20): 193. 1936.—Type: "Arabia." 1762–63, *Forsskål s.n.* (holotype: C, fide Baijnath, 1975).

Caespitose perennial, 30–150 cm tall. Rhizomes 1–2 cm in diameter, indurate, producing culms (0.5–) 1–3 cm apart. Culms 3.5–6 mm in diameter, trigonous, smooth. Leaves lacking blades, the lower culm with 1–4 blunt sheaths. Inflorescence bracts (4–) 18–22, 15–27 cm long (approximately equal in length), (1.5–) 8–12 mm wide, flat, the margins and keel scabrellate, borne approximately horizontally. Rays (14–) 20–22, (2–) 5–12 (–20) cm long; secondary rays 0.3–3 (–4) cm long; tertiary rays sometimes also present, 0.3–2.5 cm long. Heads digitate, 15–30 (–36) mm wide. Spikelets 8–20, 5–25 mm long, 1.5–2.0 mm wide, ovate to linear-lanceolate, compressed; rachilla persistent, ca. 0.3 (–0.4) mm wide, ca. 0.3 mm thick, straight, quadrate, stramineous, often with reddish spots, successive scale scars 0.6–1.0 mm

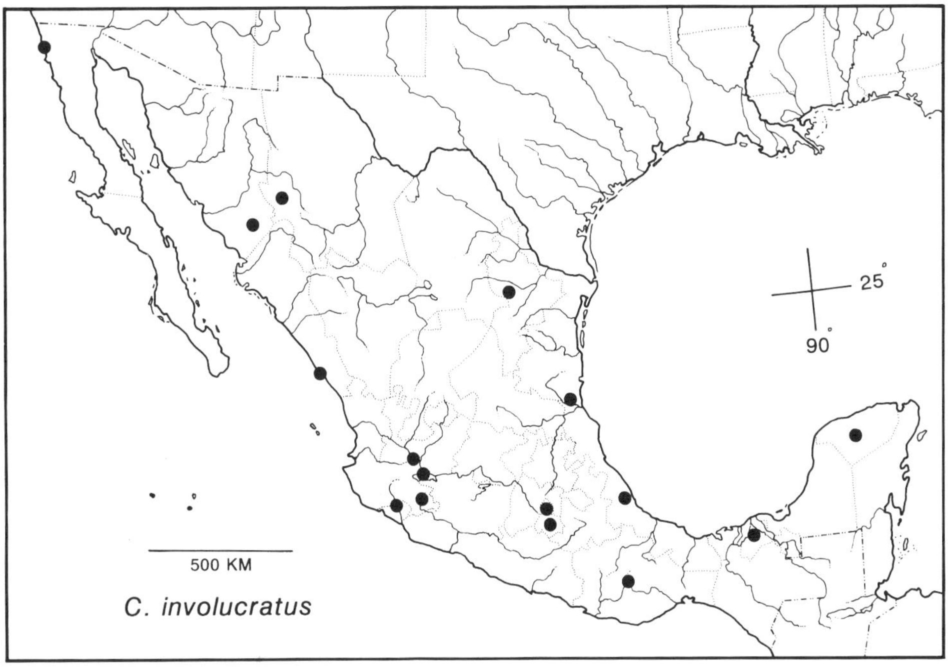

FIG. 3. Distribution of *Cyperus involucratus* in Mexico.

apart, wings absent. Scales deciduous, 8–28, 1.6–2.4 mm long, (1.0–) 1.2–1.5 (–1.7) mm wide, ovate-deltate, apex acute, muticous, laterally nerveless, whitish or light brown, more or less translucent (often appearing to have a central brown spot owing to the partial visibility of the achene through the side of the scale), medially 3-nerved, light brown, bicarinate basally. Stamens 3; filaments 1.6–2.8 mm long; anthers 0.7–1.0 mm long, linear, the connective apex 0.1–0.2 mm long, subulate, reddish. Style 0.5–1.0 mm long; stigmas 3, 0.6–1.0 mm long. Achenes 0.6–0.8 mm long, 0.4–0.6 mm wide, planoconvex to subtrigonous, broadly ellipsoid, apex obtuse, apiculate, sessile or with a stipitate base up to 0.1 mm long, the adaxial face nearly flat, the abaxial faces contiguous, broadly rounded, the surface puncticulate, brown.

Phenology. Flowering and fruiting primarily from November to May, less commonly during the remaining months.

Distribution. Native to East Africa (Baijnath 1975); introduced in the New World: Southeastern U.S.A. in Florida to Louisiana and also in central and southern California, the West Indies, Mexico, Central America, and Venezuela. Infrequent in Mexico (Fig. 3) but occurring nearly throughout the country; marshes, lake shores, and riverbanks; sea level to ca. 2000 m.

REPRESENTATIVE SPECIMENS. **Mexico.** BAJA CALIFORNIA: Baja Malibu near Tijuana, *Moran 28027* (CAS, UCR).—CHIAPAS: Tapachula, *Matuda 17706* (F).—CHIHUAHUA: Mpio. Batopilas, Barranca de Batopilas, *Bye 3294* (ECON).—COLIMA: 22 km NNW of Colima, *Sanders et al. 10401* (NYS, UCR).—DISTRITO FEDERAL: Villa Obregón, *Moore 6386* (BH, DUKE).—JALISCO: Chapala, Country Club on Laguna de Chapala, *Iltis 29163* (DUKE, MICH, WIS).—MORELOS: Mpio. Huitzilac, Las Huertas, 20 Feb 1984, *Cecilia s.n.* (IBUG); Mpio. Cuautla, 3 km W of Cuautla, *Rojas H. 98* (ARIZ, ASU, CAS, CHAPA, ENCB, IBUG).—NUEVO LEÓN: Monterrey, *Abbon 6412* (US).—OAXACA: Oaxaca, *Reko*

4262 (US).—SINALOA: Mazatlán, *Rose 14090* (NY, US).—SONORA: 8 mi SE of Alamos, *Van Devender et al. 84–180* (ARIZ).—TABASCO: Mpio. Nacajuaca, Masateupa, *Ortiz & Cowan 1866* (CAS).—TAMAULIPAS: Pánuco–Tamosí drainage, Arroyo Encino, *Darnell 15* (WIS).—VERACRUZ: Río Actopan, 3 km from Actopan, *Lot 1039* (F, NY).—YUCATÁN: Aguada Xkaxek, *Gaumer 23237* (F, G, GH, MO, MSC, NY, UC).

Cyperus involucratus can be identified by its conspicuous inflorescence bracts, ca. 20 in number. These are borne at nearly right angles to the culm. The species is similar to the native *C. canus*, but can be distinguished by its perfect flowers and glossy scales, which are laterally nerveless. *Cyperus canus* is dioecious, and its scales are laterally 5–7-nerved.

2. **Cyperus canus** J. Presl & K. Presl in K. Presl, Reliq. Haenk. 1: 179. 1828.—TYPE: MEXICO. [Guerrero:] Acapulco, 1790–91, *Haenke s.n.* (holotype: PR!; isotype: PRC!).

Caespitose perennial, 50–100 (–150) cm tall. Rhizomes ca. 1 cm in diameter, indurate, producing culms ca. 1–2 cm apart. Culms 2.5–5.5 mm in diameter, trigonous, smooth. Leaves lacking blades, the culm with 1–3 blunt sheaths basally. Inflorescence bracts 10–18, 22–26 (–36) cm long (approximately equal in length), (3–) 6–12 mm wide, flat, the margins and keel scabrous, ascendent at 30–60°. Rays 10–18, 3–6 cm long; secondary rays often present, 3–20 mm long; tertiary rays occasionally present, 1–6 mm long. Heads digitate, 4–10 (–20) mm wide. Spikelets 3–20 (–24), (2–) 5–10 (–22) mm long, 1.0–2.0 (–2.4) mm wide, ovate to linear-oblong, flatly ellipsoidal in cross section, *dioecious*; rachilla persistent, ca. 0.4 mm wide, ca. 0.2–0.3 mm thick, straight, quadrate, reddish brown, with a depression fitting the base of each achene, wings absent. Scales deciduous, 6–14 (–28), 1.6–2.0 mm long, 1.0–1.8 mm wide, ovate-deltate, apex broadly rounded to subacute, muticous, laterally 5–7-nerved nearly to the margins, reddish brown to brown, medially 2–3-nerved, the abaxial side indented proximally, reddish brown or green, distally scabrellate or smooth. Stamens 3; filaments 1.6–2.8 mm long; anthers 0.7–0.9 mm long, linear, the connective apex 0.1–0.3 mm long, linear, red. Style (0.2–) 0.5–1.2 mm long; stigmas 3, 1.0–1.5 mm long. Achenes 0.4–0.6 mm long, 0.2–0.3 mm wide, roundly trigonous, ellipsoid, apex obtuse, slightly apiculate, base stipitate, 0.1–0.2 mm long, the faces convex, the surface puncticulate, brown.

Phenology. Flowering and fruiting from January to June.

Distribution. Southern Tamaulipas and southern Sonora, Mexico, (Fig. 4) south to Colombia; occasional in Mexico, along streams, in marshes, on lake shores, and on wet shaded ledges; sea level to ca. 1500 m.

REPRESENTATIVE SPECIMENS. **Mexico.** CHIAPAS: Honduras near Siltepec, *Matuda 4412* (F, GH, LL, MICH, MO, NY, US); on Hwy 211, ca. 1 km N of Belisario Domínguez, *Tucker 2689* (CAS, DUKE, ENCB, MICH).—COLIMA: Colima, *Palmer 1304* in 1891 (F, G, GH, MICH, NY, US).—GUERRERO: Dtto. Montes de Oca, Vallecitos, *Hinton 11747* (DS, GH, LCU, MICH, NY, US).—HIDALGO: 3 km SE of San Bartolo Tutotepec, *Gimate 943* (ARIZ, ASU, ENCB).—JALISCO: Santa Cruz de Vallarta, *Mexia 1282* (CAS, DS, G, GH, MO, NY); Mpio. Acatlán de Juárez, Balneario Chorros de Santana, *Delgado Benítez 27* (IBUG).—MÉXICO: Dtto. Temascaltepec, Ixtapán, "in a salt well," *Hinton 8885* (DS, F, G, GH, LCU, MICH, NY, UC, US).—MICHOACÁN: Dtto. Coalcomán, Aquila, *Hinton 15832*

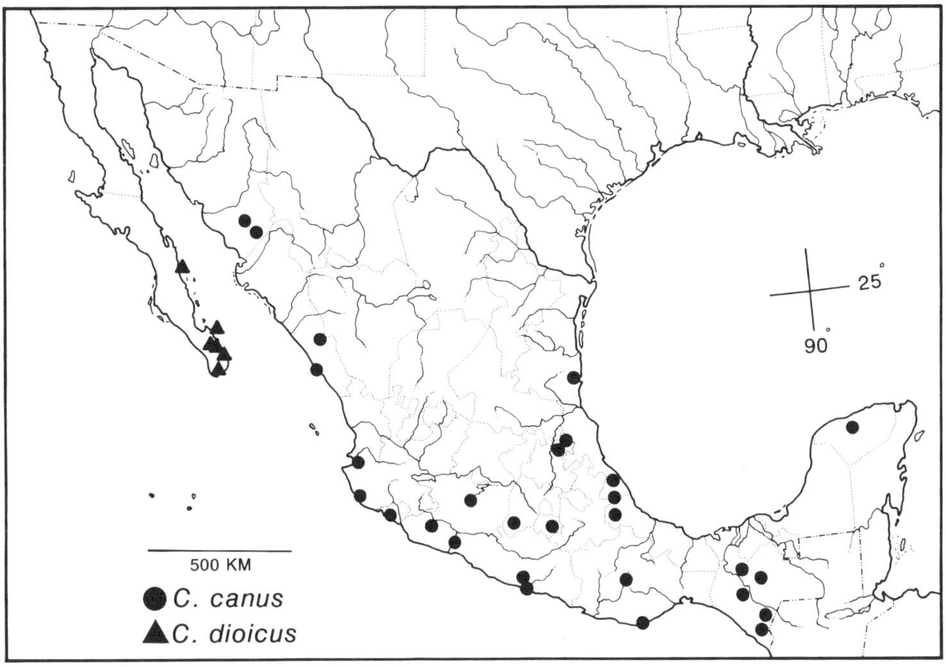

FIG. 4. Distribution of *Cyperus canus* and *C. dioicus* in Mexico.

(MICH, US).—MORELOS: Oastepec, *Matuda 26033* (NY).—OAXACA: Playa de S. Agustín, *Liebmann 14420* (C, MICH); near Oaxaca, *Pringle 6044* (AC, BKL, CAS, E, F, G, GH, LL, M, MICH, MO, MSC, MTMG, NY, PH, UC, VT, Z).—SINALOA: Mazatlán, *Ortega 5103* (US).—SONORA: Río Mayo, Arroyo Gochico, *Gentry 3626* (ARIZ, LCU); Jorinabo, *Yetman et al. s.n.* (ARIZ 302762).—TAMAULIPAS: inter Tampico et Soto La Marina, *Berlandier 3220* (NY); Hwy 85, 29 mi S of Tropic of Cancer, *Ahshapanek 123* (WIS).—VERACRUZ: Valle de Córdoba, *Bourgeau 2165* (GH, LCU, MICH, NY); Colipa, *Liebmann 14422* (C, HAL); Mpio. Cempoala, circa Mata Verde, *Cházaro & Hdz. de Cházaro 4258* (IBUG).—YUCATÁN: Izamal, *Gaumer 483* (C, E, F, G, GH, NY).

Cyperus canus and *C. dioicus* (no. 3) are the only dioecious species of *Cyperus* in the New World. They are distinguishable morphologically as follows.

1. Leaf blades absent; inflorescence bracts 10–18, ascendent at 30–60°; scales 6–14 (–28) per spikelet, with lateral nerves; anthers 0.7–0.9 mm long. *C. canus*.
1. Leaf blades present (2–30 cm long); inflorescence bracts 3–5, horizontal to reflexed; scales 10–46 per spikelet, without lateral nerves; anthers 1.0–1.8 mm long. *C. dioicus*.

Cyperus canus apparently does not set seed. All of nearly 80 collections examined from throughout its range consist of either pistillate or staminate plants, none were mixed. In *C. dioicus*, about half of the collections seen included plants of both sexes, and seeds were plentiful.

3. Cyperus dioicus I. M. Johnston, Proc. Calif. Acad. Sci., ser. 4, 12: 988. 1924.—TYPE: MEXICO. Baja California Sur: Isla del Carmen, 17 Jun 1921, *Johnston 4145* (holotype: CAS!, photo: LCU!; isotypes: C! US!).

Caespitose perennial, 10–110 cm tall. Rhizomes 3–6 mm in diameter, indurate, producing culms 2–10 mm apart. Culms 1.5–4.5 mm in diameter, trigonous, smooth. Leaves 2–5, 2–30 cm long, 3–6 mm wide, v-shaped, the margins and keel scabrous. Inflorescence bracts 3–5, 0.5–17 cm long, 0.3–6 mm wide, v-shaped, the margins and keel scabrous, borne approximately horizontally, or more often reflexed downward parallel to the culm. Rays 0–4, 0.5–2.0 cm long; secondary rays occasionally present, 3–10 mm long; rays absent in many plants, the inflorescence a single rather dense cluster of 20–35 spikelets. Heads digitate, 15–40 mm wide, loosely ovoid. Spikelets 2–8, (5–) 10–22 mm long, 2.0–2.4 (–3.2) mm wide, oblong to linear, flattened, *dioecious*; rachilla persistent, ca. 0.4 mm wide, ca. 0.2 mm thick, rectangular, stramineous, successive scale scars 0.6–0.8 mm apart, wings absent. Scales deciduous, 10–46, (1.6–) 2.0–2.3 (–2.8) mm long, 1.4–2.1 mm wide, broadly ovate to ovate-deltate, apex mucronulate, laterally nerveless, reddish to reddish brown, medially strongly 3-nerved, greenish to brownish. Stamens 3; filaments 2.0–2.8 mm long; anthers 1.0–1.8 mm long, linear, the connective apex 0.1–0.2 mm long, linear to deltate, reddish. Style 0.4–1.1 mm long; stigmas 3, 1.0–1.5 mm long. Achenes ca. 0.8 mm long, (0.5–) 0.6 mm wide, roundly trigonous, ovoid-ellipsoid, apex broadly rounded and apiculate, base stipitate, ca. 0.1 mm long, the faces convex, the surface punctate, brown.

Phenology. Flowering and fruiting from March to June.

Distribution (Fig. 4). Endemic to Baja California, Mexico, along the coast of the Gulf of California from the vicinity of Isla del Carmen south to Cabo de San Lucas; muddy or wet sandy soils, springs, permanent pools, and moist shaded arroyos; sea level to ca. 1000 m.

Representative Specimens. **Mexico.** Baja California Sur: San Raimundo, 1888, *Brandegee s.n.* (UC); Cape region, Corral de Piedra, 10 Sep 1893, *Brandegee s.n.* (UC); Sierra de la Laguna, 26 Jan 1893, *Brandegee s.n.* (UC); San Juanico, *Gentry 4309* (ARIZ, GH, MO, UC); Isla Cerralvo, *Moran 9524* (CAS, MICH, RSA, SD, UC); below Santiago along stream, *Wiggins 4659A* (CAS, GH, MICH, UC).

Cyperus dioicus and *C. canus* are the only dioecious species in the genus in the New World. Their differences are summarized in the discussion of *C. canus* (no. 2). The two species are allopatric; *C. canus* occurs on the mainland of Mexico and in Central America, and *C. dioicus* is endemic to Baja California. According to Benedict Ayers (1946), the scales of the staminate flowers are longer than those of the carpellate flowers, 3 mm versus 2 mm, respectively. Measurements made in the present study did not confirm such a difference, and indeed no plants were found to have scales longer than 2.8 mm.

4. Cyperus acuminatus Torrey & Hooker, Ann. Lyceum Nat. Hist. New York 3: 435. 1836.—Type: U.S.A. Missouri: St. Louis, *Drummond s.n.* (lectotype, designated by Denton, 1978: K; isolectotype: NY!).

Cyperus cyrtolepis Torrey & Hooker, Ann. Lyceum Nat. Hist. New York 3: 436. 1836. *Cyperus acuminatus* var. *cyrtolepis* (Torrey & Hooker) Kükenthal, Pflanzenreich 4(20): 178. 1936.—Type: U.S.A. Texas: Rio Brazos, *Drummond 450* (lectotype, designated by Denton, 1978: NY!; isolectotype: GH!).

Cyperus rufescens var. *denticarinatus* Britton, Bull. Torrey Bot. Club 11: 85. 1885. *Cyperus cyrtolepis* var. *denticarinatus* (Britton) Britton, Bull. Torrey

Bot. Club 13: 209. 1886.—TYPE: MEXICO. Valley of the Lower Río Grande, 1879–83, *Buckley s.n.* (holotype: NY!).

Small caespitose annual, (10–) 20–30 (45) cm tall. Rhizomes absent. Culms 0.4–1.2 mm in diameter, roundly trigonous, smooth or infrequently scabrous in the upper half. Leaves 2–6 (–8), 8–12 (–16) cm long, 1–2 (–4) mm wide, v-shaped or flat, the margins and keel scabrellate or smooth. Inflorescence bracts 3–6, (1.5–) 3–12 (–25) cm long, 1.0–2.5 (–3.5) mm wide, flat to v-shaped, the margins and keel sparsely scabrellate, the longest bract erect or nearly so, appearing as a continuation of the culm, the remaining bracts horizontal to ascendent at 30–60°. Rays (0–) 1–3 (–5), 0.5–2.0 cm long; secondary rays absent. Heads digitate to glomerulate, 7–12 (–15) mm wide. Spikelets (15–) 25–50 (–75), 4–7 mm long, 2–3 mm wide, ovate, compressed; rachilla persistent, ca. 0.4 mm wide, ca. 0.1–0.2 mm thick, straight, stramineous with reddish to reddish brown splotches on the faces bearing scales, successive scale scars ca. 0.5 mm apart, wings absent. Scales deciduous, 8–20 (–35), 1–2 mm long, 1.0–1.4 mm wide, ovate-lanceolate, apex acute, mucronate, the mucro ca. 0.3 mm long, straight to excurved, smooth or sparsely scabrellate, laterally yellowish, yellowish green, or light reddish brown, subtly 1-nerved, medially bicarinate, light brown or greenish brown. Stamen 1; filament 1.0–1.5 mm long; anther ca. 0.5 mm long, ellipsoid (ca. 2.5 times longer than wide), the connective apex not prolonged. Style 0.8–1.1 mm long; stigmas 3, ca. 0.5 mm long. Achenes 0.8–1.1 mm long, 0.3–0.4 mm wide, trigonous, broadly ellipsoid, apex acute, tapering into a slender beak, base abruptly contracted to a conspicuous stipe ca. 0.1 mm long, the angles rounded, the faces convex, the surface papillose, brown.

Phenology. Flowering and fruiting from July to October.

Distribution. Endemic to North America, from southwestern Virginia, Minnesota, and Washington south into Alabama, Texas, and southern California, U.S.A., and in northeastern Mexico; around water holes (tinajas); near sea level to ca. 1500 m.

ADDITIONAL SPECIMENS EXAMINED. **Mexico**. COAHUILA: Sierra de Santa Rosa, trail from S end of Hillcoat Mesa to Buena Vista Ranch headquarters, *Marsh 1491* (GH); Mesa Grande, 40 km NW of Hda. de la Encantada, *Stewart 1638* (F, GH).—TAMAULIPAS: Hwy 101 S from Matamoros at Km 256, *Tucker 1929* (DUKE, ENCB); Hwy 101, 20 km S of Santa Teresa, *Tucker 2344* (CAS, DUKE, ENCB, F, GH, MICH).

Cyperus acuminatus is distinguished by its small size and annual duration from other Mexican species of *Cyperus* with glomerulate clusters of spikelets. In addition, the erect leading inflorescence bract and the stipitate base and beaked apex of the achene are conspicuous diagnostic features of this species.

5. **Cyperus reflexus** Vahl, Enum. pl. 2: 299. 1805.—TYPE: URUGUAY. Montevideo, *Commerson s.n.* (holotype: C, not located). [Denton (1978) incorrectly cited as type: "U.S.A. Texas, *Drummond 1* (holotype: C!; isotype: NY!)".]

Cyperus fraternus Kunth, Enum. pl. 2: 42. 1837. *Cyperus reflexus* var. *fraternus* (Kunth) Kuntze, Rev. gen. pl. 3(2): 334. 1898.—TYPE: BRAZIL. *Sello s.n.* (holotype: B!).

Cyperus arenicola Steudel, Syn. pl. glum. 2: 51. 1854. *Cyperus cyrtolepis* var. *arenicola* (Steudel) Boeckeler, Linnaea 35: 558. 1868. *Cyperus pseudovegetus* var. *arenicola* (Steudel) Kükenthal, Pflanzenreich 4(20): 176. 1936.—TYPE: U.S.A. Texas: Rusk Co., *Vinzent 26* (holotype: P; isotype: BR!).

Cyperus brazas Steudel, Syn. pl. glum. 2: 316. 1854, ('*baazas*').—TYPE: U.S.A. Texas, *Drummond s.n.* (holotype: P; isotype: NY!).

Cyperus surinamensis var. *strictus* Kükenthal, Pflanzenreich 4(20): 175. 1936.—TYPE: PARAGUAY. Río Apa, *Hassler 8195* (holotype: B, destroyed; isotype: MICH).

Caespitose perennial, 30–80 cm tall. Rhizomes 2–3 mm in diameter, scaly, producing culms 1–2 (–10) mm apart. Culms 1.0–1.6 mm in diameter, roundly trigonous to terete, smooth. Leaves 1–6, 10–40 cm long, 2–8 mm wide, margins and keel glabrous or essentially so. Inflorescence bracts 3–4, 4–17 cm long, 1.5–4 mm wide, the margins scabrellate, the longest erect (or nearly so) and appearing to be a continuation of the culm, the others horizontal to ascendent. Rays (0–) 3–6, 0.5–4.0 cm long; secondary rays absent. Heads 9–15 mm wide, ovoid to pyramidal. Spikelets 25–60, 5–6 (–15) mm long, 1.5–2.5 mm wide, oblong-lanceolate, reddish with contrasting yellowish or greenish edges, strongly compressed; rachilla persistent, straight, ca. 0.2 mm wide, 0.1 mm thick, stramineous with reddish striations, successive scale scars ca. 0.5 mm apart, wings absent. Scales deciduous, 10–22, 1.4–2.2 mm long, 0.8–1.5 mm wide, ovate-elliptic, apex acute, muticous, laterally rather strongly 1-nerved, reddish, medially weakly 1-nerved, yellowish green, proximally grooved, apically scabrellate (sometimes smooth). Stamen 1; filament 1.0–1.5 mm long; anther 0.6–1.0 mm long, ellipsoid, the connective apex not prolonged. Style 0.3–0.5 mm long; stigmas 3, 0.3–0.6 mm long. Achenes 0.8–0.9 mm long, 0.3–0.4 mm wide, trigonous, ovoid-ellipsoid, apex acute to acuminate, the stipitate base ca. 0.1–0.2 mm long, the faces flat to slightly concave, the surface reticulate or subtly rugulose, reddish brown. Fig. 5e–k.

Phenology. Flowering and fruiting from May to October.

Distribution. Western Florida, Louisiana, eastern Oklahoma, and Texas, U.S.A., and the uplands of central Mexico, also in Costa Rica and temperate South America. Uncommon in Mexico; drainage ditches, marshes, pastures, and *Sphagnum*-dominated wetlands; 1500–2600 m.

REPRESENTATIVE SPECIMENS. **Mexico.** DISTRITO FEDERAL: Valle de México, *Schaffner 9* (NY).—GUANAJUATO: 16 km N of San Felipe, *Rzedowski 43690* (IEB, cited by Tucker & McVaugh, 1993).—HIDALGO: between Atotonilco and Zoquital, *Moore & Wood 4236* (A, BH).—JALISCO: Guadalajara, *Pringle 11724* (C, CAS, CU, F, K, MICH, MO, US, VT); 2 mi S of Ojuelos de Jalisco, *McVaugh 12780* (MICH, SMU); Mpio. Lagos de Moreno, *Flores Nava 1* (IBUG).—MÉXICO: San Felipe, *Mita Stein 112* (ENCB, MT); 2 km SE of Tepozotlán, *Rzedowski 36301* (ENCB, G, SD).—MICHOACÁN: Morelia, *Arsène 6654* (US).—SAN LUIS POTOSÍ: Morales, *Schaffner 564* in 1877 (GH, K).

Cyperus reflexus is recognized by its reddish scales with contrasting greenish or yellow keels and by the laterally 1-nerved scales. The lateral nerve is most often situated where the reddish and yellowish parts of the scale meet. The coloration of the scales, combined with the ovoid-ellipsoid achenes, smooth culms, and glomerulate spikelets, separate *C. reflexus* from other Mexican species of *Cyperus*.

Denton (1978) treated *C. arenicola* as a synonym of the North American *C.*

pseudovegetus Steudel; however, the isotype I saw had achenes only ca. 3 times longer than wide, reddish yellow scales with green keels, and the longest inflorescence bract erect. These characters fit *C. reflexus* rather than *C. pseudovegetus*.

6. Cyperus surinamensis Rottbøll, Descr. icon. pl. rar. 35. 1773.—Type: Suriname. *Rolander s.n.* (holotype: not located). Application of name follows Kükenthal (1935–36) and Denton (1978).

Cyperus denticulatus Schrader ex Schultes in Roemer & Schultes, Syst. veg. Mantissa 2: 104. 1824.—Type: Brazil. *Wied-Neuwied s.n.* (holotype: not located; isotype: C!).

Cyperus bipontini Boeckeler, Flora 40: 33. 1857.—Type: Mexico. Veracruz: near Mirador, *Sartorius s.n.* (holotype: B!; isotypes: GH! W, photo of W isotype: LCU! MICH!).

Cyperus surinamensis var. *lutescens* Boeckeler, Linnaea 35: 555. 1868.—Type: [Brazil. Minas Gerais:] Lagôa Santa, Nov, *Warming s.n.* (neotype, designated by Denton, 1978: C!).

Cyperus barrancae M. E. Jones, Contr. W. Bot. 18: 25. 1933.—Type: Mexico. Jalisco: Guadalajara, La Barranca, 23 Nov 1930, *Jones 27601-A* (holotype: POM!).

Caespitose annual or short-lived perennial, (10–) 35–80 cm tall. Rhizomes absent. Culms 0.5–2.0 (–4.0) mm in diameter, trigonous, usually scabrellate with retrorse prickles, infrequently smooth or rarely with some extrorse prickles also. Leaves 3–9, (12–) 25–45 (–65) cm long, 2–10 mm wide, v-shaped, the margins and keel scabrellate or smooth. Inflorescence bracts 3–8, 2–15 (–34) cm long, 1–4 (–5) mm wide, v-shaped to flat, the margins and keel scabrellate, borne approximately horizontally to ascendent at 30°. Rays 4–12, 1–6 (–9) cm long, scabrellate with retrorse prickles; secondary rays often present, 1–2 (–3) cm long; tertiary rays rarely present, 0.5–1.5 cm long. Heads umbellate, 10–20 (–30) mm wide, dense. Spikelets (6–) 15–40 (–65), (3–) 4–12 (–15) mm long, 1.5–2.5 mm wide, linear to linear-oblong, compressed; rachilla persistent, ca. 0.2 mm wide, ca. 0.1 mm thick, straight, quadrate, stramineous, successive scale scars ca. 0.5 mm apart, wings absent. Scales deciduous, 10–50 (–65), 1.0–1.5 mm long, 0.8–0.9 mm wide, lanceolate, apex mucronulate, laterally nerveless but distinctly reticulate, pale yellow, light brown, or reddish brown, medially 3-nerved, bicarinate basally, often scabrellate near the apex. Stamen 1; filament 0.9–1.3 mm long; anther ca. 0.5 mm long, oblong-ellipsoid, the connective apex not prolonged. Style 0.8–1.1 mm long; stigmas 3, ca. 0.5 mm long. Achenes 0.7–0.9 mm long, 0.2–0.4 mm wide, trigonous, narrowly ellipsoid, apex apiculate, base slightly stipitate, the faces slightly convex, the surface papillate or obscurely reticulate to rugulose, brown to reddish brown. Fig. 5a–d.

Phenology. Flowering and fruiting throughout the year.

Distribution. Neotropical and warm-temperate regions; South Carolina, central Alabama, and eastern Kansas, U.S.A., the West Indies, and Mexico to Argentina. Frequent to common in Mexico (Fig. 6); in moist sunny disturbed soils; sea level to ca. 1000 (rarely to 1600) m.

Representative Specimens. **Mexico.** Aguascalientes: 16 km N of Aguascalientes, *Rzedowski & McVaugh 728* (ENCB, MICH).—Baja California Sur: San José del Cabo, *Brandegee 602* (UC).—

FIG. 5. *Cyperus surinamensis*, *C. reflexus*, and *C. entrerianus*. a–d, *C. surinamensis*. a. Habit (×0.5). b. Spikelet (×5). c. Scale (×15). d. Achene (×15). e–k, *C. reflexus*. e. Inflorescence (×0.5). f, i. Spikelets (×5). g, j. Scales (×15). h, k. Achenes (×15). l–o, *C. entrerianus*. l. Inflorescence (×0.5). m. Spikelet (×5). n. Scale (×15). o. Achene (×15). (Based on: a–d, *Rzedowski & McVaugh 728;* e–h, *McVaugh 12780;* i–k, *Pringle 11724;* l–o, *Díaz Luna*). Drawn by Karin Douthit. Reproduced from Denton (1978) with the permission of Melinda F. Denton and the University of Michigan Herbarium.

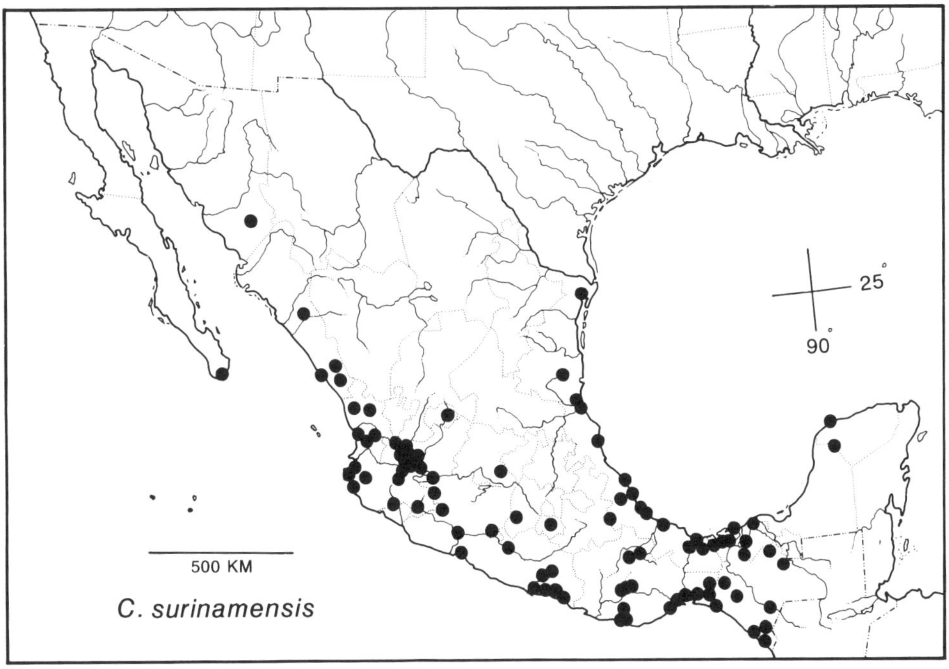

FIG. 6. Distribution of *Cyperus surinamensis* in Mexico.

CHIAPAS: Mpio. Palenque, 3–5 km N of Palenque, *Breedlove 26627* (DS); 3 km from Escuintla on rd to El Triunfo, *Tucker 2232* (DUKE, ENCB).—COLIMA: S edge of Comala, *Langman 3200* (PH).—GUANAJUATO: San Martín, SW of La Cantera, *Brown 81–23* (ARIZ).—GUERRERO: Dtto. Mina, Mesa Frijolar, *Hinton 9689* (GH, LCU); Marquelia, ca. 140 km E of Acapulco on Hwy 200, *Tucker 2926* (DUKE, ENCB).—JALISCO: near Guadalajara, *Pringle 1786* (AC, ARIZ, E, F, G, MSC, PH); Unión de Tula, *Tucker 2994* (DUKE, ENCB).—MÉXICO: Dtto. Temascaltepec, Ixtapán, *Hinton 8884* (GH, LCU).—MICHOACÁN: 22 km S of Uruapan, *King & Soderstrom 4874* (MICH).—MORELOS: 1 km S of Cuautla, *Gómez M. 50* (ARIZ, ENCB).—NAYARIT: vic. Acaponeta, *Rose et al. 14235* (US).—OAXACA: Quitepec, *Liebmann 14456* (C); 10 km N of Zaachila, *Tucker 2830* (DUKE, ENCB); 28 mi E of Juchitán, *Wilbur 35988* (DUKE, ENCB).—SINALOA: Mpio. Sinaloa y Leyva, mouth of Cañón de Tarahuares, *Breedlove & Thorne 17954* (CAS); Sierra Madre Occidental along Hwy 40 above La Guayanera, *Sanders et al. 4539* (ARIZ, UCR).—SONORA: 7 km SW of Alamos, *Van Devender et al. 92–765* (ARIZ, NYS)).—TABASCO: 30 km W of Cárdenas on Hwy 180, *Tucker 2562* (DUKE, ENCB); 15 km S of Villahermosa on Hwy 195, *Tucker 2567* (DUKE, ENCB).—TAMAULIPAS: vic. of Tampico, *Palmer 176* in 1910 (G, US); 60 km S of Matamoros on Hwy 101, *Tucker 2341* (DUKE).—VERACRUZ: Valle de Córdoba, *Bourgeau 2291* (F, G); Morro de la Mancha, *Judziewicz 3506–1/2* (WIS); 10 mi E of Jalapa, *Tucker 2141* (DUKE, ENCB); 3 km N of Lerdo de Tejada, *Tucker 2510* (DUKE, ENCB); 20 km E of Coatzacoalcos, *Tucker 2535* (DUKE, ENCB).—YUCATÁN: near Progreso, *Swallen 2962* (F).

Cyperus surinamensis is readily distinguished from all other species of the genus in the New World by the presence of retrorse prickles on the culms and rays.

7. Cyperus virens Michaux, Fl. bor.-amer. 1: 28. 1803.—TYPE: U.S.A. "Carolina," 1786–96, *Michaux s.n.* (holotype: P, photo: GH!).

Scirpus reticulatus Lamarck, Tabl. encycl. 1: 142. 1791, non *Cyperus reticulatus* (Heer) Schimper, 1853.—TYPE: U.S.A. "Carolina," ca. 1785, *Fraser s.n.* (P-LA, photo: GH!, IDC microfiche!).

Cyperus drummondii Torrey & Hooker, Ann. Lyceum Nat. Hist. New York 3: 437. 1836. *Cyperus virens* var. *drummondii* (Torrey & Hooker) Kükenthal, Pflanzenreich 4(20): 181. 1936. *Cyperus virens* subsp. *drummondii* (Torrey & Hooker) T. Koyama, Madroño 20: 254. 1970.—TYPE: U.S.A. Texas, *Drummond 449* (lectotype, designated by Denton, 1978: K; isolectotypes: B, GH! NY!).

Cyperus montanus Boeckeler, Beitr. Cyper. 2: 4. 1890.—TYPE: ARGENTINA. Cordillera de Misiones, 29 Oct 1886, *Niederlein 2116* (holotype: B; isotype: NY!).

Cyperus minarum Boeckeler, Beitr. Cyper. 2: 5. 1890. *Cyperus consanguineus* var. *minarum* (Boeckeler) Kükenthal, Pflanzenreich 4(20): 172. 1936. *Cyperus virens* var. *minarum* (Boeckeler) Denton, Contr. Univ. Michigan Herb. 11: 264. 1978.—TYPE: BRAZIL. Minas Gerais, 1845, *Widgren s.n.* (holotype: C!).

Caespitose perennial, (20–) 60–120 (–170) cm tall. Rhizomes 0.5–2.0 cm long, often absent (or not collected), indurate. Culms 2–7 (–10) mm in diameter, triquetrous, scabrous on the angles only, the crystaline prickles extrorse, antrorse, or retrorse (all three types occasionally occurring on the same culm). Leaves 4–12, 20–80 cm long, 4–14 mm wide, with conspicuous cross veins especially noticeable on the abaxial surface, flat to v-shaped, the margins and keel scabrous. Inflorescence bracts 4–11, (1.5–) 3–50 (–65) cm long, 0.5–12 mm wide, with conspicuous cross veins like the leaves, v-shaped, the margins and keel scabrous, borne horizontally to ascendent at 30°; secondary bracts 3–10 mm long, 1–2 mm wide, scabrellate especially apically. Rays 6–12, 1–9 (–14) cm long; secondary rays often present, (0.5–) 1–3 cm long; tertiary rays occasionally present, 5–12 mm long. Heads digitate, 1–3.5 cm wide. Spikelets 10–40 (–50), 5–18 mm long, 2.0–3.2 mm wide, oblong to linear-lanceolate, compressed; rachilla persistent, 0.25–0.4 mm wide, (0.1–) 0.2 mm thick, straight, reddish to stramineous, successive scale scars 0.5–0.8 mm apart, wings absent. Scales deciduous, 10–40, 1.5–2.3 (–3.0) mm long, 0.9–1.4 mm wide, ovate-deltate, apex acute, muticous or sometimes mucronulate, laterally weakly to distinctly 1-nerved, glossy with a pebbled appearance, pale reddish, brownish, or greenish, medially grooved, 2-nerved in the proximal half only, greenish or brownish, apically smooth or occasionally scabrellate. Stamens 1–2; filaments 0.8–2.0 mm long; anthers 0.8–1.4 mm long, narrowly oblong, the connective apex not prolonged. Style 0.8–1.0 mm long; stigmas 3, 0.6–0.8 mm long. Achenes 1.0–1.5 mm long, 0.3–0.6 (–0.7) mm wide, trigonous, ellipsoid, apex with a beak 0.1–0.5 mm long, base slightly to distinctly stipitate, the faces flat or slightly concave to slightly convex, the surface puncticulate, brown. Fig. 7.

Phenology. Flowering and fruiting from March to October, occasionally in the remaining months.

Distribution. Endemic to the New World; in the U.S.A. from North Carolina to Texas (once collected as an adventive in central California), the West Indies, Mexico, and Central America to central Argentina. Frequent in the uplands of Mexico (Fig. 8), especially in Chiapas, uncommon in the lowlands; wet pastures, marshes, roadside ditches, and seepage slopes in pine or oak woods; (0–) 1200–2500 (–2800) m.

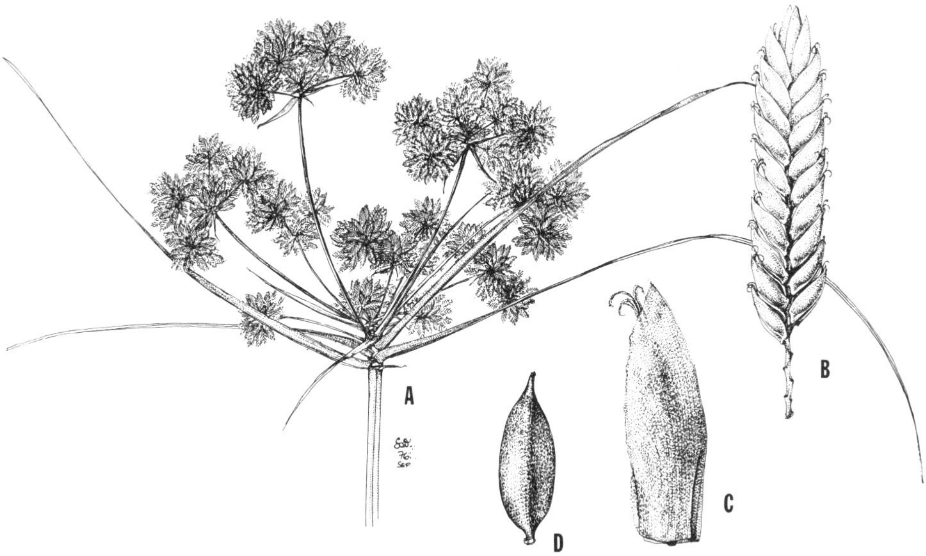

FIG. 7. *Cyperus virens* (*Moore 3468*). A. Habit (×0.5). B. Spikelet (×5). C. Scale (×17.5). D. Achene (×17.5). Drawn by Karin Douthit. Reproduced with modification from Denton (1978) with the permission of Melinda F. Denton and the University of Michigan Herbarium.

REPRESENTATIVE SPECIMENS. **Mexico.** CHIAPAS: N edge of San Cristóbal, *Breedlove 6782* (DS); Amatenango del Valle, *Breedlove 14678* (DS, DUKE); Mpio. Tenejapa, paraje of Matsab, *Ton 1295* (DS, DUKE, F); Carmelito on Hwy 190, E of San Cristóbal, *Tucker 2667* (DUKE, ENCB).—DISTRITO FEDERAL: El Rosario, *MacDaniels 628* (BH, CU, F); Valley of Mexico, *Pringle 6314* (BM, E, F, G, GH, MSC, MTMG, PH, Z).—HIDALGO: Dtto. Molango, Lake Atexca, *Moore 3468* (BH, MICH).—JALISCO: 10–20 km E of Arandas, *McVaugh 24358* (MICH); Mpio. Zapopan, nacimiento del Río Atlicolte, *Puga 4891* (IBUG).—MÉXICO: Cuautitlán, *Rzedowski 16759* (ARIZ, ENCB, MSC).—MICHOACÁN: Morelia, *Arsène 2400* (C).—OAXACA: Mpio. Ixtlán, 3 mi E of Ixtlán de Juárez, *Tucker 2171* (DUKE, ENCB); 8.5 km W of Niltepec, *Tucker 2759* (DUKE, ENCB).—QUERÉTARO: camino a la Venta, *Arguelles 368* (CAS).—TABASCO: 30 km E of Río Tonalá on Hwy 180, *Tucker 2557* (DUKE, ENCB).—VERACRUZ: Jalapa, *Galeotti 5793* (BR); Jicaltepec, *Liebmann 14451* (C, G).

Cyperus virens is readily distinguished from other species of subg. *Pycnostachys* by its triquetrous culms, which are scabrous on the angles, and its leaves and inflorescence bracts, which are conspicuously septate owing to numerous cross veins between the main veins.

The type of *Scirpus reticulatus* is clearly a plant of *Cyperus virens* (photo at GH!). The glomerulate clusters of spikelets and conspicuously septate inflorescence bracts characteristic of the species are readily visible. Combined with Lamarck's phrase "culmo aspero" in the protologue, the application of the name to *Cyperus virens* is certain. A. E. Schuyler, a student of the genus *Scirpus*, examined the type of *S. reticulatus* and annotated it "*Cyperus* sp."

Denton (1978) recognized four varieties of *C. virens*. In the study of the Central America representatives of *Cyperus* (Tucker 1983) I was not able to confirm the distinctness of these varieties. Additional study of Mexican specimens and ca. 300 collections from the southeastern United States and South America indicated that the "varieties" intergrade extensively and do not merit recognition.

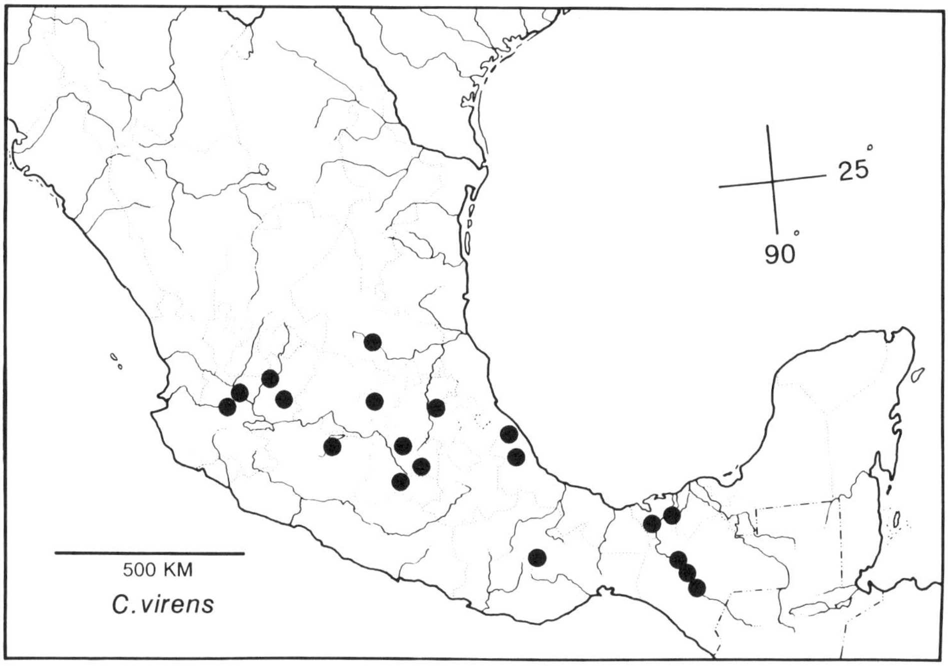

FIG. 8. Distribution of *Cyperus virens* in Mexico.

8. Cyperus ochraceus Vahl, Enum. pl. 2: 325. 1805.—TYPE: VIRGIN ISLANDS. Saint Croix, *West 15* (holotype: C-herb. Vahl!; isotype: C-herb. Schum.!).

Cyperus ochraceus var. *excelsior* Kükenthal, Pflanzenreich 4(20): 182. 1936.—TYPE: BOLIVIA. Prov. Tarija, *Pflanz 651* (lectotype, designated by Denton, 1978: B).

Caespitose perennial, 20–50 (–80) cm tall. Rhizomes 1–3 cm long, 0.3–1.5 cm in diameter, indurate. Culms 1–3 mm in diameter, roundly trigonous, smooth. Leaves 3–11, 12–40 cm long, 1.5–4 mm wide, v-shaped, the margins smooth or sparsely scabrous distally. Inflorescence bracts 4–8, 10–36 cm long, 1.5–4.0 mm wide, v-shaped, the margins smooth, approximately horizontally to ascendent at 30°. Rays 5–10, 2–12 cm long; secondary rays occasionally present, 5–25 mm long. Heads glomerulate, 15–30 mm, loosely to rather densely hemispherical. Spikelets 10–30, (5–) 7–10 (–35) mm long, 2–3 mm wide, oblong to linear-oblong, compressed; rachilla persistent, 0.4–0.5 mm wide, 0.1–0.2 mm thick, rectangular, greenish to reddish brown, successive scale scars 0.6–0.8 mm apart, wings absent. Scales deciduous, 10–25 (–85), (1.2–) 1.5–2.0 mm long, 1.2–2.0 mm wide, broadly ovate (widest in the lower half), apex obtuse, muticous, laterally nerveless, yellowish brown, medially 2-nerved with a groove between the nerves, yellowish or brownish, not overlapping, the margin of a scale just reaching the edge of the next, widely spreading from the rachilla at an angle of 60–90°. Stamen 1; filament 1.2–1.5 mm long; anther 0.8–1.2 mm long, oblong, the connective apex not prolonged. Style 0.6–1.2 mm long, basally thickened, confluent with the apex of the achene; stigmas 3, 0.3–0.6 mm long. Achenes 1.0–1.5 mm long, 0.4–0.6 (–0.7) mm wide, roundly

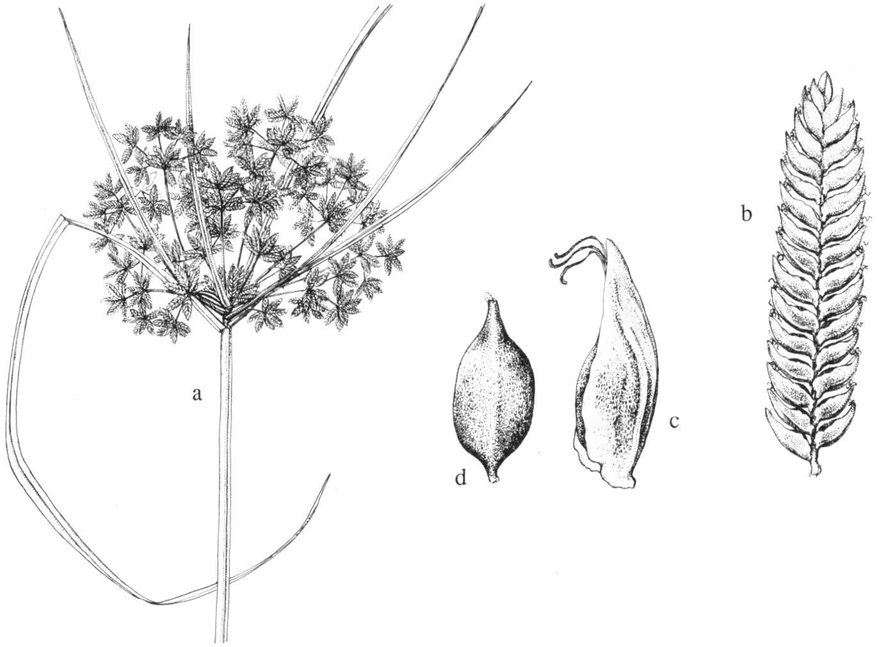

FIG. 9. *Cyperus ochraceus* (*Dieterle 4256*). a. Inflorescence (×0.5). b. Spikelet (×5); scales are oriented nearly perpendicular to the rachilla. c. Scale enclosing achene (×17.5). d. Achene, abaxial view (×17.5). Drawn by Karin Douthit. Reproduced with modification from Denton (1978) with the permission of Melinda F. Denton and the University of Michigan Herbarium.

trigonous, broadly ellipsoid, apex acute, with a beak 0.1–0.5 mm long, base cuneate to substipitate, 0.1–0.2 mm long, the faces convex to flat, the surface reticulate with a network of low ridges, reddish brown to dark purplish brown, dull or more often iridescent. Fig. 9.

Phenology. Flowering and fruiting throughout the year but especially from July to September.

Distribution. Southern U.S.A., throughout the West Indies, Mexico, Central America to Venezuela, Ecuador, Bolivia, and northern Argentina. Frequent to common on the Gulf Coast of Mexico, less frequent on the Pacific Coast, occasionally inland (Fig. 10); in various open wet situations, such as riverbanks, pastures, ditches, and edges of marshes; sea level to 2000 m.

Representative Specimens. **Mexico.** Campeche: Tuxpeña, *Lundell 803* (LL, MO).—Chiapas: Tuxtla Gutiérrez, *Miranda 7464* (MEXU); just N of Pichucalco, *Tucker 2603* (DUKE, ENCB).—Coahuila: Múzquiz Swamp, *Marsh 889* (GH, TEX).—Colima: 2.5 km W of Orteaga [Cerro de Ortega], *Tucker 2959* (DUKE, ENCB).—Guerrero: ca. 35 km S of Chilpacingo on Hwy 95, *Tucker 2270* (DUKE, ENCB).—Hidalgo: 4 km SE of San Bartolo Tutotepec, *Gimate 639* (ARIZ).—Jalisco: 1 mi E of Ayo el Chico, *McVaugh 17221* (G, MICH); Guadalajara, 15 Nov 1980, *Silva s.n.* (IBUG); Santa Rosa, *Vázquez 9* (IBUG).—México: 4 km S of Ixtapán, *Dunn et al. 20443* (DS, TEX).—Michoacán: 11–13 km WSW of Apatzingan, *Dieterle 4256* (MICH).—Morelos: Puente de Ixtla, *Sharp 441351-B* (TENN).—Nayarit: Mpio. Tuxpan, Sur Tuxpan, 21 Feb 1979, *Ranchero s.n.* (IBUG).—Nuevo León: 20 mi S of Monterrey, *Roberts & Keil 10446* (ASU, PH).—Oaxaca: 12 km W of Juchitán, *Tucker 2763*, *Wilbur 36138* (DUKE, ENCB).—Puebla: 44 km SW of Poza Rica on rd to Xicotepec, *Koch 747* (CAS).—Quintana Roo: Laguna de Paitoro, *Novelo 320* (MEXU, MO).—San Luis Potosí: Las

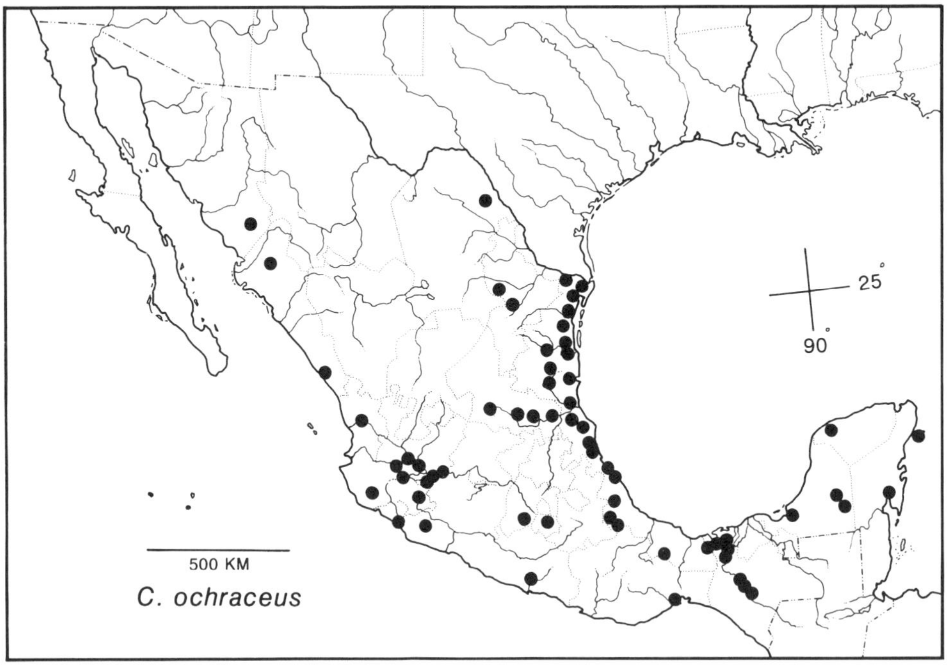

FIG. 10. Distribution of *Cyperus ochraceus* in Mexico.

Canoas, *Pringle 3724* (AC, ARIZ, E, F, G, GH, MICH, MSC, MTMG, PH, US, VT, Z).—SINALOA: Mpio. Sinaloa y Vela, El Alamo, *Breedlove & Thorne 17956* (CAS).—SONORA: 11.5 mi E of Alamos, *Sanders et al. 8813* (NYS, UCR); ca. 0.6 mi NE of turnoff to Bacobampo, *Sanders et al. 8940* (NYS, UCR).—TABASCO: 15 km S of Villahermosa, *Tucker 2566* (DUKE, ENCB).—TAMAULIPAS: 10 mi N of San Fernando on Hwy 101, *Tucker 1936* (DUKE, ENCB); ca. 20 mi N of Cd. Mante at Cinco de Mayo, *Tucker 1992* (DUKE, ENCB).—VERACRUZ: Esperanza, E of Jalapa, *Tucker 2140* (DUKE, ENCB); 30 km S of Pánuco on Hwy 127, *Tucker 2422* (DUKE, ENCB); 10 km N of Riachuelos on Hwy 180, *Tucker 2467* (DUKE, ENCB).—YUCATÁN: *Gaumer 388* (C, F, G, US).

The combination of glomerulate spikelets, scales strongly spreading from the rachilla (forming an angle of 60–90°), and broadly ellipsoid achenes distinguishes *C. ochraceus* from other species of the genus in Mexico. Brown-scaled collections of *C. niger* and *C. lanceolatus* occasionally have been misidentified as *C. ochraceus*, probably because of the similar color and shape of the spikelets. Examination of the achenes readily resolves this confusion: *C. ochraceus* has trigonous achenes and three stigmas; *C. niger* and *C. lanceolatus* have lenticular achenes and only two stigmas.

9. Cyperus entrerianus Boeckeler, Flora 61: 139. 1878. *Cyperus luzulae* var. *entrerianus* (Boeckeler) Barros, Anales Mus. Argent. Ci. Nat. "Bernardino Rivadavia" 39: 309. 1938.—TYPE: ARGENTINA. Concepción del Uruguay, Apr 1876, *Lorentz s.n.* (holotype: B).

Cyperus tucumanensis Boeckeler, Bot. Jahrb. Syst. 7: 274. 1886. *Cyperus luzulae* var. *tucumanensis* (Boeckeler) C. B. Clarke ex Chodat & Hassler, Bull. Herb. Boissier, sér. 2, 3: 1007. 1903.—TYPE: ARGENTINA. Tucumán, 29 Dec 1872, *Lorentz & Hieronymus 1086* (holotype: B).

Coarse perennial, (30–) 40–65 (–95) cm tall. Rhizomes [seldom collected] 5–12 mm in diameter, oblique, indurate, clothed with fibrous brown scales, 1–5 cm long. Culms (1–) 2–3 mm in diameter, trigonous to roundly trigonous, smooth. Leaves (1–) 5–10, (10–) 30–45 (–70) cm long, 3–7 mm wide, v-shaped, the margins and keel scabrous, the sheaths greenish or purplish. Inflorescence bracts (5–) 6–8 (–10), (5–) 15–40 (–55) cm long, (1–) 3–7 mm wide, v-shaped, the margins and keel scabrous, borne ascendent at 45–60° (–75°). Rays 6–10 (–12), 1–8 (–12) cm long, ascendent at 45–60° (–75°); secondary rays 1–4, 5–15 (–23) mm long. Heads (6–) 10–12 (–20) mm, loosely to densely globose-ovoid, off-white to dull grayish green or light brown. Spikelets (1–) 30–50 (–65), 4.0–6.5 mm long, 1.8–3.2 mm wide, broadly elliptic, flattened; rachilla persistent, 0.2–0.3 mm wide, 0.1 (–0.2) mm thick, straight, quadrate, dark red (stramineous at the nodes), successive scale scars 0.4–0.5 mm apart, wings absent. Scales deciduous, (10–) 16–20 (–26), (1.2–) 1.4–1.6 (–1.8) mm long, (0.7–) 0.8–1.0 mm wide, oblong-lanceolate, apex acute, mucronulate, laterally weakly 1-nerved, hyaline, pale green, off-white, or light brown, medially 2-nerved, bicarinate basally, smooth or scabrellate distally, green to light brown. Stamen 1; filament 1.2–2.0 mm long; anther 0.6–0.7 mm long, ellipsoid, the connective apex 0.1–0.2 mm long, acute, reddish brown. Style 0.2–0.4 mm long; stigmas 3, 0.4–0.6 mm long. Achenes 0.9–1.1 mm long, 0.3–0.4 mm wide, roundly trigonous, ellipsoid to narrowly ellipsoid, distally gradually tapered to an acute apex, proximally broadly rounded or contracted into a stipitate base ca. 0.1 mm long, the faces convex, the surface finely reticulate, brown. Fig. 5l–o.

Phenology. Flowering and fruiting from July to February.

Distribution. Amphitropically disjunct; in North America, from Georgia to Missouri and Texas, U.S.A., west to southern Sonora and south to the Isthmus of Tehuantepec in Mexico, in Nicaragua and northwestern Costa Rica, in Cuba and Jamaica, in southern Brazil, Paraguay, and northern Argentina. Common in Mexico (Fig. 11); moist pastures, roadside ditches, marshes, riverbanks, and muddy or sandy lake shores; sea level to ca. 1800 m.

Representative Specimens. **Mexico.** Chiapas: 3.5 mi NW of Arriaga, *Tucker 2721*, *Wilbur 35945* (DUKE, ENCB).—Guanajuato: San Miguel Allende, *Kenoyer 2360* (GH).—Jalisco: Mpio. Zapopan, San Isidro, 30 May 1979, *Espino s.n.* (IBUG); Mpio. Tlajomulco, El Zapote, 15 July 1978, *García Villalobos s.n.* (IBUG); Mpio. Ocotlán, E of Ejido San Andrés, 20 Oct 1979, *Rodríguez s.n.* (IBUG).—Michoacán: 1 km E of Tuxpan, *Iltis & Doebley 33*, *38* (WIS); Parque Nacional de Uruapan, *Nieves & Vázquez 67* (IBUG); vicinité de Morelia, *Arsène 3117* (MO, US).—Nayarit: 12 mi SE of Tepic, *Feddema 590* (DUKE, ENCB).—Oaxaca: 5 mi E of Temascal, 23 July 1964, *Janzen s.n.* (UC).—San Luis Potosí: Rascón, *Palmer 651* in 1905 (F, MO, US); Las Canoas, *Pringle 3806* (ARIZ, ASU, DUKE, G, MTMG, US, VT, Z); Mpio. Alaquines, 2 km SSE of Ojo de Agua, *Rzedowski 9669* (ENCB, MT).—Sinaloa: Culiacán, *Palmer 1540* in 1891 (F, US).—Sonora: Mpio. Huatabampo, 5 mi N of Estación Don, *Breedlove & Thorne 18640* (CAS).—Veracruz: 10 km E of Jalapa on the truck route to Coatepec, *Tucker 2148* (DUKE, ENCB); ca. 30 km S of Pánuco on Hwy 127, *Tucker 2423*, *2430* (DUKE, ENCB).

Cyperus entrerianus has been confused with *C. luzulae* (no. 10). Kükenthal (1935–36) accepted Boeckeler's *C. entrerianus*. Yet, Barros (1940), Benedict Ayers (1946), and Pedersen (1970) treated *C. entrerianus* as a variety of *C. luzulae*, whereas Denton (1978) considered *C. entrerianus* a synonym of *C. luzulae*. She stated that the features of these two taxa merged so completely throughout their geographic ranges that *C. entrerianus* did not merit specific status.

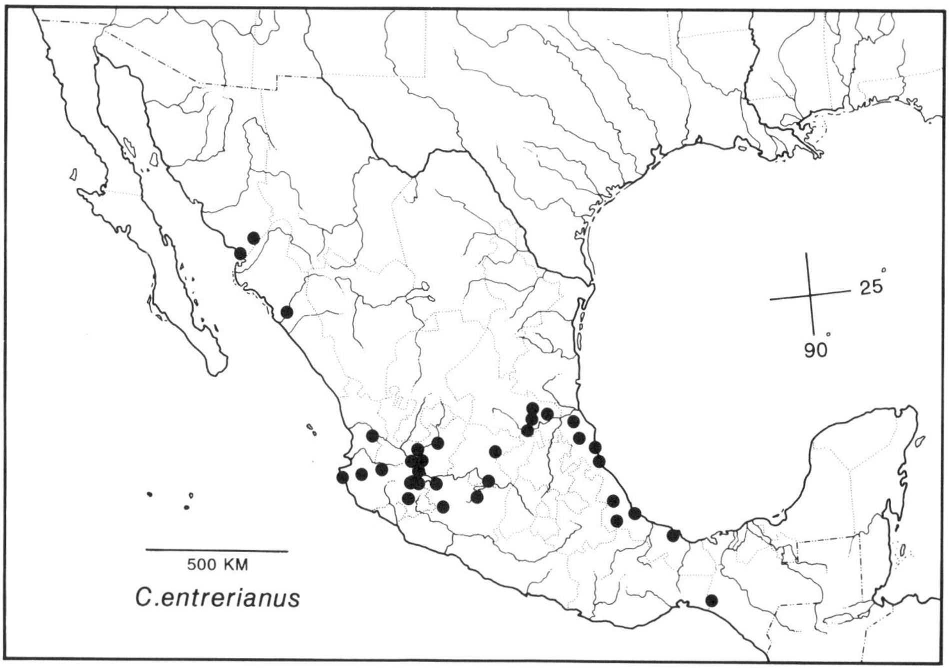

FIG. 11. Distribution of *Cyperus entrerianus* in Mexico.

In the field in Mexico the two species differ strikingly: the greenish ovoid spikes and ascendent inflorescence bracts of *C. entrerianus* contrast with the whitish oblong to pyramidal spikes and horizontal bracts of *C. luzulae*. The two species are allopatric in Mexico: *C. luzulae* occurs east of the region of the Isthmus of Tehuantepec, whereas *C. entrerianus* is found north and west of the Isthmus (Figs. 11, 12). Although the two species are readily distinguished in the field, the distinctive colors of the spikelets are frequently altered in drying; however, the shape of the heads remains intact and is readily observable. An additional difference is the color of the rachilla at anthesis: those of *C. luzulae* are green, those of *C. entrerianus* dark red. This difference in color is not affected by drying. The achenes of the two species are identical, but the scales differ in color and shape. Examination of some 550 specimens from throughout the ranges of both species revealed that the taxa do not intergrade as Denton stated. Morphological differences between the two species can be summarized as follows.

1. Culms (30–) 40–65 (–95) cm tall; inflorescence bracts ascendent at 45–60° (–75°); primary rays ascendent at 45–60° (–75°); secondary rays present; heads of spikelets loosely globose-ovoid, light greenish white to golden-brown; rachilla dark red with stramineous scale scars; scale apices acute. *C. entrerianus*.
1. Culms (18–) 25–40 (–55) cm tall; inflorescence bracts approximately horizontal to ascendent at 30°; primary rays ascendent at 15–30°; secondary rays absent; heads of spikelets densely oblong to pyramidal, bright to dull white; rachilla green throughout; scale apices rounded. *C. luzulae*.

10. Cyperus luzulae (L.) Rottbøll ex Retzius, Observ. bot. 4: 11. 1786. *Scirpus luzulae* L., Sp. pl. 2: 75. 1762, ('luculae').—TYPE: WEST INDIES OR SURINAME. Collector unknown (holotype: LINN no. 71.45, IDC microfiche!). [See Tucker and McVaugh (1993) for discussion concerning the publication of the combination.]

Cyperus globulosus Aublet, Hist. pl. Guiane 1: 47. 1775.—TYPE: FRENCH GUIANA. *Aublet s.n.* (holotype: P-Herb. Rousseau-Aublet, IDC microfiche!).

Cyperus sphaerostachys Link, Jahrb. Gewächsk. 1(3): 89. 1820.—TYPE: "America meridionale," *Humboldt & Bonpland s.n.* (B-W no. 1412, IDC microfiche!).

Cyperus trinitatis Steudel, Syn. pl. glum. 2: 26. 1854.—TYPE: TRINIDAD. *Sieber fl. Trin. no. 1* (holotype: P?; isotypes: HAL! M!).

Cyperus luzulae var. *minor* Boeckeler, Linnaea 35: 561. 1868.—TYPE: SURINAME. *Hostmann s.n.* (holotype: B).

Cyperus pseudosurinamensis Boeckeler, Allg. Bot. Zeitschr. Syst. 1: 201. 1895.—TYPE: MARTINIQUE. *Hahn s.n.* (holotype: B, fide Denton, 1978).

Cyperus bangianus Gandoger, Bull. Soc. Bot. France 66: 297. 1920.—TYPE: BOLIVIA. Yungas, 1890, *Bang 531* (holotype: not located).

Cyperus guatemalensis Gandoger, Bull. Soc. Bot. France 66: 297. 1920.—TYPE: GUATEMALA. Alta Verapaz, *von Türckheim II–128* (holotype: LY, fide Denton, 1978; isotypes: BR! E! F! G! Z!).

Cyperus luzulae f. *pallidiflorens* Kükenthal, Pflanzenreich 4(20): 171. 1936.—TYPE: DOMINICAN REPUBLIC. Santo Domingo, Llano Costero, Cuenca, 3 Jan 1929, *Ekman 10975* (holotype: B; isotypes: GH! S).

Caespitose perennial, (18–) 25–40 (–55) cm tall. Rhizomes (infrequently collected) 3–10 mm in diameter, oblique to horizontal, indurate, clothed with brown scales or their fibrous remnants. Culms 1–2 (–2.5) mm in diameter, trigonous to roundly trigonous, smooth. Leaves 3–10, (10–) 15–30 (–45) cm long, 1.5–5 mm wide, v-shaped, the margins and keel scabrellate or smooth. Inflorescence bracts (4–) 7–12, (3–) 15–25 (–40) cm long, (0.5–) 1.5–4.0 mm wide, v-shaped, the margins and keel usually scabrellate, occasionally smooth, borne approximately horizontally to ascendent at 30°. Rays 4–10, (0.5–) 1–3 (–5) cm long, ascendent at 15–30°; secondary rays absent. Heads 7–15 (–17) mm wide, densely oblong to pyramidal, rarely proliferous, bright to dull white. Spikelets 40–100, (1.5–) 3.0–4.5 mm long, 1.0–2.2 mm wide, ovate to broadly ellipsoid, flattened, white to reddish or brownish white; rachilla persistent, 0.1–0.2 mm wide, 0.1 mm thick, flat, green or rarely white at anthesis, becoming light brown after achenes fall, successive scale scars ca. 0.4 mm apart, wings absent. Scales deciduous, 6–18, 1.0–1.6 (–1.8) mm long, 0.6–0.8 (–1.0) mm wide, oblong-ovate, apex obtuse to broadly rounded, mucronulate, laterally weakly 1-nerved or essentially nerveless, whitish or reddish white, medially white or green, proximally grooved, smooth or occasionally scabrellate near the apex. Stamen 1; filament 0.5–1.8 mm long; anther 0.5–0.7 mm long, ellipsoid to oblong-ellipsoid, the connective apex 0.1–0.2 mm long, acute, reddish brown. Style 0.2–0.4 mm long; stigmas 3, 0.3–0.4 mm long. Achenes 0.9–1.1 mm long, (0.2–) 0.3–0.4 mm wide, trigonous, narrowly ellipsoid, distally gradually tapered to an acute apex, proximally cuneate or contracted into a stipitate base

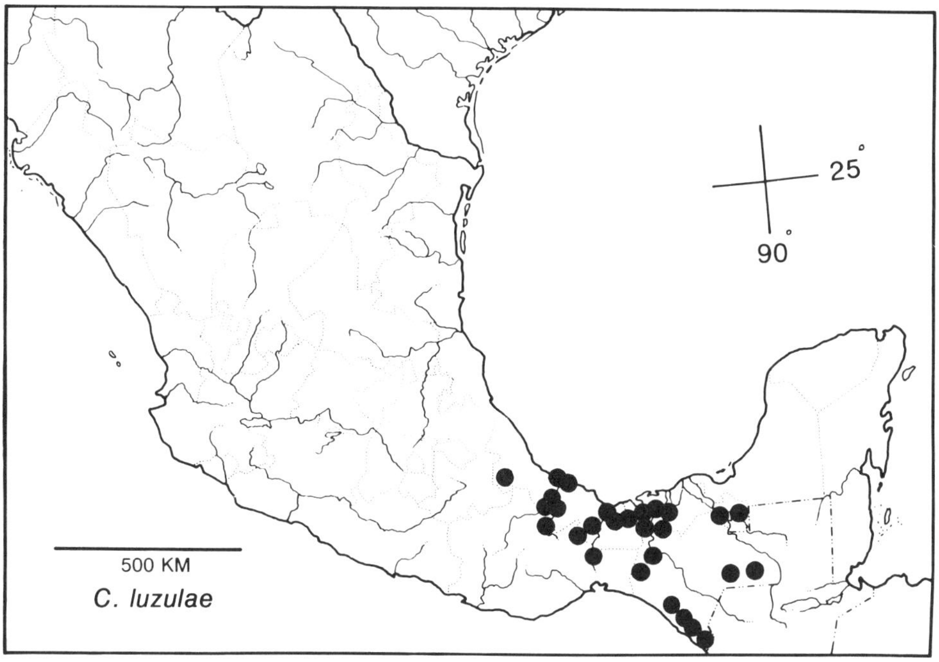

FIG. 12. Distribution of *Cyperus luzulae* in Mexico.

ca. 0.1 mm long, the angles obtuse, the faces flat or slightly convex, the surface finely reticulate with a network of ridges forming cells, brown.

Phenology. Flowering and fruiting from June to September and infrequently during the remaining months.

Distribution. Widespread in the neotropics; from southern Mexico through Central America to Bolivia and east-central Brazil, and from the Virgin Islands through the Lesser Antilles. In Mexico frequent east of the Isthmus of Tehuantepec (Fig. 12); in various moist open situations: stream banks, moist pastures, wet meadows, and roadside ditches; sea level to ca. 1000 (–2000) m.

REPRESENTATIVE SPECIMENS. **Mexico.** CHIAPAS: Mpio. Ostuhuacán, Presa de Mal Paso, 21 Jul 1969, *Arreguín s.n.* (DS, MSC); Palenque, *Beristain 7772* (F, US); Santa Anita entre Huixtla y Tapachula, *Boege 1102* (DS); Mpio. Tenejapa, paraje de Mahosik, *Breedlove 14841* (CAS, F, LL, MICH); Mpio. Ocosingo, 6–8 km N of Ocosingo along rd to Bachajón, *Breedlove 27883* (DS, MICH, NY); Mpio. Arriaga, 15 km NW of Arriaga, *Breedlove 36789* (DS); La Florida between Oxchuc and Ocosingo, *Tucker 2227* (DUKE, ENCB); between Escuintla and El Triunfo, *Tucker 2242* (DUKE, ENCB); 15 km NE of Pichucalco near El Azufre, *Tucker 2586* (DUKE).—OAXACA: SW of Valle Nacional on Hwy 175, *Elliot 359* (WIS); Lacoba at 3000 ft, *Galeotti 5866* (BR, F, G, US); Sierra San Pedro Nolasco, *Jurgensen 633* (G, US); 10 km S of Matías Romero, *King 855* (MICH); Mpio. Tuxtepec, Chiltepec, *Martínez Calderón 100* (CAS, MEXU, MO, NY, SD); Dtto. Choapam, Yaveo, Arroyo de las Culebras, *Mexia 9129* (G, LCU).—TABASCO: ca. 21 mi W of Cárdenas, *Conrad 2958* (GH, MO); ca. 30 km E of Río Tonalá on Hwy 180, *Tucker 2554* (DUKE, ENCB).—VERACRUZ: Mpio. Minatitlán, along Río Tancochoapan just below San José del Carmen, *Gilly & Hernández X. 9* (MICH, MSC); 16 km S of Playa Vicente, *Hill 1725* (A); Canal de la salida de la Laguna de Sontecomapan, *Novelo & Arce 478* (MEXU); 3 km N of Lerdo de Tejada, *Tucker 2502* (DUKE); 20 km E of Coatzacoalcos on Hwy 180, *Tucker 2534* (DUKE, ENCB).

Cyperus luzulae can be readily recognized by its dense whitish heads of spikelets. It has been confused with *C. entrerianus* (no. 9); see comments under that species. Several collections of *C. luzulae* from Mexico and Central America lack rays. *Cyperus guatemalensis* and *C. trinitatis* are based upon such variants, which otherwise are typical representatives of *C. luzulae* and do not merit taxonomic status.

The type of *Cyperus globulosus* Aublet is a plant of *C. luzulae*. The name *C. globulosus* has been misapplied by Kükenthal and Fernald to *C. croceus* Vahl, a species closely related to *C. retrorsus* (no. 88). *Cyperus croceus* occurs in the southeastern U.S.A. (Tucker 1987), the West Indies, Central America (cited as *C. globulosus* in Tucker, 1983b), and northern South America but has never been reliably reported from Mexico (see Doubtful and Excluded Names).

11. Cyperus megalanthus (Kükenthal) G. C. Tucker, comb. nov. *Cyperus pseudovegetus* var. *megalanthus* Kükenthal, Pflanzenreich 4(20): 176. 1936.—Type: Mexico. San Luis Potosí: Las Canoas, banks of streams, 17 Jun 1891, *Pringle 3716* (lectotype, designated by Denton, 1978: B!; isolectotypes: BM! BR! E! G! MSC! MTMG! NY! PH! VT! Z!).

Caespitose perennial, 30–55 cm tall. Rhizomes 5–8 mm in diameter, short, upright, indurate. Culms 1.8–2.5 mm in diameter, trigonous, smooth. Leaves 4–7 (–10), (20–) 40–60 cm long, 4.5–8.0 mm wide, v-shaped, the margins and keel scabrellate. Inflorescence bracts (5–) 10–15, 6–30 (–40) cm long, 5–8 (–10) mm wide, v-shaped, the margins and keel scabrellate, borne horizontally to ascendent at 30° (–45°). Rays 3–6, 1–30 mm long; or rays absent, the inflorescence a dense irregularly lobate head 2.0–3.5 cm wide; secondary rays infrequently present, 2–5 mm long. Heads glomerulate, 7–15 (–18) mm wide, dense, hemispherical to pyramidal, ordinarily proliferous, green, becoming light brown as the achenes mature. Spikelets 60–100, 3.5–5.0 mm long, 3.0–4.0 mm wide, orbiculate-ovate, flattened; rachilla persistent, ca. 0.3 mm wide, 0.1–0.2 mm thick, quadrate, reddish brown, yellow at the nodes, successive scale scars ca. 0.4 mm apart, wings absent. Scales deciduous, (8–) 12–16, 2.0–2.5 mm long, (0.7–) 1.0–1.2 mm wide, oblong-spatulate (widest at or above the middle), apex mucronulate, uniformly light greenish to light brown, laterally 1-nerved, the nerve impressed on the adaxial surface, medially indistinctly 3-nerved, bicarinate basally, apically scabrellate or infrequently smooth. Stamen 1; filament ca. 2 mm long; anther 0.7–1.0 mm long, oblong, the connective apex ca. 0.1–0.2 mm, subulate, light brown. Style 0.5–0.8 mm long, thick, confluent with the apex of and persistent on the achene; stigmas 3, 0.6–1.0 mm long. Achenes 1.2–1.4 mm long, 0.2 (–0.3) mm wide, trigonous, linear and slightly curved, apex acute, base stipitate, 0.1–0.2 mm long, ca. 0.2 mm wide, whitish, spongy, the faces flat, the surface papillose, brown. Fig. 13E–H.

Phenology. Flowering and fruiting from February to July (–August).

Distribution. Eastern Mexico and eastern Guatemala (Fig. 14); uncommon or occasional, stream banks, roadside ditches, and marshy pastures; sea level to ca. 400 m.

Additional Specimens Examined. **Mexico**. Quintana Roo: El Paso, *Lundell 1606* (DS, US).—San Luis Potosí: along the rd to Jalpan ca. 12 mi NE of Xilitla, *King 4382* (MICH); Las Canoas, *Pringle 3716* (B, BM, BR, E, G, MSC, MTMG, NY, PH, VT, Z,).—Tamaulipas: near Villagran, Río Pilón, 3 km E of Est. Garza Valdes, *Rodríguez & Carranza 806* (ARIZ, CHAPA); ca. 20 mi N of Cd. Victoria

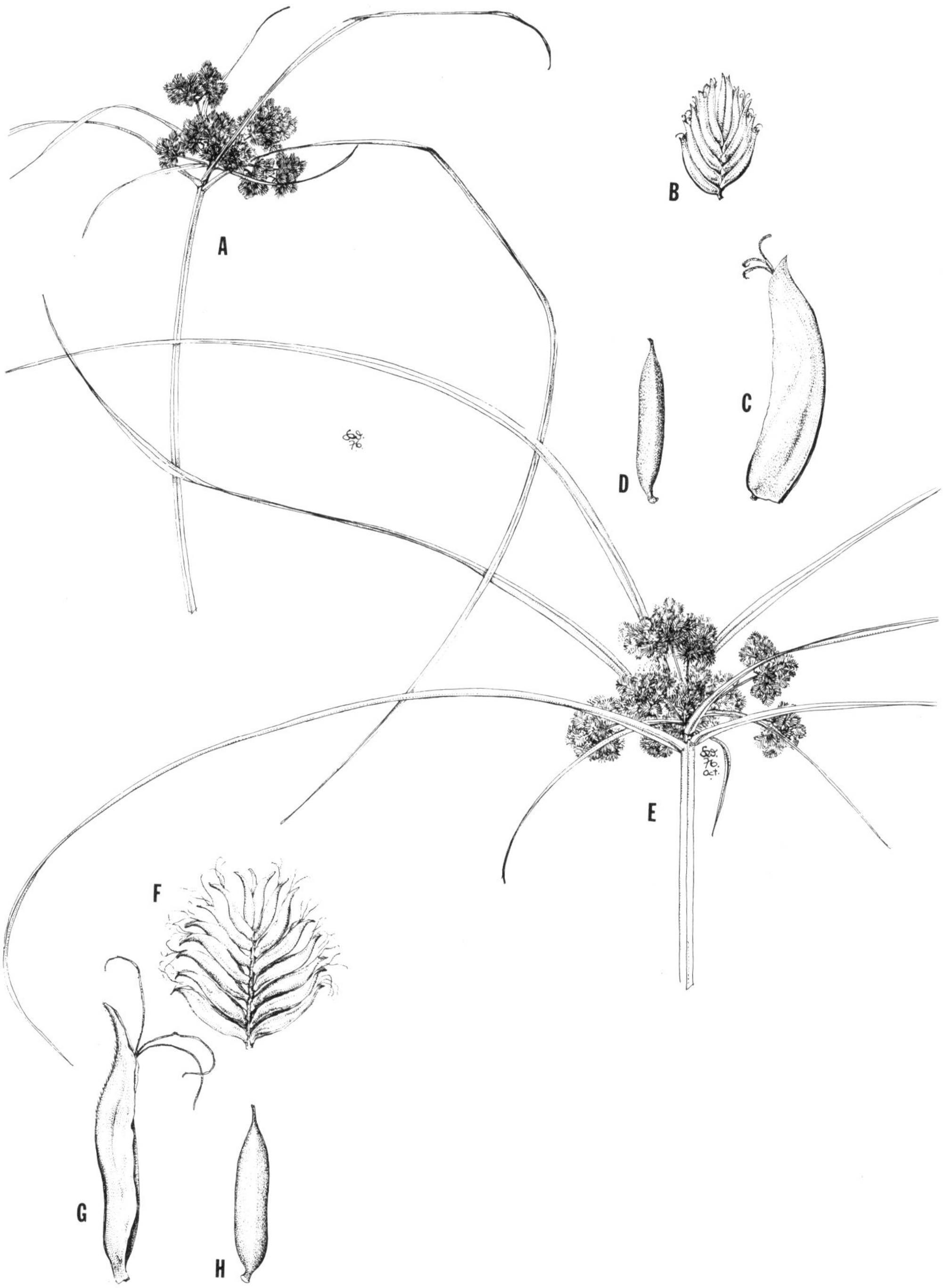

FIG. 13. *Cyperus pseudovegetus* and *C. megalanthus*. A–D, *C. pseudovegetus*. A. Inflorescence (×0.5). B. Spikelet (×5). C. Scale (×17.5). D. Achene (×17.5). E–H, *C. megalanthus*. E. Inflorescence (×0.5). F. Spikelet (×5). G. Scale (×17.5). H. Achene (×17.5). (Based on: A, *Fisher 372*; B–D, *Demaree 3437*; E–H, *Lundell 2658*.) Drawn by Karin Douthit. Reproduced with modification from Denton (1978) with the permission of Melinda F. Denton and the University of Michigan Herbarium.

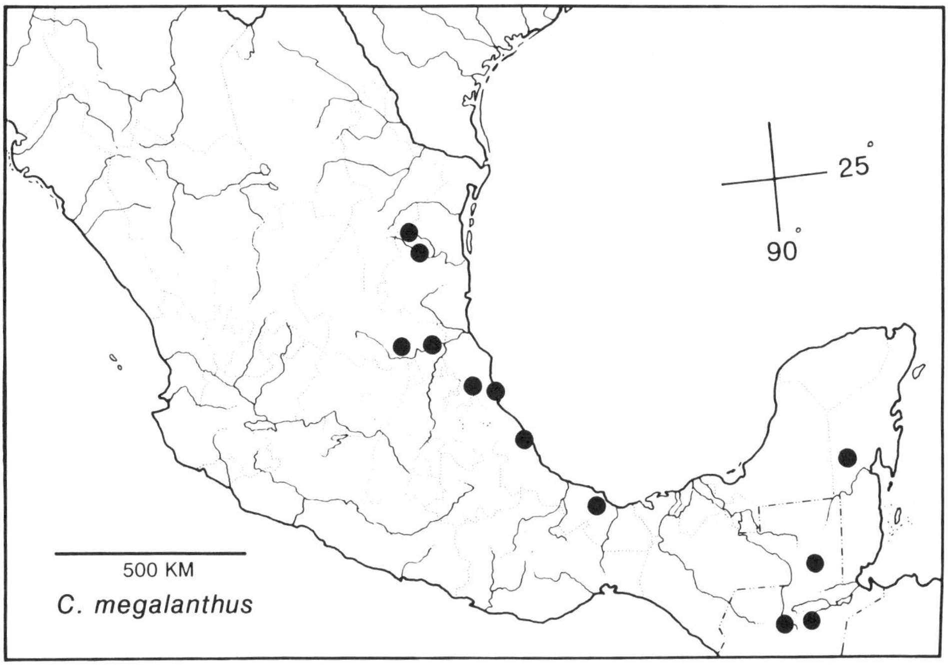

FIG. 14. Distribution of *Cyperus megalanthus*.

near Ejido Cruciata, *Tucker 1953* (CAS, DUKE, ENCB).—VERACRUZ: Mpio. Cosamaloapan, Los Dobles, *Hernández & Trios 1078* (F); Colipa, *Liebmann 14434* (C); Sanborn, *Orcutt 3243* (BM); ca. 6 mi S of Tamiahua on the rd to Tuxpan, *Tucker 2062* (DUKE); ca. 2–4 km W of San Sebastián on Hwy 127, *Tucker 2446* (CAS, DUKE, ENCB, MICH).

The combination of pyramidal, greenish heads, orbiculate-ovate, flattened spikelets, and linear achenes distinguishes *C. megalanthus* from other species in Mexico and Central America. The inflorescences of this species are usually proliferous, with one or more of the heads bearing leafy vegetative shoots. As the inflorescences mature, the culms bend downward into the moist or muddy soils in which this species grows. The vegetative shoots thus take root, increasing the number of plants in the population. Proliferous inflorescences are a character shared with *C. luzulae*.

Denton (1978), examining herbarium material only, noted the narrower achenes and greater number of inflorescence bracts of *C. megalanthus* but treated this taxon as a variety of *C. pseudovegetus* (Fig. 13A–D), a North American species ranging from New Jersey to Missouri south to Florida and eastern Texas. In the field, the two species do not, however, greatly resemble each other; *C. pseudovegetus*, with its hemispherical heads, resembles *C. entrerianus* and *C. surinamensis* more than *C. megalanthus*. Morphological differences (including those noted by Denton) of the allopatric *C. megalanthus* and *C. pseudovegetus* are summarized below.

1. Inflorescence bracts (5–) 10–15, 5–8 (–10) mm wide; secondary rays infrequently present, up to 5 mm long; heads ordinarily proliferous; spikelets orbiculate-ovate, 3.0–4.0 mm wide; scales 2.0–2.5 mm long, (0.7–) 1.0–1.2 mm wide, the apex scabrellate (or smooth), laterally firm, glossy; achenes linear, ca. 1/2 as long as scale. *C. megalanthus*.
1. Inflorescence bracts (3–) 4–8, 3–5 mm wide; secondary rays frequently present, up to 15 mm long; heads not proliferous; spikelets broadly ellipsoid, 1.5–3.0 mm wide; scales 1.5–2.2 mm long, 0.5–1.0 mm wide, the apex smooth, laterally membranous, reticulate; achenes narrowly ellipsoid, ca. 3/4 as long as scale. *C. pseudovegetus*.

12. Cyperus laxus Lamarck, Tabl. encycl. 1: 146. 1791.—TYPE: FRENCH GUIANA. *Stoupy s.n.* (lectotype, designated by Raynal, 1978: P-LA, IDC microfiche!; isolectotype: C-Herb. Vahl!).

Cyperus tolucensis H. B. K., Nov. gen. sp. 1 [quarto]: 206. 1816. *Cyperus diffusus* var. *tolucensis* (H. B. K.), Kükenthal, Pflanzenreich 4(20): 211. 1936. *Cyperus diffusus* subsp. *tolucensis* (H. B. K.) A. Borhidi, Acta Bot. Acad. Hung. 25(1–2): 1. 1979.—TYPE: MEXICO. "Nevado de Toluca," *Humboldt & Bonpland s.n.* (holotype: P?).

Cyperus umbrosus Lindley & Nees ex Nees in Martius, Fl. bras. 2(1): 31. 1842. *Cyperus diffusus* var. *umbrosus* (Nees) Kükenthal, Pflanzenreich 4(20): 211. 1936.—TYPE: BRAZIL. Bahia, unknown collector (lectotype, here designated: CGE-Herb. Lindley!).

Caespitose perennial, 18–55 cm tall. Rhizomes 3–10 mm in diameter, upright or oblique, indurate. Culms 1.4–2.5 mm in diameter, trigonous, smooth or infrequently very sparsely scabrellate on the angles just below the bracts. Leaves 5–10, 7–25 (–40) cm long, 3–10 mm wide, v-shaped or inversely w-shaped, the margins and larger nerves on both surfaces scabrellate. Inflorescence bracts 6–12, 3–23 cm long, 2–11 mm wide, v-shaped or inversely w-shaped, the margins and veins scabrellate, borne ascendent at 30–60° (–75°). Rays 6–13, 1–17 cm long; secondary rays 1–3 (–4), 5–40 mm long; tertiary rays frequently present, 2–6 mm long. Heads digitate, 6–30 mm wide. Spikelets 1–5, 8–17 mm long, 2.5–3.6 mm wide (including the cusps of the scales), oblong-elliptic to linear-oblong, compressed; rachilla persistent, 0.4–0.6 mm wide, (0.1–) 0.2 mm wide, quadrate, rectangular, greenish to yellowish green, purplish at the nodes, successive scale scars ca. 0.1 mm high and 1.3–1.5 mm apart, the wings very narrow, hyaline, ca. 0.1 mm wide or less. Scales deciduous, 8–20, 1.6–2.4 mm long, the cusp an additional (0.4–) 0.6–1.0 mm long, 1.6–2.6 mm wide, ovate-orbiculate (widest at the middle), laterally 2–4-nerved, hyaline, stramineous, or reddish brown, medially 3–5-nerved, green, smooth or more often scabrellate, especially apically. Stamens 3; filaments 1.0–1.4 mm long; anthers 0.7–1.0 mm long, oblong-linear, the connective apex 0.1 mm long, reddish brown, setose. Style 0.1–0.2 mm long; stigmas 3, 1–2 mm long, persistent on the achene. Achenes 1.4–1.6 mm long, 1.0–1.1 (–1.3) mm wide, trigonous, obovoid to ovoid, apex obtuse to apiculate, base rounded, the stipitate portion 0.1–0.2 mm long, 0.3–0.4 mm wide, the faces concave, the surface puncticulate, stramineous while immature, becoming smooth, glossy, and brown at maturity.

Phenology. Flowering and fruiting throughout the year but primarily from June until December.

Distribution. Neotropical regions; Cuba and Mexico south to Paraguay and Bolivia. In Mexico (Fig. 15) from northern Veracruz, Morelos, and Nayarit south

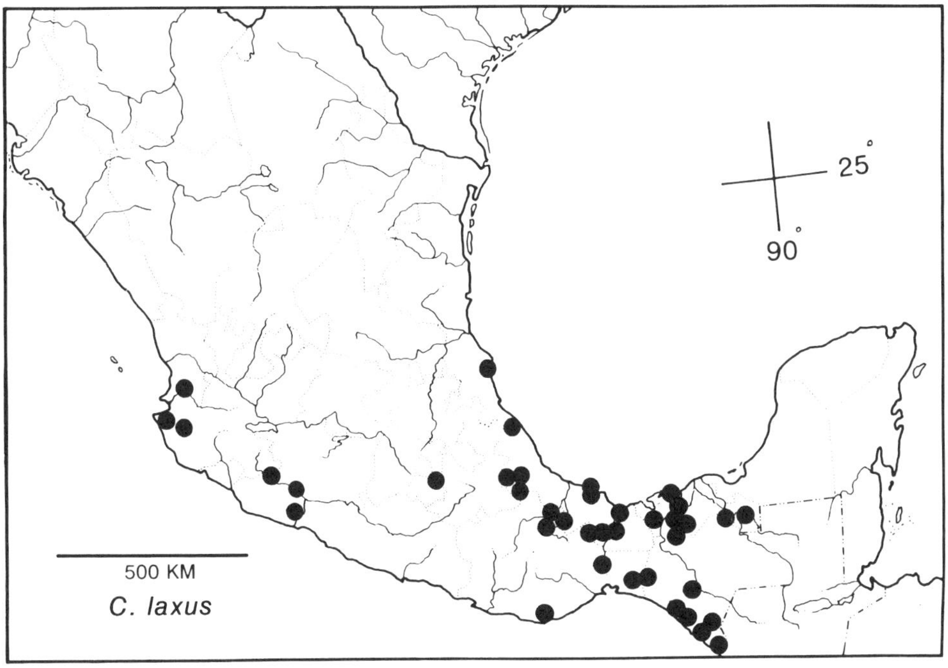

FIG. 15. Distribution of *Cyperus laxus* in Mexico.

to Chiapas, absent from most of the Yucatan Peninsula although present in Belize; in moist, shaded forests, both primary and secondary, coffee groves, along moist shaded roadsides, riverbanks, and in croplands; sea level to 1500 m.

REPRESENTATIVE SPECIMENS. **Mexico.** CHIAPAS: Tapachula, *Fisher 35278* (ARIZ, F, LCU, NY, US); Escuintla, *Matuda 301* (BM, LCU, MICH, NY, US); ca. 15 km NE of Pichucalco near El Azufre, *Tucker 2583* (DUKE, ENCB); 1 km E of Ixtacomitán, *Tucker 2608* (CAS, DUKE, ENCB); ca. 25 km N of Huixtla on Hwy 211, *Tucker 2685* (DUKE, ENCB).—GUERERRO: Dtto. Mina, *Hinton 10792* (DS, GH, LCU, MICH, NY); Dtto. Galeana, Plan del Carrizo, *Hinton 14645* (DS, GH, LCU, MICH, NY, US).—JALISCO: 2 mi N of La Cuesta, rd to Talpa Allende, *McVaugh 21133* (G, MICH, NY, TEX, US); 10–13 km SE of El Tuito, Mpio. de Cabo Corrientes, *McVaugh 25375* (MICH).—MICHOACÁN: environs de Tibor, *Langlassé 302* (G).—MORELOS: NE of Cuernavaca, *Ramírez 102* (NY).—NAYARIT: Mpio. Valle de Banderas, 1 km N of El Cuatante, *Rzedowski 17873* (LL, MICH, MSC).—OAXACA: Trapiche de la Concepción, *Liebmann 14427* (C, HAL); Hwy 175, between Pochutla and Miahuatlán, 1.5 km S of Puente San Juan, *Tucker 2251* (DUKE, ENCB).—TABASCO: ca. 15 km S of Villahermosa, *Tucker 2570* (DUKE, ENCB); 40 km S of Villahermosa, *Tucker 2574* (DUKE, ENCB).—VERACRUZ: Córdoba, *Fisher 35348* (CAS, F, TEX, US); Mpio. Jesús Carranza, above Jesús Carranza, *Gilly & Hernández 3* (MSC, TEX); Est. Biol. Trop. "Los Tuxtlas" 33.5 km NE of Catemaco, *Iltis 27252* (WIS); Laguna de Tamiahua, *LeSueur 14* (ARIZ, F, GH, TEX); 15 km E of Coatzacoalcos, *Tucker 2532* (DUKE, ENCB).

Cyperus laxus can be separated from all other Mesoamerican species of *Cyperus* with digitate spikelets by its very wide scales (1.6–2.6 mm wide). It is the only species of the region in which the body of the scale (excluding the cusp) is always wider than long. The cuspidate scales, the obovoid to ovoid achenes with concave sides, and the minute style help to distinguish *C. laxus* from other New World species of subg. *Pychnostachys*.

Kükenthal (1935–36) and O'Neill (1940) included *C. laxus* in *C. diffusus* Vahl, an Old World species. The name *C. chalaranthus* J. Presl & K. Presl has been applied to *C. laxus* (Tucker 1983), although the two species differ in scale shape (*C. chalaranthus* longer than wide, *C. laxus* wider than long). Also, *C. chalaranthus* is restricted to middle elevations of western South America (Colombia to Paraguay).

13. Cyperus matudae G. C. Tucker, Rhodora 88: 503. 1986.—TYPE: MEXICO. Chiapas: Mt. Ovando, pine land, 1250–2370 m, Jul 1938, *Matuda 2553* (holotype: MICH!; isotypes: F! GH! LL! NY! VT!).

Perennial, 50–75 cm tall. Rhizomes 5–12 mm in diameter, oblique, indurate, clothed with fibrous leaf bases. Culms 1.3–2.5 mm in diameter, triquetrous, scabrellate on the angles only, especially distally. Leaves 3–10, 40–70 cm long, 5–15 mm wide, flat to slightly inversely w-shaped, the margins and keels scabrellate. Inflorescence bracts 6–8, 3–30 cm long, 2–10 mm wide, inversely w-shaped, erect to ascendent at (60–) 75°. Rays 10–11, (1–) 3–15 cm long; secondary rays 10–40 mm long; tertiary rays 3–15 mm long. Heads digitate, 5–10 mm wide. Spikelets 1–3, 7–15 mm long, 2.3–3.5 mm wide (including the apices of the scales), linear-lanceolate, slightly compressed; rachilla persistent, 0.4–0.6 mm wide, 0.1–0.2 mm thick, slightly flexuous, greenish to purplish brown, successive scale scars (1.6–) 1.8–2.1 mm apart, wings absent. Scales 8–16, 2.2–2.4 mm long, the excurved cusp 0.4–0.8 mm long, 1.4–1.6 mm wide, oblong-elliptic, laterally 3–4-nerved, reddish to brownish green, medially 3–7-nerved, the midvein green, smooth or scabrellate distally. Stamens 3; filaments (2.0–) 2.8–3.8 mm long; anthers 0.9–1.2 mm long, linear-oblong, the connective apex 0.1–0.2 mm long, reddish brown, setose. Style 0.6–0.8 mm long; stigmas 3, (1.2–) 1.6–2.6 mm long. Achenes 1.8–2.0 mm long, ca. 1.2 mm wide, trigonous, ellipsoid, apex apiculate, base sessile, the faces slightly convex, the surface puncticulate, brown. Fig. 1A–C, G, H.

Phenology. Plants with mature achenes have been collected in July and August.

Distribution. Chiapas, Mexico; known from only two collections from near the Guatemalan border.

ADDITIONAL SPECIMENS EXAMINED. **Mexico**. CHIAPAS: Escuintla, Finca Fuárez, 12 Aug 1937, *Matuda 1937* (GH, MICH).

Cyperus matudae finds its place in subg. *Pycnostachys* because of its trigonous achenes and digitately arranged spikelets. Both collections of this species had been identified previously as "*C. diffusus* Vahl" [=*C. laxus*], a widespread neotropical species. *Cyperus matudae* differs strikingly from *C. laxus* in features of the achenes and scales (see key to species) and in its more highly branched inflorescences. *Cyperus laxus* has only primary and secondary rays, whereas all available specimens of *C. matudae* have tertiary rays as well.

14. Cyperus breedlovei G. C. Tucker, Rhodora 88: 505. 1986.—TYPE: MEXICO. Chiapas: on Hwy 211 ca. 25 km N of Huixtla, 28 Jul 1984, *Tucker & Schwartz 2686* (holotype: DUKE!; isotypes: BH! CAS! ENCB! F! GH! K! MEXU! MICH! MO! NY! US! WIS!).

Perennial, 25–50 cm tall. Rhizomes ca. 2 mm in diameter, indurate. Culms 0.8–1.2 mm in diameter, trigonous and smooth below, triquetrous and scabrellate or smooth above. Leaves 1–5, 10–40 cm long, 2–3.5 mm wide, v-shaped, the margins and keels scabrellate. Inflorescence bracts 3–5, 2–12 cm long, 1–2.5 mm wide, v-shaped, the margins and keels scabrellate, the longest erect or strongly ascendent, appearing as a continuation of the culm, the others ascendent. Rays 2–8, 0.5–4 cm long, scabrellate; secondary rays absent or present, 0.3–1 cm long. Heads digitate, 8–27 mm wide. Spikelets 1–5 (–7), 4–15 mm long, 1.8–2.4 mm wide, linear-oblong, compressed; rachilla persistent, 0.3–0.4 mm wide, 0.1–0.2 mm thick, straight, purplish brown, successive scales 1.0–1.2 mm apart, wings absent. Scales deciduous, 7–16, 1.6–2.1 mm long (excluding the mucronate apex of 0.2–0.4 mm), 1.0–1.2 mm wide, oblong-lanceolate, laterally 2–3-nerved, reddish brown, the margins sometimes sparsely ciliate-scabrellate, medially 3–5-nerved, green, the midvein scabrellate. Stamens 3; filaments 1.4–1.7 mm long, anthers ca. 1.0 mm long, narrowly ellipsoid, the connective apex ca. 0.2 mm long, reddish brown, setose. Style 0.2–0.4 (–0.6) mm long; stigmas 3, 2.3–2.5 mm long. Achenes 1.2–1.4 mm long, ca. 0.6 mm wide, trigonous, ellipsoid, apex apiculate, base cuneate to substipitate, the faces slightly convex, the surface puncticulate, brown.

Phenology. Flowering and fruiting in July and October.

Distribution. Chiapas, Mexico, known from only three collections at the type locality; in crevices of steep moist ledges of granodiorite in moist tropical forest; ca. 700 m.

ADDITIONAL SPECIMENS EXAMINED. **Mexico**. CHIAPAS: Mpio. Motozintla, 25–27 km NE of Huixtla on rd to Motozintla [Hwy 211], 7 Oct 1972, *Breedlove 28603* (DS); 15 mi N of Huixtla on Rte 211, 28 Jul 1984, *Wilbur 35843* (DUKE).

Cyperus breedlovei is distinguished from all other Mexican and Central American species of subg. *Pycnostachys* by the combination of linear-oblong spikelets, reddish brown, mucronate scales, and ellipsoid, apiculate achenes. *Cyperus breedlovei* shares phenetic similarities with *C. chalaranthus* of South America. Both species have oblong-lanceolate scales nerved nearly to the margins, scales with straight mucronate apices, and ellipsoid achene. *Cyperus breedlovei* differs from *C. chalaranthus* in having the longest inflorescence bract erect to ascendent (at 60°), and the scales reddish brown and scabrellate along the midvein; in *C. chalaranthus,* the bracts are horizontal to ascendent at 30°, and the scales light green and smooth.

15. Cyperus simplex H. B. K., Nov. gen. sp. 1 [quarto]: 207. 1816.—TYPE: COLOMBIA. [Tolima: Ibague] "Ibagna," *Humboldt & Bonpland s.n.* (holotype: P?; isotype: B-W no. 1301, IDC microfiche!).

Caespitose perennial, 7–20 cm tall. Rhizomes (seldom collected) 1–3 mm in diameter, indurate, producing tufts of 1–10 culms at intervals of 1–3 cm. Culms 0.5–1.2 mm in diameter, trigonous, smooth. Leaves 1–10, 10–40 cm long, 2.2–5 mm wide, v-shaped to inversely w-shaped, the margins and keel scabrellate. Inflorescence bracts 5–10, 1–35 cm long, 1.5–7.0 mm wide, v-shaped, the margins and upper and lower surfaces scabrellate, ascendent at 45–60°; secondary bracts 2–

6 cm long, 1–2 mm wide. Rays 3–10, 1–30 cm long, the longest equalling or exceeding the culm in length, scabrellate or rarely smooth on the angles; secondary rays infrequently present, 0.5–7.0 cm long, scabrellate. Heads digitate, 7–32 mm wide. Spikelets 1–5 (–10), 8–35 mm long, (1.5–) 2.0–2.3 (–2.5) mm wide, linear, ellipsoid in cross section; rachilla persistent, 0.7–0.9 mm wide, 0.3–0.5 mm thick, zigzag, greenish, stramineous or reddish brown, successive scale scars 1.0–1.6 mm apart, wings absent. Scales deciduous, (6–) 20–36 (–56), 2.0–2.6 mm long, 1.2–2.0 mm wide, ovate-lanceolate, the apical portion 0.6–0.8 mm long and incurved parallel to the rachilla, laterally weakly 1–4-nerved, whitish or translucent, often somewhat reddish or greenish tinged, the margins especially so apically and often the entire abaxial scale surface, scabrellate, the medial part of the scale nerveless or faintly 1–3-nerved. Stamen 1; filament (1.6–) 2.0–2.6 mm long; anther 0.6–0.8 mm long, oblong, the connective apex not prolonged. Style 0.7–1.3 mm long; stigmas 3, 0.5–0.9 mm long. Achenes 1.0–1.2 mm long, 0.8 (–0.9) mm wide, trigonous, obovoid, the apex obtuse or emarginate, apex apiculate, base broadly rounded to cuneate, the faces barely concave, flat or slightly convex especially apically, the surface divided by fine grooves into rather regular hexagonal isodiametric sections, each papillate in its center, white or sometimes brownish or reddish brown, the color differing even among adjacent achenes in a spikelet.

Phenology. Flowering and fruiting from (March–) June to October.

Distribution. Trinidad, southern Mexico south to Bolivia and southern Brazil. In Mexico known only from Oaxaca and Chiapas; riverbanks, trailsides, and moist shaded soils of lowland forests, north of Panama uncommon and local; sea level to 1000 m. Recorded as an epiphyte [e.g., *Hill 1662* (A, NY, VT) from Oaxaca] by Kükenthal (1935–36) and Tucker (1983).

Representative Specimens. **Mexico.** Chiapas: Mpio. Solosuchiapa, 3–5 km above Solosuchiapa, *Breedlove 26442* (DS, DUKE); Escuintla, *Matuda 305* (F, GH, LCU, MICH, NY, UC, US); Mt. Madre Vieja, *Matuda 2518* (GH, LCU, MICH).—Oaxaca: Mpio. Juquila, 5 km N of San Gabriel Mixtepec, *Hill 1662* (A, NY, VT); Dtto. Choapam, Yaveo, *Mexia 9201* (ARIZ, B, BH, GH, LL, MICH, MO, NY, UC, US); between Pochutla and Miahuatlán, 1.5 km S of Puente San Juan, *Tucker 2255* (CAS, DUKE, ENCB, MICH).

Cyperus simplex is the only neotropical species in which the longest rays of the inflorescence equal or exceed the length of the culm. Kükenthal (1935–36), Svenson (1943), and Benedict Ayers (1946) stated that the rays are longer than the culms, but this is misleading. In *C. simplex*, as in nearly all species of the genus, the rays vary considerably in length within any one inflorescence. As noted in the description, it is the longest ray of an inflorescence that equals or exceeds the culm in length. Often the longest rays are 2–4 times longer than the culm. Although the longest ray within an inflorescence is very long, the shortest ray is nearly always less than 1 cm long, so that its spikelets appear sessile, nestled among the base of the inflorescence bracts.

In about half of the specimens examined from Mexico and Central America, adventitious roots occur at the base of the spikelets. Presumably, as the inflorescences mature and the rays bend downward to the soil, the plant forms a root and shoot at the junction of the ray and the spikelet. Kunth noted this, remarking in the original description, "Variat rarissime spicis basi viviparis." Several other Mexican species of *Cyperus* have also been noted to form viviparous propagules in the spikelets, viz., *C. dioicus*, *C. megalanthus*, *C. luzulae*, *C. tenerrimus*, and *C. elegans*.

16. Cyperus lundellii O'Neill, Carnegie Publ. Bot. 522: 276. 1940.—TYPE: GUATEMALA. Petén: Subin River, Monte Santa Teresa, 12 Apr 1933, *Lundell 2657* (holotype: MICH!; isotypes: CAS! LCU!).

Perennial, 40–100 cm tall. Rhizomes 1–3 mm in diameter, short, indurate. Culms 1.7–4.0 mm in diameter, trigonous to triquetrous, smooth or sparsely scabrellate on the angles. Leaves 3–8, 15–60 (–90) cm long, 3–6 mm wide, v-shaped, the margins and keel scabrellate. Inflorescence bracts 3–5, 4–45 cm long, 2–5 mm wide, v-shaped, the margins and keel scabrellate, erect to ascendent at ca. 45°; secondary bracts lanceolate, 2–4 mm long. Rays 5–7, 1–6 cm long; secondary rays 5–8, 0.2–2 cm long. Heads digitate, 4–15 mm wide. Spikelets (1–) 3–6, 6–12 mm long, 1.5–2 mm wide, linear-oblong to linear-lanceolate, subcylindric at anthesis, becoming flatly elliptic in cross section as the achenes mature; rachilla persistent, ca. 0.4 mm wide, 0.2 mm thick, rectangular, slightly geniculate, stramineous, the narrow reddish wings ca. 0.1 mm wide. Scales deciduous, (12–) 16–36, 1.5–1.8 (–2) mm long, 1.4–1.7 mm wide, ovate to suborbiculate, widest near the middle, apex mucronate, laterally nerveless, light reddish brown with hyaline edges especially near the apex, medially weakly 3-nerved, distinctly green, sharply contrasting with the sides of the scales, overlapping but loosely imbricate. Stamens 3; filaments 1.0–1.2 mm long, the bases forming a white hypogynous ring; anthers 0.4–0.5 mm long, ellipsoid, the connective apex ca. 0.1 mm long, obtuse, reddish. Style 0.5–1 mm long; stigmas 3, 0.5–1 mm long. Achenes ca. 1.0 mm long, (0.5–) 0.6 mm wide, trigonous but dorsiventrally flattened, ellipsoid, apex acute and slightly apiculate, base slightly to distinctly stipitate, whitish, the faces convex, the surface coarsely papillose or granular, reddish brown.

Phenology. Flowering and fruiting from January to March.

Distribution. Southeastern Tamaulipas and southeastern Chiapas, Mexico, to northern Guatemala and Belize, also in Venezuela; riverbanks and waterholes in lowland forest; sea level to 1400 m.

ADDITIONAL SPECIMENS EXAMINED. **Mexico**. CHIAPAS: Mpio. La Trinitaria, along stream canal at ruins of Ch'encultik, near Monte Bello National Park, 1400 m, 22 Jan 1973, *Breedlove & Smith 32092* (DS).—TAMAULIPAS: River at Altamira, 6 Mar 1939, *LeSueur 39* (ARIZ, GH, TEX).

Cyperus lundellii can be distinguished from all others of the genus in Mexico and Central America by its digitate spikelets, broadly rounded scale apices, and granular, dorsiventrally flattened achenes. *Cyperus haspan* and *C. simplex* also have granular achenes, but theirs are not flattened.

O'Neill (1940) believed *C. lundellii* to be most closely related to *C. miliifolius* Poepp. & Kunth, a species known from Costa Rica and northern South America (Tucker 1983). Observation of the specimens cited here, and four others from Guatemala, leads me to agree with O'Neill's placement. The dark spikelets and the granular to papillose achenes confirm a phenetic similarity between *C. miliifolius* and related South American species (sect. *Incurvi* as delimited by Kükenthal, 1935–36); however, *C. lundellii* differs from *C. miliifolius* in having closely imbricate scales and achenes that are slightly dorsiventrally compressed. Also, *C. lundellii* lacks long acuminate scale apices, an apparently specialized character that marks *C. miliifolius* and its close relatives. Satisfactory appraisal of the relationships of *C. lundellii* must await further study of *C. miliifolius* and its South American relatives.

Cyperus lundellii, described from Guatemala, is here reported from Mexico for the first time. It also occurs in Venezuela: Est. Carabobo, 6 km S of Valenzia, *Koyama 13654* (NY); Est. Portuguesa, 14 km NE of Guanare, *Aymard 4990* (MO).

17. Cyperus haspan L., Sp. pl. 1: 45. 1753.—TYPE: "India." Hermann Herbarium [Fl. Zeyl. 37] (lectotype: BM, fide Trimen, 1888, and McGivney, 1938).

Cyperus juncoides Lamarck, Tabl. encycl. 1: 147. 1791. *Cyperus haspan* subsp. *juncoides* (Lamarck) Kükenthal, Repert. Spec. Nov. Regni Veg. 23: 184. 1926.—TYPE: "North America," collector unknown (holotype: P-LA, IDC microfiche!).

Cyperus aphyllus Vahl, Eclog. amer. 2: 6. 1798.—TYPE: TRINIDAD. *Von Rohr s.n.* (holotype: C!).

Cyperus autumnalis Vahl, Enum. pl. 2: 318. 1805. *Cyperus haspan* var. *americanus* Boeckeler, Linnaea 35: 575. 1886.—TYPE: U.S.A. "Carolina," *Bosc s.n.* (holotype: C!).

Cyperus efoliatus Boeckeler, Allg. Bot. Zeitschr. Syst. 1: 226. 1895.—TYPE: COSTA RICA. Cartago: Turrialba, Sep 1893, *Tonduz 8304* (holotype: B; isotype: US!).

Perennial, (2–) 25–60 (–100) cm tall. Rhizomes 2–6 mm in diameter, soft or indurate, producing culms 1–4 mm apart. Culms 1–4 mm in diameter, trigonous (soft, flattened in drying), smooth. Leaves bladeless, represented by 1–4 reddish sheaths at the base of the culm; blades occasionally present, (3–) 10–30 cm long, (1–) 2.5–5 mm wide, flat to v-shaped, smooth. Inflorescence bracts 2 (–3), (0.5–) 6–18 cm long, 1–4 mm wide, flat to slightly v-shaped, smooth, horizontal to ascendent at 30–60°. Rays (5–) 10–12 (–15), 1–15 (–20) cm long; secondary rays usually present, (1–) 6–12 (–25) mm long; tertiary rays sometimes present, 1–6 mm long. Heads digitate, 4–15 (–30) mm wide. Spikelets 1–15, 3–18 mm long, 1.0–1.6 mm wide, linear-lanceolate, subquadrate-compressed; rachilla persistent, 0.3–0.4 mm wide, 0.2–0.3 mm thick, straight, quadrate, stramineous with purplish sides, successive scale scars 0.7–1.2 mm apart, wings absent. Scales deciduous, 10–20 (–40), (1.0–) 1.4–1.9 mm long, 0.8–1.1 mm wide, oblong to obovate, apex mucronulate, laterally 1-nerved, reddish to greenish brown, dull, often hyaline-edged, medially 1-nerved, greenish, smooth except for a cluster of crystalline prickles at the apex. Stamens 3; filaments (0.6–) 1.5–2.0 mm long; anthers 0.3–0.5 mm long, ellipsoid, the connective apex not prolonged but often bearing a stellate tuft of trichomes. Style 0.4–0.9 mm long; stigmas 3, 0.5–0.8 (–1.3) mm long. Achenes 0.5–0.6 (–0.7) mm long, 0.3–0.4 (–0.5) mm wide, trigonous to subterete, globose to obovoid or broadly ellipsoid, apex obtuse to acute, apiculate, or entire, base stipitate to subcuneate, ca. 0.1 mm long, 0.1–0.2 mm wide, the angles rounded, the faces convex, the surface granular to papillose, white or reddish brown.

Phenology. Flowering and fruiting throughout the year, but primarily from September to January.

Distribution. Pantropic to warm temperate regions; in the New World in the

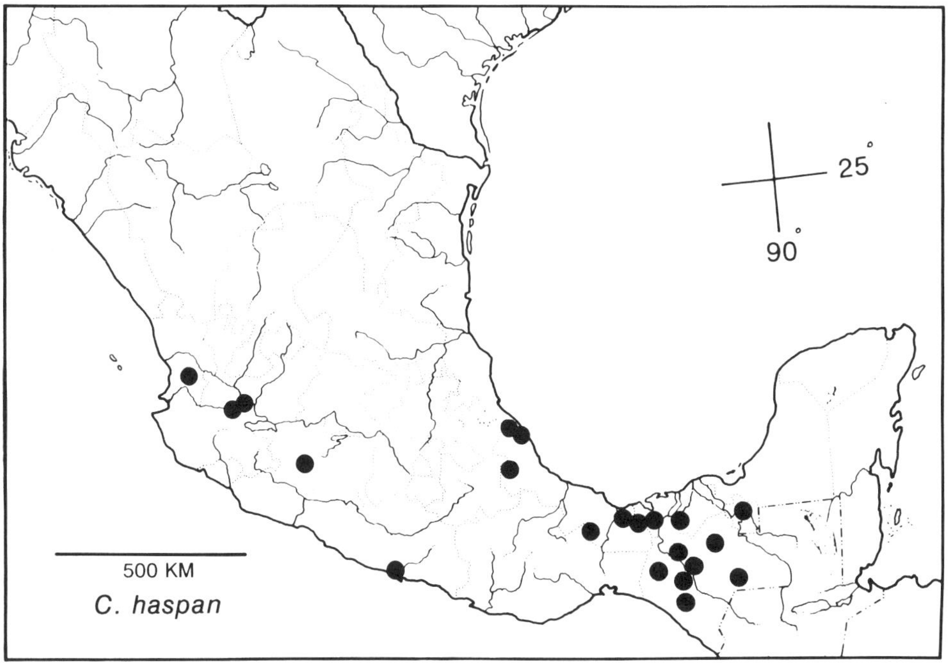

FIG. 16. Distribution of *Cyperus haspan* in Mexico.

West Indies and from Virginia to Texas, U.S.A., south through Mexico and Central America to northern Argentina. In Mexico (Fig. 16) from Veracruz and Nayarit to Tabasco and Chiapas; uncommon, in wet, muddy, often disturbed soils and shallow water; sea level to 1900 m.

REPRESENTATIVE SPECIMENS. **Mexico.** CHIAPAS: Mpio. La Trinitaria, Lagunas de Monte Bello Nat. Park, *Breedlove 37025* (DS, DUKE); Mpio. Ocozocoautla de Espinosa, 14 km N of Ocozocoautla, *Breedlove 38256* (DS).—GUERRERO: vicinity of Acapulco, *Palmer 118* in 1894 (E, F, GH, NY, UC).—JALISCO: near Guadalajara, *Pringle 3283* (AC, ARIZ, CAS, F, G, GH, M, MICH, MSC, MTMG, NY, UC, US, VT, WRSL, Z).—MICHOACÁN: 10 km S of Ario de Rosales, *McVaugh 22508* (MICH).—NAYARIT: 4–5 mi N of Compostela, *McVaugh 19327* (G, MICH, TEX).—TABASCO: 20 km N of Balancán, *Rzedowski 30122* (ENCB, MICH, SD).—VERACRUZ: Mirador, *Liebmann 14428* (C, G, UC); 15 km E of Coatzacoalcos, *Tucker 2530* (DUKE, ENCB); 10 km N of the turnoff to Riachuelos on Hwy 180, *Tucker 2471* (DUKE).

Cyperus haspan can be distinguished from other Mexican species with digitate spikelets by its globose-ovoid, whitish granular achenes and its usually leafless culms. This species exhibits considerable variation in the length of the inflorescence rays, in the presence and absence of leaf blades, and in the shape and color of the achenes.

Plants with reddish brown achenes account for about half of the collections of this species seen from Chiapas, whereas only one other collection from elsewhere in Mexico has such achenes. Also, several collections from Chiapas have ellipsoid rather than ovoid achenes. The color and achene shape are not correlated; numerous intermediates in achene shape between the typical ovoid to subglobose form and the ellipsoid form occur. Apparently, the Chiapas plants simply represent extremes in the shape of the achene not previously noted and do not merit taxonomic recognition.

Kern (1974) suggested that the spelling of the epithet "haspan" ought to be treated as an orthographic error and corrected to "halpan." Wilson (1991) has discussed this and believes that the spelling "haspan" should be retained. She noted that Linnaeus in his *Flora Zeylanica* (1747) cited Hermann in J. Burman's *Thesaurus Zeylanicus* (1737), in which the spelling is "haspan," and also Hermann's *Musaeum Zeylanicum* (1726), in which the spelling is "halpan." Linnaeus in his *Species plantarum* (1753) made a deliberate choice of "haspan," and this selection should be followed.

18. Cyperus difformis L., Cent. pl. 2: 6. 1756.—TYPE: "India." Collector unknown (holotype: LINN-70.10, IDC microfiche!).

Cyperus lateriflorus Torrey, Rep. U.S. Mex. Bound. 2: 226. 1859.—TYPE: MEXICO. Sonora: along mountain torrents E of Santa Cruz, 1851–52, *Wright 1950* (holotype: NY!; isotypes: GH! YU!).

Caespitose annual, 7–30 cm tall. Rhizomes absent. Culms 1.2–2.5 mm in diameter, trigonous (flattened in pressing), smooth. Leaves 2–7, (2–) 7–22 cm long, 2.2–4 mm wide, flat, the margins and keel sparingly scabrellate. Inflorescence bracts 2–4, 1–22 cm long, 0.5–3.5 mm wide, flat, the margins and keel scabrellate, the longest bract erect or nearly so, appearing as a continuation of the culm, the other bracts horizontal to ascendent. Rays 1–5, 2–32 mm long, when short, the heads sessile or nearly so and the inflorescence a dense irregularly lobate head, 12–35 mm wide; secondary rays absent. Heads glomerulate, 7–17 mm wide, dense. Spikelets 30–120, (2–) 3–5 (–6) mm long, 0.8–1.2 mm wide, oblong-elliptic, compressed; rachilla persistent, ca. 0.2 mm wide, ca. 0.1 mm thick, quadrate, purplish brown, successive scale scars 0.4–0.6 mm apart, wings absent. Scales deciduous, (6–) 12–20 (–30), 0.6–0.8 mm long, 0.6–0.8 mm wide, obovate to orbiculate, apex mucronulate, laterally nerveless, hyaline-margined, stramineous to deep purple, medially 3-nerved, greenish, stramineous, or purplish. Stamens 1 or 2; filaments 0.3–0.4 mm long; anthers ca. 0.1 mm long, ovate-ellipsoid, the connective apex not prolonged. Style up to 0.1 mm long; stigmas 3, 0.1–0.3 mm long. Achenes 0.6–0.8 mm long, 0.3–0.4 mm wide, trigonous, obovoid-ellipsoid, apex obtuse, apiculate, base cuneate, the faces flat, the surface finely reticulate with papillae, light brown.

Phenology. Flowering and fruiting in August, September, and October.

Distribution. Native in the Old World, from southern Europe to southern Africa, and eastward to southeast Asia and Australia (Lipscomb 1980); in the New World in the southern half of the U.S.A. (Tucker 1992), and in Mexico, Nicaragua, Panama, and also in Puerto Rico and northern South America. In Mexico in Sonora, Sinaloa, Guanajuato, Jalisco, and Michoacán; disturbed, wet, muddy soils and shallow waters; near sea level to ca. 2000 m.

ADDITIONAL SPECIMENS EXAMINED. **Mexico**. GUANAJUATO: Salamanca, *Waterfall 13901* (F, TEX, US).—JALISCO: al este de El Salto, carr. Guadalajara–Chapala, *Amante 1-A* (IBUG).—MICHOACÁN: Morelia, *Arsène 3092* (GH, MO, NY, US), *3173* (F, MO, NY, US); valley of the Lerma at La Barca, *Pringle 3261* (AC, ARIZ, BKL, E, F, G, GH, LL, M, MICH, MSC, MTMG, NY, PH, US, VT, WRSL, Z).—SINALOA: 26 Feb 1967, *Rios s.n.* (CHAPA).

The very short, distinctly 3-nerved, usually purple scales distinguish *C. difformis* from any other species with densely capitate heads. Rayless plants of *C. manimae* may suggest *C. difformis* by their reddish purple, dense inflorescences, but can be readily distinguished by their long (1.3–1.6 mm) oblong-elliptic scales. Several collections of *C. difformis* that I examined had been misidentified as *C. haspan*, perhaps because both have 3-nerved scales. *Cyperus haspan* has digitate, rather than capitate, clusters of spikelets, and whitish, globose achenes, rather than light brown trigonous achenes.

Lipscomb (1980) has provided an interesting account of the spread of this species in North America. It was first collected in the New World in Sonora in 1851–52 [*Wright 1950*]. It is a significant weed of rice fields in California but has not yet become a problem in the rice producing states of the southern United States (Bryson 1984). By inference, its limited and sporadic occurrence in Mexico suggests it is not a significant weed in Mexico. In contrast to the other weedy species of the genus (e.g., *C. esculentus* L.), *C. difformis* is an annual that is capable of completing its life cycle in only one month; a single plant can produce thousands of achenes. The species is adapted to frequent flooding, such as occurs in rice fields. The seeds germinate best in shallow water (McIntire 1985).

19. Cyperus humilis Kunth, Enum. pl. 2: 23. 1837.—Type: Cuba. Nov 1824, *Poeppig s.n.* (holotype: B!; isotypes: BR! HAL!).

Cyperus humilis var. *elatior* Britton, Bull. Torrey Bot. Club 15: 99. 1888.—Type: Guatemala. Alta Verapaz: Cobán, Aug 1885, *von Türckheim 705* (holotype: NY!; isotypes: BKL! F! GH! MICH! TEX!).

Caespitose annual, 1–45 cm tall. Rhizomes absent. Culms 0.2–0.9 mm in diameter, trigonous, smooth. Leaves 1–3, (0.5–) 5–18 cm long, (0.8–) 1.2–2.5 mm wide, flat or v-shaped, the margins and keel smooth, except scabrellate apically. Inflorescence bracts 3–6, 1–18 cm long, (0.3–) 2–7 mm wide, flat to slightly v-shaped, the margins and keel sparsely scabrellate near the tip, borne approximately horizontally. Rays absent, the inflorescence a dense spherical to hemispherical head (4–) 6–18 (–23) mm in diameter. Spikelets (10–) 30–100 (–120), 3–6 (–10) mm long, 1.2–2.0 mm wide, oblong to oblong-lanceolate, compressed; rachilla persistent, (0.2–) 0.3–0.4 (–0.7) mm wide, 0.1–0.2 mm thick, straight, rectangular, stramineous to greenish brown, spongy, successive scale scars ca. 0.2 m apart, wings absent. Scales deciduous, (6–) 12–38, 1.2–1.6 mm long, 0.8–1.0 (–1.4) mm wide, oblong-ovate, apex acute, minutely mucronulate, laterally nerveless, hyaline to pale greenish, medially prominently 3-nerved, green or light brown. Stamen 1; filament 0.8–1.4 mm long; anther ca. 0.2 mm long, oblong-ellipsoid, the connective apex not prolonged. Style 0.2–0.4 (–0.6) mm long; stigmas 3, 0.3–0.6 mm long. Achenes (0.7–) 0.8 (–0.9) mm long, (0.3–) 0.4–0.5 mm wide, plano-convex, ovoid-deltoid, apex acute, base broadly rounded, sessile, the adaxial face flat, the abaxial faces confluent, broadly rounded, the surface finely pitted, light brown.

Phenology. Flowering and fruiting throughout the year.

Distribution. West Indies, and the Caribbean lowlands of Mexico and Central America: from the vicinity of Victoria, Tamaulipas, south to Heredia Province, Costa Rica; also in northwestern South America. Frequent in Mexico (Fig. 17); in

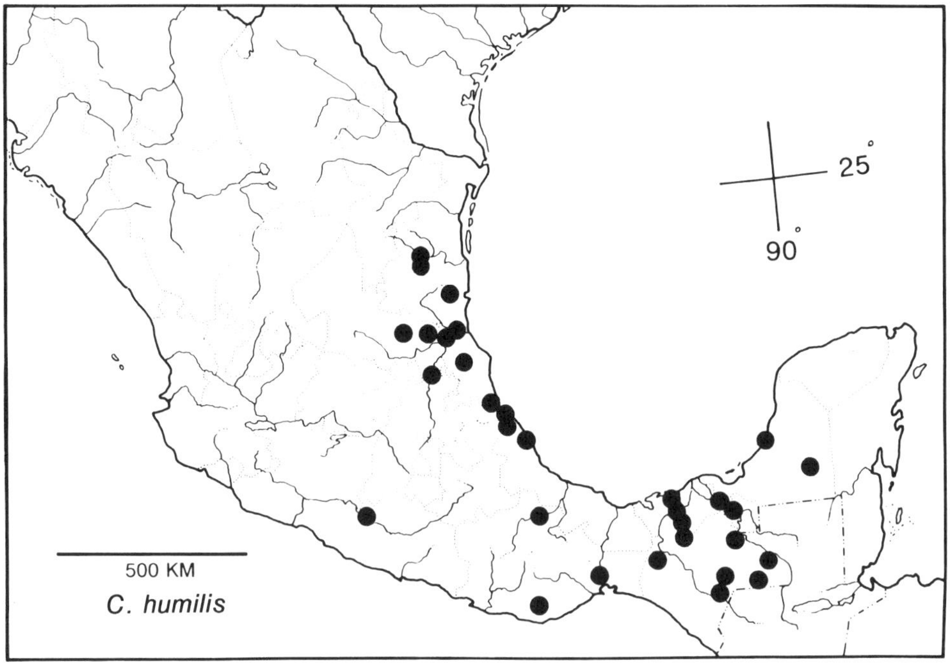

FIG. 17. Distribution of *Cyperus humilis* in Mexico.

moist, sunny, disturbed sites, such as river banks, sand bars, burrow pits, lake shores, and edges of puddles; sea level to ca. 1000 m.

REPRESENTATIVE SPECIMENS. **Mexico.** CAMPECHE: Tuxpeña, *Lundell 1191* (ARIZ, DS, GH, LL, MICH, MO, NY, UC).—CHIAPAS: Mpio. La Trinitaria, near ruins of Lagarteros, *Breedlove 47091* (CAS); Mpio. Ocosingo, near ruins of Bonampak, *Breedlove & Almeda 58025* (CAS); ca. 8 km SE of Ixtacomitán, *Tucker 2620* (DUKE, ENCB).—GUERRERO: Coyuca, *Hinton 5536* (G, GH, LCU, LL, MO, NY, PH).—OAXACA: ca. 55 km N of Pochutla, *Tucker 2887* (DUKE, ENCB).—QUINTANA ROO: Mpio. Chetumal, 6.5 7 km N of Tomás Garrido, *Sanders 9942* (NYS, UCR).—SAN LUIS POTOSÍ: Las Canoas, *Pringle 3917* (AC, ARIZ, BKL, C, CAS, E, G, GH, M, MICH, MO, MSC, MTMG, NY, PH, TEX, UC, VT); 8.2 mi E of Tamuín, *Tucker 2001* (DUKE, ENCB).—TABASCO: just S of Villahermosa, *Tucker 2565* (DUKE, ENCB); 40 km S of Villahermosa, *Tucker 2571* (DUKE, ENCB).—TAMAULIPAS: Arroyo Presa, 7 mi N of Victoria on Hwy 85 toward Monterrey, *Tucker 1943* (CAS, DUKE, ENCB, F, MICH).—VERACRUZ: Pital ad ripas Río Nautla, *Liebmann 14447* (C); Hwy 127, 10 km E of San Sebastián, *Tucker 2453* (DUKE, ENCB).

Cyperus humilis can be distinguished from all other annual species of the genus in the New World by its dense, capitate, greenish inflorescences and its compressed, plano-convex achenes with their unusual ovoid-deltoid shape.

20. Cyperus seslerioides H. B. K., Nov. gen. sp. 1 [quarto]: 209. 1816.—TYPE: VENEZUELA. "In ripa fluminis Orinoci prope Atures," May 1800, *Humboldt & Bonpland s.n.* (holotype: B-W no. 1310, IDC microfiche!; isotype: P!). [See note by Tucker and McVaugh, 1993.]

Cyperus perpusillus Boeckeler, Linnaea 35: 499. 1868.—TYPE: MEXICO. *Coulter 1611* pro parte (holotype: B?). [The type gathering was evidently a mixed collection. Kükenthal (1935–36) also cited *Coulter 1611* under *C. spectabilis*; *Coulter 1611* (GH) is strictly *C. spectabilis*.]

Cyperus seslerioides f. *minor* C. B. Clarke in Donnell Smith, Enum. pl. guatemal. 3: 87. 1893. *Cyperus seslerioides* var. *minor* (C. B. Clarke) Kükenthal, Pflanzenreich 4(20): 310. 1936.—TYPE: GUATEMALA. Santa Rosa, *Heyde & Lux 3546* (holotype: K; isotypes: G! M!).

Cyperus andinus Palla ex Kükenthal, Pflanzenreich 4(20): 310. 1936.—TYPE: BOLIVIA. *Bang 995* (lectotype, here designated: GH!; isolectotype, GH!).

Caespitose perennial, (0.5–) 4–30 cm tall. Rhizomes 2–5 mm in diameter, much branched, scaly. Culms 0.4–1.1 mm in diameter, roundly trigonous, smooth but clothed at the base with dark fibrous remnants of old leaf sheaths. Leaves 1–5, 3–18 cm long, 1.0–2.5 (–3.0) mm wide, flat, involute toward the apex, the margins and keel scabrous, but scabrellate apically. Rays absent, the inflorescence (3–) 6–15 (–24) mm in diameter, densely hemispherical to spherical, whitish, greenish, or brownish. Inflorescence bracts 2–5, 1–13 cm long, 1–3 mm wide, flat, margins and keel sparsely distally scabrellate, horizontal to reflexed parallel to the culm. Spikelets (5–) 20–60 (–100), 3–7 mm long, 2–4 mm wide, ovate to oblong-lanceolate, compressed; rachilla persistent 0.2–0.5 (–0.8) mm wide, 0.1–0.4 mm thick, quadrate, zigzag, becoming spongy in age, whitish to brownish, often reddish spotted, successive scale scars 0.2–0.5 mm apart, wings absent. Scales ± deciduous, (8–) 10–12 (–24), 1.6–3.0 (–3.4) mm long, 0.8–1.3 mm wide, ovate to ovate-lanceolate, apex acuminate, mucronulate, laterally nerveless, milky white, yellowish, pale greenish, brown or bronze, medially green or white, 3-nerved. Stamens 3; filaments 1.3–2.4 mm long; anthers (0.4–) 0.6–0.8 (–1.1) mm long, linear, the connective apex not prolonged. Style 0.8–1.8 mm long; stigmas 3, (0.5–) 1.0–1.5 (–2.0) mm long. Achene 0.5–1.0 mm long, 0.3–0.75 mm wide, trigonous, broadly obovoid, apex apiculate, base subcuneate to rounded, slightly stipitate, faces flat or concave, finely papillose, brown to blackish. Fig. 1D, E.

Phenology. Flowering and fruiting from July to October, very few collections in the remaining months.

Distribution. Western Texas, U.S.A., to Sonora, Mexico, and south to Honduras (Fig. 18), also in the mountains of South America. Common in the uplands of Mexico; forest clearings, meadows, pastures, seepages, trailsides, and disturbed soils; (1200–) 1600–2700 (–3700) m.

REPRESENTATIVE SPECIMENS. **Mexico.** AGUASCALIENTES: Aguascalientes, *Hartweg 256* (BM, E).—CHIAPAS: Mpio. San Cristóbal de Las Casas, *Breedlove 37284* (DS).—CHIHUAHUA: Mpio. Guachochic, NW of Norogachic, *Bye 8756a* (ARIZ, MICH, TEX, UC, US); Sierra Charuco, Río Fuerte, *Gentry 1525* (ARIZ, GH, LCU, MO, UC); 15 mi E of San Antonio, *Muller 3362* (GH, LCU, LL, UC); near Chihuahua, *Pringle 806* (CAS, E, G, LL, MICH, MO, UC); Sierra del Nido, 7.5 mi W of Bella Vista, *Worthington 7338* (UTEP).—DISTRITO FEDERAL: Deleg. Coyoacán, *Mexia 2663* (CAS, E, G, MICH, UC, Z); Tlalpam, lava fields, *Pringle 11206* (CU, GH, MO, US, Z).—DURANGO: Mpio. Santiago Papasquiaro, crest of sierra, *Díaz 242* (DUKE, UTEP); Otinapa, *Palmer 337* (GH, UC, US).—GUANAJUATO: Between Guanajuato and Santa Rosa, *Moore 1348* (BH, GH).—GUERRERO: 8 km SE de Chichihualco, *Rzedowski 22726* (ENCB, MICH).—HIDALGO: Dtto. Pachuca, Parque Nac. El Chico, *Moore & Wood 4115* (A, BH, BM), *López 124* (CAS, DS, MICH, TEX).—JALISCO: Cerro Viejo, above Zapotitan Hidalgo, *Gregory & Eiten 226* (C, GH, MICH, MSC); Cerro de Talcozagua, 2 km E of Tagalpa, *Iltis et al. 770* (MICH, WIS); Guadalajara, *Palmer 290* (DS, G, GH, MICH).—MÉXICO: Dtto. Temascaltepec, Timbres, *Hinton 1229* (CAS, G, LCU, MO); Mpio. Ecatepec, Sierra de Guadalupe,

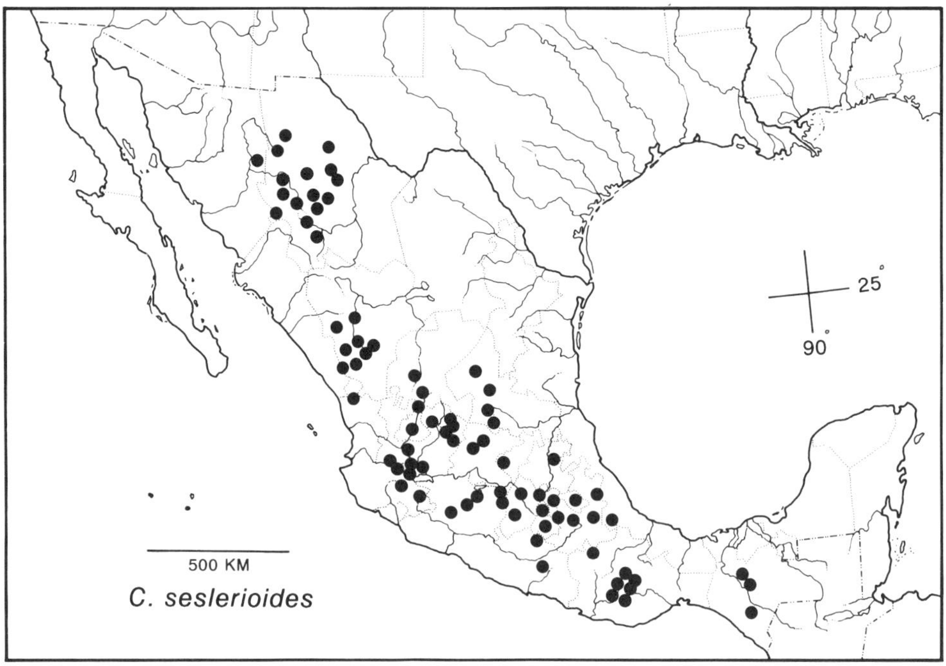

FIG. 18. Distribution of *Cyperus seslerioides* in Mexico.

Moreno 292 (ENCB, MICH, SD).—MICHOACÁN: Summit of Cerro San Andrés, *Beaman 4301* (GH, MSC).—MORELOS: Barranca W of Cuernavaca, *Crespo 330* (DS, MICH, MSC).—NAYARIT: Mpio. Acaponeta, W of Jesús María, *Norris & Taranto 14169* (MICH).—OAXACA: 10 km SW of Guelatao de Juárez, *Solheim & Powers 819* (WIS); vicinity of Oaxaca near San Felipe, 23 Jun 1966, *Te Strake s.n.* (DUKE); 43 km from Mitla on rd to Ayutla, *Tucker 2187* (DUKE); ca. 15 km S of Ejutla, *Tucker 2821A* (DUKE).—PUEBLA: pr. Cerro León, *Liebmann 14397* (C).—QUERÉTARO: camino a San Miguel, frontera con Guanajuato, *Arguelles 1652* (MEXU).—SAN LUIS POTOSÍ: 12 mi W of San Luis Potosí, *Templeton 8736* (CAS, MICH); Charcas, *Whiting 818* (DS, LCU, MICH).—SINALOA: NE of Pánuco, *Pennell 20067* (GH, LCU).—SONORA: Maicoba, *Pennington 212* (TEX).—TLAXCALA: Mt. Malinche, *Balls 4870* (BM).—VERACRUZ: Mpio. Las Vigas, Encino Gacho, *Ventura A. 8674* (ARIZ, MICH).—ZACATECAS: 7 mi S of Pinos, *Shreve 8671* (ARIZ, LCU); 10 km S of Zapoqui on Hwy 54, *Tucker 3002* (DUKE, ENCB).

Cyperus seslerioides includes several variants with regard to the color of the scales: white, yellowish, pale greenish or brownish, or bronze. White-scaled plants account for about 50% of the Mexican material of this species. Critical study of these color forms, based on some 400 Mexican collections, failed to reveal correlation of this character with any other morphological features, such as dimensions of the scales or achenes, or with geography, elevation, flowering time, or habitat. Further, the collections seen exhibit a continuum of scale colors from white to brown. Evidently, color of the scales in *C. seslerioides* is a variable trait, which does not merit taxonomic recognition.

21. Cyperus arsenei O'Neill & Benedict Ayers, Leafl. W. Bot. 4: 33. 1944, ("arsenii").—TYPE: MEXICO. Michoacán: San Miguel del Monte, vicinité de Morelia, 8 Aug. 1912, *Arsène 7411* (holotype: MO!; isotypes: GH! LCU!).

Perennial, 2–10 cm tall. Rhizomes 0.5–1 mm in diameter, slender, clothed with lanceolate scales, the internodes 5–15 mm long, producing 1 or 2 culms at the tip. Culms 0.2–0.3 mm in diameter, trigonous, sparsely scabrellate on the angles, especially just below the inflorescence. Leaves 1–3, 1.5–12 cm long, 0.3–0.8 mm wide, flat or v-shaped, margins and keel scabrellate. Rays absent, the inflorescence 4–8 mm in diameter, loosely digitate. Inflorescence bracts 2, 2.5–30 mm long, 0.5–1 mm wide, flat, margins and keel scabrellate, the longer bract ± erect (vertical), the shorter bract approximately horizontal. Spikelets 2–6 (–12), 3.5–5 mm long, 1.3–1.7 mm wide, oblong-lanceolate, somewhat compressed; rachilla persistent, 0.3–0.4 mm wide, 0.2–0.3 mm thick, straight, stramineous, successive scales 0.2–0.3 mm apart, wings absent. Scales deciduous, 6–10, 1.2–1.4 mm long, 1–1.2 mm wide, ovate, apex mucronulate, laterally nerveless, light brown, bronze, or whitish, medially 3-nerved, green, often red-speckled. Stamens 1 or 2; filaments ca. 1.2 mm long; anthers ca. 0.3 mm long, oblong, the connective apex not prolonged. Style 0.4–0.5 mm long; stigmas 3, 0.4–0.5 mm long. Achenes 0.8–1.0 mm long, 0.5–0.6 mm wide, dorsiventrally compressed, weakly trigonous, obovoid to ovoid, apex apiculate, base slightly to decidedly stipitate up to 0.1 mm long, the adaxial face widest, flat or concave, the abaxial faces narrower, flat or slightly concave, the surface papillose, brown, the stipitate base whitish.

Phenology. Flowering and fruiting in August and September.

Distribution. Western Mexico (Baja California Sur, Nayarit, Michoacán) and also the Distrito Federal; rare, lava fields and montane forests, especially oak forests; ca. 500–2500 (–3000) m.

REPRESENTATIVE SPECIMENS. **Mexico.** BAJA CALIFORNIA SUR: Cape Region, El Taste, 14 Sep 1893, *Brandegee s.n.* (UC).—DISTRITO FEDERAL: Valley of México, lava fields, *Pringle 7182* (GH, LCU, MICH, US, VT); Tlalpan, at Pedregal Peña Pobre, *Sharp 44275* (TENN).—MICHOACÁN: vicinité de Morelia, près San Miguel del Monte, *Arsène 7411* (GH, LCU, MO); Cerro Sta. María, 8–10 km SW of Jiquilpan, *Feddema 241A* (DUKE, MICH).—NAYARIT: Mpio. San Blas, 3 mi E of Jalcocotán, *Norris & Taranto 14982* (MICH).

Cyperus arsenei is readily distinguished from other species of the genus in Mexico by its small inflorescence with only two bracts and its dorsiventrally flattened achenes.

22. Cyperus perennis (M. E. Jones) O'Neill, Contr. U.S. Natl. Herb. 29: 93. 1944. *Cyperus aristatus* var. *perennis* M. E. Jones, Contr. W. Bot. 18: 25. 1933.—TYPE: MEXICO. Baja California: Laguna Mountains, 22 Sep 1930, *Jones 27595* (holotype: POM!; isotype: UC!).

Caespitose perennial, 0.5–8.5 cm tall. Rhizomes 0.5–2 cm long, knotted. Culms 0.4–0.6 mm in diameter, trigonous, smooth but densely clothed basally with a thick layer of persistent fibers from the preceding years' leaf sheaths. Leaves 1–3, 3–10 cm long, 0.5–2 mm wide, v-shaped, the margins sparsely scabrellate. Inflorescence bracts 1–4, 1–8 cm long, 1–2 mm wide, v-shaped, margins scabrellate, horizontal to reflexed. Rays absent, the inflorescence 5–15 mm in diameter, densely spherical to hemispherical. Spikelets (5–) 10–25 (–60), 3–5 (–7) mm long, 1.6–1.7 mm wide, linear-oblong, compressed; rachilla persistent, 0.4–0.6 mm wide, 0.3–0.4

mm thick, slightly flexuous, spongy, yellowish, usually red-spotted, successive scale scars 0.4 (–0.5) mm apart. Scales deciduous, 10–24 (–34), 0.6–1.3 mm long, the recurved uncinate cusp an additional 0.5–0.8 mm long, 0.8–1.2 (–1.3) mm wide, oblong-ovate, subtruncate, laterally nerveless or weakly 1-nerved, stramineous, reddish brown lineolate, medially strongly 3-nerved, greenish or stramineous. Stamens 1, 2, or 3; filaments 1.0–1.2 mm long; anthers (0.4–) 0.5 (–0.6) mm long, oblong, the connective apex less than 0.1 mm long, reddish. Style 0.4–0.5 mm long; stigmas 3, 0.4–1.0 mm long. Achene 0.4–0.5 (–0.6) mm long, (0.3–) 0.4 mm wide, obscurely trigonous, globose-obovoid, apex truncate, broadly rounded, apiculate, base substipitate, the faces very broadly convex, papillose, light to dark brown, iridescent.

Phenology. Flowering and fruiting from August to February.

Distribution. Mexico, endemic to northwestern Sonora and Baja California; occasional or frequent, edges of streams, arroyos, temporary pools, and wet spots in low desert flats; near sea level to ca. 2000 m.

ADDITIONAL SPECIMENS EXAMINED. **Mexico**. BAJA CALIFORNIA: N side of La Laguna Meadow, Sierra de la Victoria, *Chambers 909* (DS, UC); Santa Genoviva Peak ca. 20 mi SE of El Pescadero, *Hammerly 250* (CAS); above beach at SW base of Los Frailes, *Porter 309* (DS).—BAJA CALIFORNIA SUR: valley S of Pico La Aguja, *Breedlove & Axelrod 43342* (CAS); San José del Cabo, *Dawson 1164* (LCU, MICH); La Paz, *Jones 24063* (ARIZ, F, LCU, POM, UC); 10 mi S of La Paz, *Keil et al. 16750* (UCR); 3 km N of Cabo San Lucas, *Moran 7038* (CAS, DS).—SONORA: Camahuiroa between Agiabampo and Las Bocas on the Gulf of California, *Van Devender et al. 93–278* (NYS); Cerro Terucuchi, 3 km N of Teachive de Masiaca, *Van Devender et al. 93–896* (NYS); junction of rd to Horcasitas with old hwy on railroad 6 mi S of Carbo, *Wiggins 7271* (US).

Cyperus perennis is distinguished from all other species of the genus in the New World by its recurved, uncinate scale apices. Jones first described this species as a variety of *C. aristatus* [=*C. squarrosus*], but *C. perennis* differs from *C. squarrosus* in several features, and O'Neill quite rightly elevated it to specific rank. *Cyperus squarrosus* is an annual with spicate spikelets, deciduous rachillas, and laterally 2–3 (–4) -nerved scales; *C. perennis*, although small, is a perennial with digitate spikelets, persistent rachillas, and laterally nerveless (or weakly 1-nerved) scales.

23. Cyperus uncinulatus Schrader ex Nees in Martius, Fl. bras. 2(1): 23. 1842.—TYPE: BRAZIL. Piaui: in campsis, *Martius s.n.* (holotype: M!, photo: F!).

Caespitose annual, (1–) 4–8 (–18) cm tall. Rhizomes absent. Culms 0.3–0.5 mm diameter, trigonous, smooth. Leaves 2–3 (–4), 2–5 (–8) cm long, 1–1.5 mm wide, v-shaped, the margins and keel scabrellate. Inflorescence bracts 3–4, (1–) 3–6 (–8.5) cm long, 0.5–1.5 mm wide, v-shaped, the margins and keel scabrellate especially distally, horizontal to reflexed up to 30°. Rays absent, the inflorescence 6–15 (–18) mm in diameter, hemispherical to subspherical. Spikelets (3–) 10–25 (–45), (3.5–) 5–8 (–9) mm long, 2.4–3.2 mm wide, linear-oblong, compressed; rachilla persistent, 0.2–0.4 mm wide, 0.2–0.3 mm thick, straight, becoming spongy in maturity, stramineous, densely red-dotted, becoming light brown in maturity, successive scale scars 0.4–0.6 mm apart, wings absent. Scales deciduous, (10–) 18–34, 0.8–1.1 (–1.2) mm long, the straight or slightly excurved cusp an additional (0.4–) 0.6–0.8 (–1.1) mm long, 0.8–1.0 mm wide, broadly ovate, apex rounded, laterally nerveless, light brown to stramineous, with reddish speckles, medially strongly 3-nerved, green at anthesis, becoming brown throughout at maturity, bicarinate basally. Stamen 1; filament 0.5–1.2 mm long; anther 0.4–0.5 (–0.6) mm long, narrowly oblong, the connective apex less than 0.1 mm long, red. Style 0.3–0.4 (–0.5) mm long; stigmas 3, 0.4–0.6 mm long. Achenes 0.5 (–0.6) mm long, 0.3–0.4 mm wide, roundly trigonous, ovoid to globose-ovoid, broadly rounded apically with a conspicuous apiculus, abruptly contracted into a stipitate base ca. 0.1 mm long, the faces convex, the surface finely papillose, golden-brown to dark brown, the apiculus and base contrastingly reddish purple.

Phenology. Flowering and fruiting in June and July.

Distribution. Primarily in South America: Venezuela, southern Brazil, Uruguay, and northern Argentina, also on the Pacific Coast of southern Mexico (Chiapas and Oaxaca) and Guatemala; pastures, brushy slopes, sandy roadsides, and thorn-scrub savannas; sea level to ca. 150 m.

ADDITIONAL SPECIMENS EXAMINED. **Mexico**. CHIAPAS: Monserrate, *Purpus 10267* (LCU, M, UC).—OAXACA: along hwy to Ixtepec, 12 km NW of Juchitán, *King 777* (DUKE, MICH, TEX, WIS); on Hwy 200, ca. 13 km E of Niltepec on the Istmo de Tehuantepec, *Tucker 2748* (DUKE, ENCB) [with *Cyperus microbrunneus*, *Tucker 2743* and the putative hybrid, *C. microbrunneus* × *C. uncinulatus*, *Tucker & Schwartz 2745*].

Cyperus uncinulatus is distinguished from other species in the New World by the following combination of characters: small size, digitate spikelets, and ovate scales with a conspicuous, straight or slightly excurved, cuspidate apex. This species may be a recent introduction from South America. Kükenthal (1935–36), Standley (1931), and Benedict Ayers (1946) did not record *C. uncinulatus* as occurring in Mexico or Central America.

Cyperus uncinulatus hybridizes with *C. microbrunneus* (no. 24). These hybrids occur in the area of sympatry, from the Isthmus of Tehuantepec to southeastern Guatemala. Table 1 demonstrates the morphological intermediacy of the hybrids. The two parent species set seed abundantly, but in the putative hybrids only ca. 0.1% of the gynoecia mature to form shrivelled achenes lacking endosperm; the spikelets of the hybrid plants are sufficiently mature that the scales have begun to disarticulate, at which point of development similarly mature plants of the parent species would have begun to show mature achenes.

Both *C. uncinulatus* and *C. microbrunneus*, as well as their hybrids, were found in a mixed population in southeastern Oaxaca in 1984 (*Tucker 2743*, *2745*, *2748*). Additionally, these three taxa are represented in one collection from southeastern Guatemala (*Steyermark 29038*, F).

24. Cyperus microbrunneus G. C. Tucker, Bull. Torrey Bot. Club 110: 163. 1983.—TYPE: HONDURAS. Choluteca: vicinity of Pespire, 18–27 Oct 1950, *Standley 27233* (holotype: GH!; isotype: F!).

Small tufted perennial, 4–15 (–30) cm tall. Rhizomes 1–2 cm long, knotted. Culms 0.3–0.5 (–0.7) mm in diameter, trigonous, smooth, bulbous-thickened and whitish proximally, the base clothed with fibrous remnants of preceding year's leaf sheaths. Leaves 1–4, 2–10 (–16) cm long, 0.4–1.3 mm wide, flat to v-shaped,

Table 1. Comparison of *Cyperus uncinulatus, C. microbrunneus,* and their putative hybrids.

	C. uncinulatus	*putative hybrids*	*C. microbrunneus*
Diameter of inflorescence	6–15 (–18) mm	(5–) 8–15 (–26) mm	6–12 (–18) mm
Number of inflorescence bracts	3–4	3–4	(2–) 3 (–4)
Spikelets: shape	broadly elliptic	elliptic	linear-lanceolate
length	(3.5–) 5.0–8.0 (–9.0) mm	(3.0–) 4.5–7.0 (–12) mm	2.5–4.5 (–8.0) mm
width	2.4–3.2 mm	1.2–1.6 mm	1.0–1.2 mm
Rachilla	stramineous to light brown, red-dotted throughout	stramineous with reddish splotches on the sides	whitish with deep purple stripes on the sides
Scales: shape	broadly ovate	oblong-ovate	oblong
length of cusp	0.6–0.8 mm	0.2–0.3 mm	<0.1 mm
Length of anther	0.4–0.5 (–0.6) mm	(0.3–) 0.4 mm	0.3 mm

apically triquetrous, the margins and keel distally scabrellate. Inflorescence bracts (2–) 3 (–4), (0.5–) 2–6 (–8.5) cm long, 0.5–1.6 mm wide, v-shaped, margins and keel scabrellate, horizontal to reflexed. Rays absent, the inflorescence 6–12 (–18) mm in diameter, densely hemispherical. Spikelets 20–70 (–100), 2.5–4.5 (–8.0) mm long, 1.0–1.25 mm wide, oblong-lanceolate, strongly compressed; rachilla persistent, ca. 0.2 mm wide, ca. 0.1 mm thick, straight, rectangular, deep purplish brown, with a white stripe on the center of each side bearing scales, successive scale scars ca. 0.4 mm apart, wings absent. Scales deciduous, (10–) 14–20 (–42), 0.8–0.9 (–1.0) mm long, 0.6–0.8 (–0.9) mm wide, oblong, apex broadly rounded, laterally nerveless, reddish brown to brown, glossy, medially 3-nerved, greenish distally, whitish proximally, often red-speckled as well. Stamen 1; filament 0.6–0.9 mm long; anther ca. 0.3 mm long, oblong-elliptic, the connective apex minute, much less than 0.1 mm long. Style 0.2–0.4 mm long; stigmas 3, (0.3–) 0.5–0.6 mm long. Achenes 0.6 (–0.7) mm long, 0.3 mm wide, trigonous, ellipsoid, apex subacute, apiculate, base cuneate, substipitate to stipitate, ca. 0.1 mm long, the faces slightly convex, the surface with low rounded papillae, brown, iridescent.

Phenology. Flowering and fruiting in June, July, and October.

Distribution. Southeastern Oaxaca, Mexico, to northwestern Costa Rica near the Pacific Coast. Rare in Mexico; rocky, brushy, and grassy slopes, pastured thorn-scrub woodlands, and edges of fields; near sea level to ca. 300 m.

ADDITIONAL SPECIMENS EXAMINED. **Mexico**. OAXACA: along Hwy 200 on the Istmo de Tehuantepec, ca. 27 km E of Niltepec, 31 Jul 1984, *Tucker 2743* (DUKE, ENCB, MICH); along Hwy 200, ca. 30 km E of Niltepec, 31 Jul 1984, *Tucker 2740* (CAS, DUKE, ENCB, MICH).

Cyperus microbrunneus is distinguished from other species of *Cyperus* in Mexico and Central America by its rayless inflorescence of brown to reddish brown spikelets, its short scales (0.8–1.0 mm long), and its small ellipsoid achenes (0.6–0.7 mm long). It might be confused with *C. difformis*, also a small species with rayless brownish inflorescences and small scales and achenes. The scales of *C. difformis* are obovate to orbiculate, and the achenes have acute angles, whereas in *C. microbrunneus* the scales are oblong and the achene angles are rounded.

Cyperus microbrunneus is morphologically most similar to *C. arsenei*. They have in common small size, rayless inflorescences, persistent leaf sheaths, broadly rounded scale apices, small iridescent achenes, and similar habitats in dry open situations.

25. Cyperus nayaritensis G. C. Tucker, Bull. Torrey Bot. Club 11: 161. 1983.—TYPE: MEXICO. Nayarit: Volcán Ceboruco near Ahuacatlán, 13 Aug 1959, *Feddema 418* (holotype: DUKE!; isotypes: ENCB! MICH!).

Caespitose perennial, 23–40 cm tall. Rhizomes 1–2 cm long, indurate, dark brown. Culms 0.7–1.8 mm in diameter, triquetrous, the angles densely scabrellate especially distally. Leaves 2–4, 5–36 cm long, 2.5–5.5 mm wide, flat or broadly v-shaped, the margins and keel scabrellate. Inflorescence bracts (2–) 3–4, 2–10 cm long, 1–2.5 mm wide, flat or broadly v-shaped, conspicuously whitened on the proximal 5–20 cm of both surfaces, the margins and keel scabrellate, reflexed from 30–60°. Rays absent or sometimes 1 or 2 present, 5–10 mm long; the inflorescence 6–13 mm in diameter, densely hemispherical, dull white to pale reddish brown. Spikelets 25–45, 3–5 mm long, 1.9–2.4 mm wide, ovate to oblong-elliptic, strongly compressed; rachilla persistent, ca. 0.2 mm wide, ca. 0.2 mm thick, slightly zigzag, quadrate, green, successive scale scars ca. 0.4 mm apart, wings absent. Scales deciduous, 8–16, 2.1–2.5 mm long, 0.8–1.2 mm wide, oblong-lanceolate, apex obtuse to tridentate, laterally 2–3-nerved, white or yellowish white, with numerous tiny reddish spots (dull white to reddish brown when dried), medially 3-nerved, white. Stamens 2; filaments 2.0–2.4 mm long; anthers ca. 0.4 mm long, oblong, the connective apex not prolonged. Style ca. 0.4 mm long; stigmas 3, ca. 0.8 mm long. Achenes 1.0–1.1 mm long, ca. 0.4 mm wide, trigonous, linear-oblong, apex acute, apiculate, base broadly rounded, not stipitate, the adaxial face concave, the abaxial faces convex, the adaxial angles sharp, the abaxial angle broadly rounded, the surfaces puncticulate, dark brown.

Phenology. All collections were made in August.

Distribution. Endemic to western Mexico, in the mountains of Nayarit, Jalisco, Colima, and Michoacán; in pockets of soil on moist ledges of roadcuts, shaded boulders in lava fields, and moist gravelly arroyos in volcanic soils; 800–1600 m.

ADDITIONAL SPECIMENS EXAMINED. **Mexico**. COLIMA: Mpio. Comala, Rancho El Jabalí, *Vázquez & Phillips 913* (NYS, UCR).—JALISCO: on Hwy 180, ca. 18 km S of Autlán, 12 Aug 1984, *Tucker 2982* (CAS, DUKE, ENCB, F, GH, MICH); 21 km SE of Autlán, 2 km NW of Tecopatlán, 16 Jul 1987, *Vázquez 4318* (ZEA).—MICHOACÁN: 1 mi W of Uruapan, 3 Aug 1940, *Leavenworth 2270* (F).—NAYARIT: Volcán Ceboruco near Ahuacatlán, 13 Aug 1959, *Feddema 418* (DUKE, ENCB, MICH); 10 Aug 1970, *Norris & Taranto 15222* (MICH); E slopes of Volcán Ceboruco, 4 km W of Jala, 7 Aug 1983, *Tucker 2272* (BM, CAS, DUKE, ENCB, F, GH, MICH, MO, NY, US, WIS); El Ceboruco lava flow W of Ahuacatlán on Hwy 15, 7 Aug 1983, *Tucker 2276* (DUKE).

Cyperus nayaritensis is morphologically similar to *C. tenerrimus* and *C. michoacanensis*. It can be distinguished from both of these relatives by its greater height, scabrellate, triquetrous culms, wider leaves, and two stamens per flower (one in *C. tenerrimus*, two or three in *C. michoacanensis*).

In *C. nayaritensis*, the inflorescence bracts are conspicuously whitened basally. This coloration is easily noted in living plants, but is not obvious in dried specimens. Such coloration pattern has not been noted elsewhere in *Cyperus*. Its biological importance is unknown; however, species of *Rhynchospora* sect. *Dichromena* similarly have basally colored bracts, which attract pollinators (Thomas 1984). I did not observe any pollinator activity while making three collections of this species in Nayarit and Jalisco.

26. Cyperus tenerrimus J. Presl & K. Presl in K. Presl, Reliq. Haenk. 1: 166. 1828.—TYPE: MEXICO. [Guerrero:] Acapulco, 1790–91, *Haenke s.n.* (holotype: PRC!; isotype: B!).

Cyperus cymbaeformis Liebmann, Mexic. halvgr. 20. 1850.—TYPE: NICARAGUA. Granada, 1848, *Oersted 14389* (lectotype, designated by Tucker, 1986: C!).

Cyperus wawrai Boeckeler, Linnaea 38: 363. 1874.—TYPE: COLOMBIA. Santa Marta, Soledad, *Wawra 162* (holotype: B!).

Caespitose perennial, (1–) 5–20 (–50) cm tall. Rhizomes much branched, short, thick, indurate, covered with scales persisting as fibrous remnants. Culms 0.3–1.1 mm in diameter, trigonous to roundly trigonous, smooth or rarely sparsely scabrellate just below the inflorescence. Leaves 3–6, 5–25 cm long, 1–2 mm wide, flat or v-shaped, the margins and keel scabrellate. Inflorescence bracts (4–) 6–8, 0.5–18 cm long, 0.5–1.6 mm wide, flat to v-shaped, the margins and keel scabrellate, horizontal to reflexed by 30°. Rays absent, the inflorescence (6–) 10–15 (–20) mm in diameter, densely hemispherical, white. Spikelets (3–) 15–25 (–45), 4–10 mm long, 1.4–2.5 (–3.2) mm wide, ovate to elliptic, strongly flattened; rachilla persistent, ca. 0.2 mm wide, ca. 0.2 mm thick, geniculate, quadrate, green, successive scale scars 0.4–0.6 mm apart, wings absent. Scales deciduous, 6–32, 1.6–2.2 mm long, 0.8–1.2 mm wide, oblong-elliptic, apex mucronulate, tridentate, laterally 1-nerved, white, medially sulcate, 3-nerved, white to greenish white. Stamen 1; filament (1.3–) 1.6–2.0 mm long; anther (0.5–) 0.6–0.8 (–1.0) mm long, oblong-ellipsoid, the connective apex not prolonged. Style 0.4–0.6 (–0.8) mm long; stigmas 3, 0.4–0.6 mm long. Achenes 1.3–1.5 (–1.6) mm long, (0.3–) 0.4 mm wide, trigonous distally, subterete proximally, narrowly ellipsoid, apex acute, base slightly stipitate, the faces slightly concave distally to slightly convex proximally, the surface finely reticulate with isodiametric cells, brown to black.

Phenology. Flowering and fruiting from May to October.

Distribution. Along the Pacific Coast from northwestern Mexico south to Colombia and Venezuela, also in eastern Bolivia. In Mexico (Fig. 19) from southwestern Sonora to Chiapas; sunny moist or dry soil of roadsides, pastures, rocky bluffs, and open woods; sea level to 250 m (–2000 m).

REPRESENTATIVE SPECIMENS. **Mexico.** CHIAPAS: 3 mi NW of Arriaga, *King 3116* (MICH, TEX, UC); 30 km N of Escuintla on rd to Pijijiapán, *Tucker 2699* (DUKE, ENCB).—COLIMA: Colima, *Palmer 5* in 1897 (ARIZ, MICH, UC, US).—GUERRERO: Dtto. Coyuca, Pungarabato, *Hinton 6454* (G, LCU, LL, MO); Dtto. Galeana, Moreno, *Hinton 14342* (GH, LCU, MICH, TEX, UC); Mpio. Acapulco, Cerro de Tierra Colorada, 8 Jul 1966, *Solis s.n.* (ENCB, IBUG); near La Unión on Hwy 200, *Tucker 2940* (DUKE).—JALISCO: Mpio. La Huerta, Est. Biol. Chamela, *Lott 1275* (DUKE, MEXU, MICH); 30 km S of Autlán at El Tigre, *Tucker 2979* (DUKE, ENCB, MICH).—MÉXICO: Dtto. Temascaltepec, Ixtapán, *Hinton 1060* (ARIZ, G, LCU, MO, PH, TEX).—MICHOACÁN: Dtto. Apatzingan, Tepalcatepec, *Hinton 12110* (GH, LL, MICH).—NAYARIT: Isla María Madre, *Ferris 5718* (DS); Mpio. El Nayar, Jesús María, *Colunga & Zizumbo 43* (ARIZ, MEXU); lower slopes of Volcán Ceboruco, *Tucker*

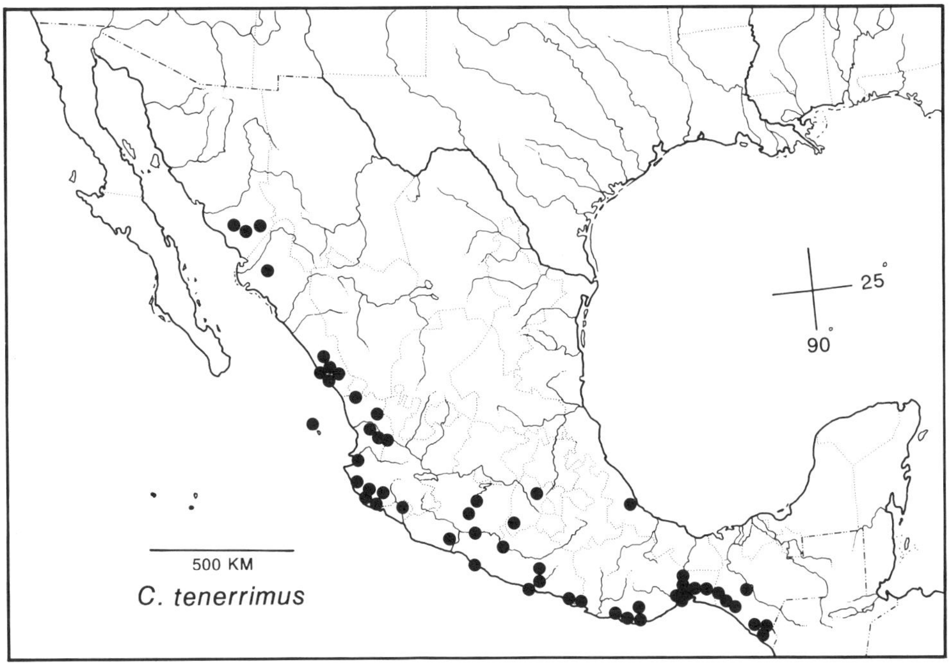

FIG. 19. Distribution of *Cyperus tenerrimus* in Mexico.

2273 (DUKE, ENCB).—OAXACA: 22 km NW of Zanatepec, *King 470* (MICH, TEX, US); 40 km W of Puerto Escondido, *Tucker 2907* (DUKE, ENCB); 15 km W of Pinotepa Nacional, *Tucker 2914* (DUKE, ENCB).—SINALOA: Mpio. La Concordia, 1–2 km N of Copala, *Breedlove 43795* (CAS).—SONORA: 1 km N of Alamos, *Collins et al. 80–463* (ARIZ); San Bernardo, Río Mayo, *Gentry 1586* (ARIZ, LCU, MO, UC).—VERACRUZ: Mpio. Puente Nacional, Conejos, *Ventura 3930* (ARIZ, ENCB).

Cyperus tenerrimus is readily identified by its dense white inflorescences, slender culms, and caespitose habit. From *C. nayaritensis*, *C. michoacanensis*, and *C. seslerioides*, it is quickly distinguished by its lowland, not montane, habitat.

27. Cyperus michoacanensis Britton ex C. B. Clarke, Bull. Misc. Inform. Kew, add. ser. 8: 5. 1908.—TYPE: MEXICO. Michoacán: Pátzcuaro, 23 Oct 1890, *Pringle 3427* (holotype: K! isotype: NY!). [Simpson (1990) indicated that the type specimen was at VT, but I found no duplicate of the type number there.]

Mat-forming perennial, 15–35 cm tall. Rhizomes much branched, the internodes 5–10 mm long, scaly. Culms 0.5–0.9 mm in diameter, roundly trigonous, smooth, basally thickened and surrounded by fibrous remnants of old leaf bases. Leaves 2–5, 5–10 (–15) cm long, 1.5–3 mm wide, flat, the margins and keel distally scabrellate, otherwise smooth. Rays absent, the inflorescence 3–8 mm in diameter, densely spherical, white. Inflorescence bracts 2 (–3), 0.3–6 cm long, 0.4–1.7 mm wide, the margins and keel scabrellate throughout but especially so distally, horizontal to reflexed downward parallel to the culm. Spikelets 10–20 (–26), 2–4 mm long, ca. 2 mm wide, oblong, flattened; rachilla persistent, 0.4–0.5 mm wide, 0.1–0.2

mm thick, flat, quadrate, white, successive scale scars 0.3–0.4 mm apart, wings absent. Scales deciduous, 8–12, 1.3–1.6 mm long, 0.7–1.1 mm wide, oblong-ovate, apex obtuse, slightly emarginate, laterally inconspicuously 1- (–2-) nerved, whitish, medially subtly 3-nerved, white. Stamens 3; filaments 1.2–1.4 mm long; anthers 0.25–0.5 mm long, oblong, the connective apex not prolonged. Style 0.4–0.6 mm long; stigmas 3, ca. 0.6 mm long. Achenes 0.8–1.1 mm long, 0.6–0.7 mm wide, plano-convex, dorsiventrally flattened, ovoid, apex broadly rounded, apiculate, base stipitate, lighter than the body of the achene, the adaxial face flat or slightly concave, the abaxial faces broadly rounded, the surface finely reticulate, dark brown to jet black.

Phenology. Flowering and fruiting in July, August, and October.

Distribution. Endemic to western Mexico, on shaded rocks of lava flows around Pátzcuaro, Michoacán, at ca. 1500 m and on rocks in hardwood-pine-fir forests southeast of Autlán, Jalisco, in the Sierra de Manantlán at 2000–2800 m; rare.

ADDITIONAL SPECIMENS EXAMINED. **Mexico**. JALISCO: 15 mi SE of Autlán, 28 Jul 1949, *Wilbur & Wilbur 1909* (MICH), 29 Jul 1949, *Wilbur & Wilbur 1951* (MICH); Sierra de Manantlán, Las Joyas, camino en el Filo de la Vaca, 28 Jul 1985, *Vázquez 3409* (ZEA); Sierra de Manantlán, 16–17 km NE de Cuautitlán, 22 Jul 1988, *Cuevas & Núñez 3173* (WIS).—MICHOACÁN: Pátzcuaro, 7 Oct 1892, *Pringle 4269* (AC, ARIZ, BM, C, CAS, DS, E, F, G, GH, MO, MSC, MTMG, NY, UC, US, VT, Z).

Cyperus michoacanensis is morphologically similar to both *C. nayaritensis* and *C. tenerrimus*; all have rayless whitish inflorescences. *Cyperus michoacanensis* is distinguished from the other two by its ovoid, rather than slenderly oblong, achenes, shorter anthers, and fewer spikelets per inflorescence. Simpson (1990) also emphasized the difference in achene morphology in his revision of *Cyperus* sect. *Leucocephali*, which includes these three species. *Cyperus michoacanensis* grows at higher elevations than *C. nayaritensis* and *C. tenerrimus*.

28. Cyperus amabilis Vahl, Enum. pl. 2: 318. 1805.—TYPE: GUINEA. 1783–86, *Thonning s.n.* (lectotype, designated by Tucker in Tucker & McVaugh, 1993: C!; isolectotypes: C!).

Cyperus microstachyus Vahl, Enum. pl. 2: 318. 1805.—TYPE: GUINEA. 1783–86, *Thonning s.n.* (lectotype, here designated: C!; isolectotypes: C!).

Cyperus aureus H. B. K., Nov. gen. sp. 1 [quarto]: 205. 1816. *Cyperus aureus* [var.] *α macrostachyus* Boeckeler, Linnaea 35: 495. 1868. *Cyperus amabilis* var. *macrostachyus* (Boeckeler) Kükenthal, Pflanzenreich 4(20): 266. 1936.—TYPE: ECUADOR. Guayaquil, *Humboldt & Bonpland s.n.* (holotype: B-W no. 1326, IDC microfiche!, photo, LCU!; isotype: C!).

Cyperus oligostachyus H. B. K., Nov. gen. sp. 1 [quarto]: 205. 1816. *Cyperus aureus* var. *oligostachyus* (H. B. K.) Boeckeler, Linnaea 35: 495. 1868. *Cyperus amabilis* var. *oligostachyus* (H. B. K.) Kükenthal, Bot. Not. 67. 1934.—TYPE: VENEZUELA. *Humboldt & Bonpland 546* (holotype: B-W no. 1300, IDC microfiche!).

Cyperus aurantiacus H. B. K., Nov. gen. sp. 1 [quarto]: 205. 1816. *Cyperus aureus* var. *aurantiacus* (H. B. K.) Boeckeler, Linnaea 35: 495. 1868.—TYPE: VENEZUELA. Near Orinoco River near Conucos de Siquita and

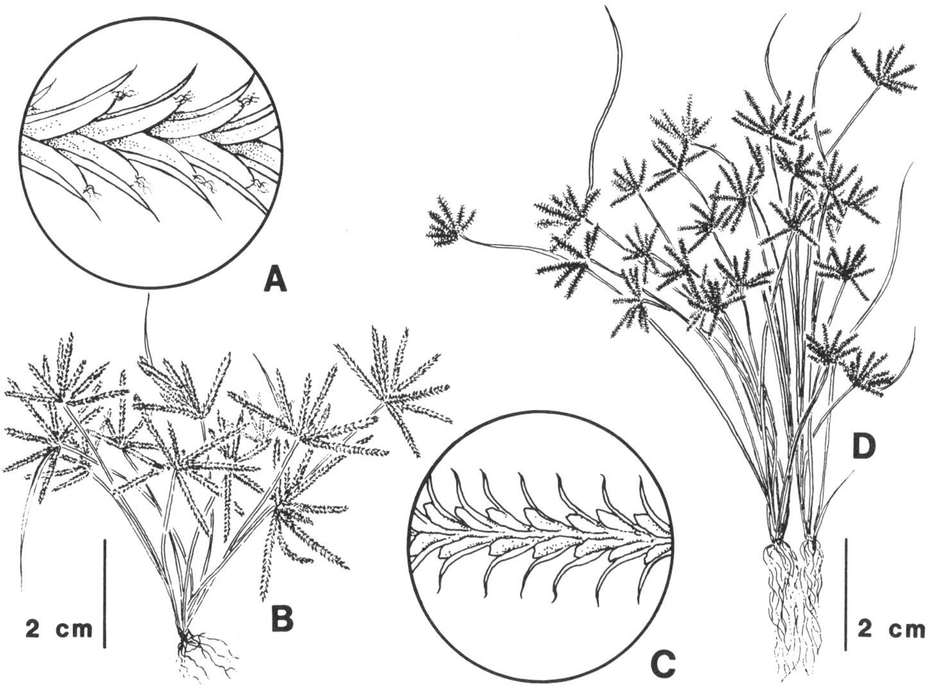

FIG. 20. *Cyperus amabilis* and *C. cuspidatus*. A, B, *C. amabilis*. A. Portion of spikelet. B. Habit. C, D, *C. cuspidatus*. C. Portion of spikelet. D. Habit. Drawn by Patricia M. Kernan.

Aricagua, *Humboldt & Bonpland s.n.* (holotype: B-W no. 1330, IDC microfiche!).

Cyperus glareosus Liebmann, Mexic. halvgr. 16. 1850.—TYPE: MEXICO. Veracruz: Potrero de Consoquitla, 1841–42, *Liebmann 14365* (lectotype, designated by Tucker, 1986: C!; isolectotypes: G! UC!).

Cyperus guadalajaranus M. E. Jones, Contr. W. Bot. 18: 25. 1933.—TYPE: MEXICO. Jalisco: Guadalajara, 26 Nov 1930, *Jones 27603* (holotype: POM!; isotypes: DS! MO! UC!).

Caespitose annual, 3–12 cm tall. Rhizomes absent. Culms 0.3–0.8 mm in diameter, roundly trigonous, smooth. Leaves 1–3 (–4), 1–6 cm long, 1.0 (–1.5) mm wide, flat to v-shaped, the margins sparsely scabrellate. Inflorescence bracts 3–6, (0.5–) 2–10 (–15) cm long, 0.5–2.0 mm wide, flat to v-shaped, the margins scabrellate, erect to ascendent. Rays generally absent, the inflorescence 10–30 (–50) mm in diameter, loosely ovoid; sometimes rays present, 1–8 cm long; secondary rays infrequently also present, 1–2 cm long. Heads digitate, 5–20 mm long, 10–30 mm wide. Spikelets 5–10 (–20), (5–) 8–12 (–35) mm long, 1.6–2.3 (–2.5) mm wide, linear to linear-lanceolate; rachilla persistent, 0.2 mm wide, 0.1 mm thick, geniculate, reddish brown, successive scales 0.6–0.8 mm apart, wings absent. Scales deciduous, (6–) 16–30 (–76), 1.6–2.0 mm long excluding the straight to slightly excurved cusp of 0.3–0.4 (–0.5) mm, 0.8–1.1 mm wide, ovate-lanceolate, truncate, emarginate, laterally nerveless, glossy reddish brown to brown, medially strongly 3-nerved, greenish or sometimes also reddish or brownish. Stamens (1–) 2–3; fila-

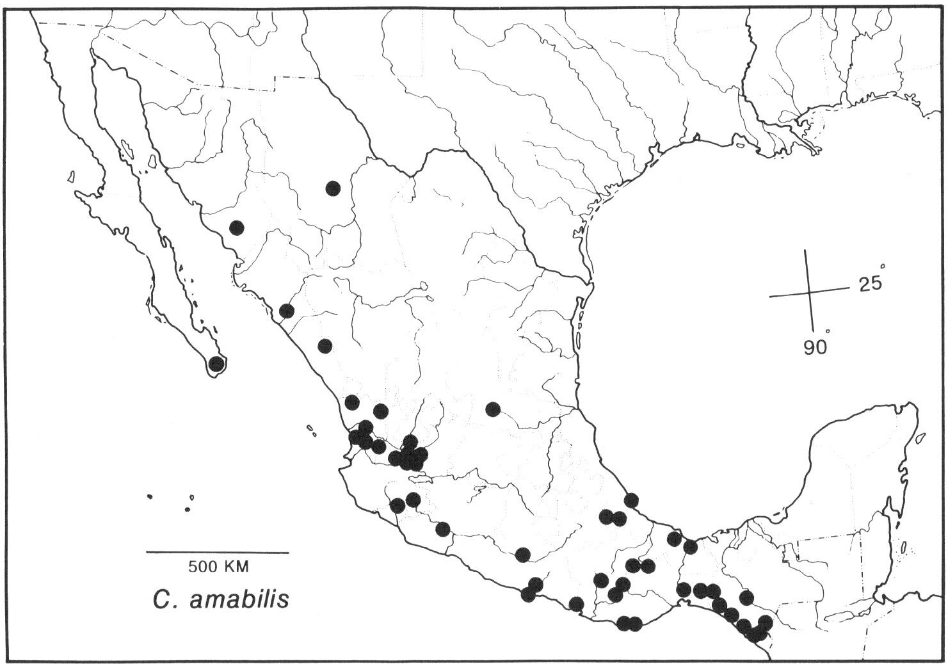

FIG. 21. Distribution of *Cyperus amabilis* in Mexico.

ments (0.7–) 1.0–1.3 mm long; anthers ca. 0.3 mm long, ellipsoid, the connective apex not prolonged. Style 0.3–0.4 mm long; stigmas 3, 0.4–0.7 mm long. Achenes 0.8–0.9 mm long, 0.4–0.5 mm wide, trigonous, oblong-obovoid, apex abruptly apiculate, base slightly stipitate, the faces flat distally, concave basally, the surface papillose, brown. Fig. 20A, B.

Phenology. Flowering and fruiting from July to October (–November).

Distribution. Pantropic regions; Australia, southern Asia, Africa; in the New World in the Lesser Antilles (Curaçao), and from southern Arizona, U.S.A., south to southern Brazil. Frequent in Mexico (Fig. 21), especially along the Pacific Coast; dry open slopes, soil pockets on wet roadcuts or lava flows, moist sandy roadsides, and sandy alluvium; sea level to ca. 2000 m.

REPRESENTATIVE SPECIMENS. **Mexico.** BAJA CALIFORNIA SUR: Cape region, El Taste, 11 Sep 1893, *Brandegee s.n.* (UC).—CHIAPAS: 27 km N of Acala, *Breedlove 36896* (DS); ca. 3 km NE of Escuintla, *Tucker 2237* (DUKE, ENCB); ca. 5 km N of Tonalá, *Tucker 2705* (DUKE, ENCB, MICH); Belisario Domínguez on Rt. 211, *Wilbur 35868* (CAS, DUKE, ENCB, MICH).—CHIHUAHUA: mountains near Chihuahua, *Pringle 811* (CAS, F, G, GH, LCU, MICH, MO, MSC, NY, PENN, PH, SMU, UC, VT).—COLIMA: Colima, *Orcutt 4543* (DS, GH, MO).—GUERRERO: Dtto. Mina, Zihuaqueo–Fila Mayor, *Hinton 9305* (BM, DS, F, GH, LCU, MICH, NY, UC); between Acapulco and Tierra Colorada, *Tucker 2263* (DUKE).—JALISCO: Mpio. Zapopan, Las Agujas, *Bernely 11* (IBUG); 2 mi NW of Tequila, *McVaugh 18640* (G, MICH, NY, US); near Guadalajara, *Pringle 11722* (BKL, C, CAS, CU, F, GH, LCU, MICH, MSC, VT); on Hwy 54 ca. 40 km N of Guadalajara, *Tucker 2999* (DUKE).—MICHOACÁN: 8 km N of Arteaga, 24 Oct 1961, *King & Soderstrom s.n.* (US).—NAYARIT: 15 km W of Ahuacatlán, *Feddema 510* (DS, MICH, TEX).—OAXACA: 2 km S of Niltepec, *King 1776* (MICH, NY, TEX, UC); ca. 10 km S of Xoxocotlán, *Tucker 2822* (DUKE, ENCB, MICH); ca. 26 km N of Arriaga, *Tucker 2728* (DUKE, ENCB); ca. 30 km E of Puerto Escondido, *Tucker 2895* (DUKE, ENCB).—SAN

Luis Potosí: region of San Luis Potosí, *Parry & Palmer 433* (BM).—Sinaloa: Culiacán, 20 Aug 1904, *Brandegee s.n.* (UC); Sierra Surutato, Mpio. Sinaloa y Vela, El Alamo, *Breedlove & Thorne 17881* (CAS).—Sonora: Canyon Sapopa, Río Mayo, *Gentry 1082* (MICH); 2 km NE of El Guayabo, *Martin & O'Rourke s.n.* (ARIZ 302747).—Veracruz: Orizaba, 10 Aug 1924, *Fisher s.n.* (DS, MO, US, WIS); Zacuapan, *Purpus 8025* (G, GH, M, MO, NY, UC, US, Z).

Cyperus amabilis is recognizable by its small size, annual habit, and its strongly 3-nerved, emarginate, cuspidate scales. It is distinguished from the morphologically similar (and possibly closely related) *C. cuspidatus* by its larger achenes and scales and its shorter scale awns.

29. Cyperus cuspidatus H. B. K., Nov. gen. sp. 1 [quarto]: 204. 1816. *Dichostylis cuspidata* (H. B. K.) Palla, Bot. Jahrb. Syst. 10: 296. 1889.—Type: Venezuela. Along Orinoco River, *Humboldt & Bonpland s.n.* (holotype: B-W no. 1299, IDC microfiche!).

Cyperus exiguus Nees in Martius, Fl. bras. 2(1): 22. 1842. *Cyperus cuspidatus* f. *exiguus* (Nees) Kükenthal, Pflanzenreich 4(20): 264. 1936.—Type: Brazil. Bahia, collector unknown, *Herb. Lindley* (holotype: CGE!).

Caespitose annual, (0.5–) 2–5 (–9) cm tall. Rhizomes absent. Culms 0.2–0.4 (–0.6) mm in diameter, trigonous to roundly trigonous, smooth. Leaves 1–6, 1–6 cm long, 0.3–1.0 mm wide, v-shaped, the margins and keel smooth or occasionally distally scabrellate. Inflorescence bracts 2–3 (–5), 2–60 mm long, 0.3–1.2 mm wide, v-shaped, the margins and keel smooth or sparsely scabrellate, horizontal to ascendent at 45° (–60°). Rays absent, or 1–6 (–8), 2.5–20 (–30) mm long. Heads digitate, 3–18 mm long, 3–26 mm wide, loosely ovoid. Spikelets (3–) 10–20 (–30), 3–11 mm long, (0.8–) 1.2–1.5 mm wide, narrowly ellipsoid to oblong; rachilla persistent, 0.2–0.3 mm wide, 0.1–0.2 mm thick, straight, reddish or yellowish brown, successive scale scars 0.5–0.7 mm apart, wings absent. Scales deciduous, (3–) 6–20 (–28), (0.6–) 0.9–1.2 mm long excluding the strongly recurved cusp (0.6–1.2 mm long), 0.8–1.0 (–1.2) mm wide, oblong-ovate, laterally nerveless, greenish to reddish brown, glossy, medially strongly 3-nerved, greenish to light reddish brown. Stamens 2–3; filaments 0.7–1.1 mm long; anthers 0.1 (–0.2) mm long, oblong-ovoid, the connective apex not prolonged. Style (0.2–) 0.4–0.6 mm long; stigmas 3, 0.3–0.5 mm long. Achenes (0.5–) 0.6 mm long, (0.3–) 0.4 mm wide, trigonous, oblong-obovoid, wide, apex blunt, base cuneate (not stipitate), the faces slightly convex, the surface papillose, brown. Fig. 20C, D.

Phenology. Flowering and fruiting from June through October.

Distribution. South Carolina to Texas, U.S.A., the West Indies, and Mexico south to Guatemala and Honduras; also in South America from Colombia to eastern Brazil. Occasional in Mexico (Fig. 22), from central Veracruz, western México, and Nayarit to Chiapas, absent from the Yucatan Peninsula; sand bars along rivers, moist sandy or gravelly roadsides, croplands, and pastures; 0–500 (–900) m; one collection from Guerrero at 1340 m.

Representative Specimens. **Mexico.** Chiapas: 3 km from Escuintla, *Tucker 2238* (DUKE); 30 km N of Escuintla on rd to Pijijiapán, *Tucker 2696* (DUKE); 5 km N of Tonalá, *Tucker 2703* (DUKE).—Guerrero: between Acapulco and Tierra Colorada, *Tucker 2267* (DUKE); 80 mi SE of Acapulco, *Wilbur 36539* (DUKE, ENCB, MICH).—Jalisco: near La Resolana, *Wilbur & Wilbur 2261* (MICH);

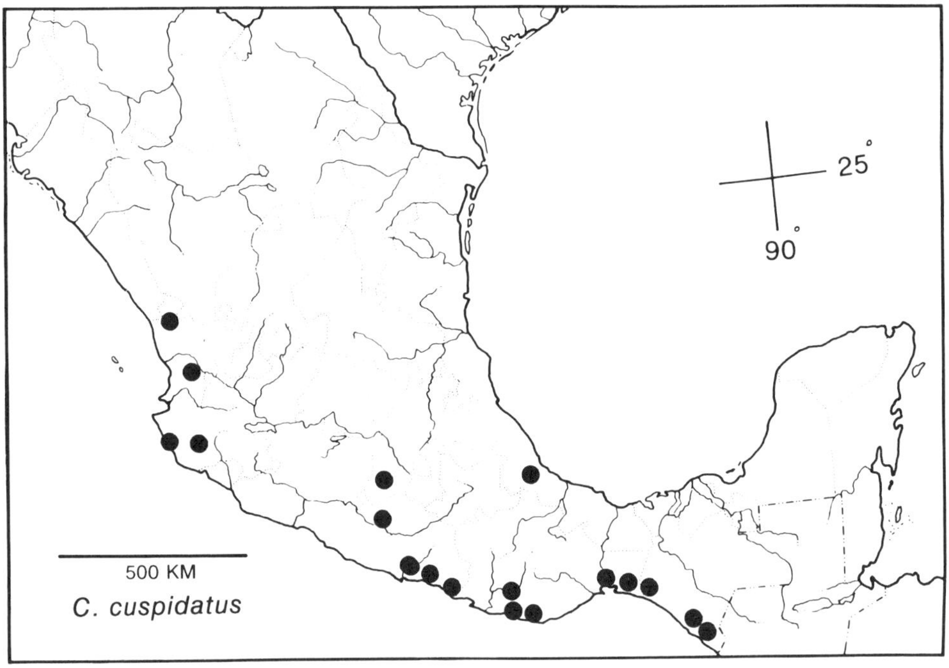

FIG. 22. Distribution of *Cyperus cuspidatus* in Mexico.

Tomatlán, *Santana M. 182* (IBUG).—MÉXICO: Dtto. Temascaltepec, Tejupilco, *Hinton 4762* (BM, G, GH, LCU, MO).—NAYARIT: 3 mi NE of Puga, *Feddema 892* (MICH).—OAXACA: Hwy 175, 25 km N of Pochutla, *Tucker 2246* (DUKE); 26 km N of Arriaga, *Tucker 2727* (DUKE, ENCB); 30 km E of Niltepec, *Tucker 2738* (DUKE); 40 km W of Puerto Escondido, *Tucker 2905* (DUKE, ENCB); 15 km W of Pinotepa Nacional, *Tucker 2916* (DUKE, ENCB).—VERACRUZ: Consoquitla, 1841–42, *Liebmann s.n.* (C, GH, HAL).

Cyperus cuspidatus is readily identified by its annual habit, small size, and conspicuously cuspidate scales. It is morphologically similar to *C. amabilis*, as indicated by the annual habit, and glossy, 3-nerved, cuspidate scales. The two species are distinguished by the shape of the achenes, length of the scales, and, especially, by the length of the cusp of the scales (see Fig. 20).

II. Cyperus subgenus **Pycreus** (Palisot de Beauvois) A. Gray, Man. Bot. ed. 1, 517. 1848. *Pycreus* Palisot de Beauvois, Fl. Oware 2: 48. 1807. *Cyperus* sect. *Pycreus* (Palisot de Beauvois) Grisebach, Spicil. Fl. Rumel. 2: 419. 1844.—TYPE: *Cyperus polystachyos* Rottbøll.

Culms and leaves with kranz (chlorocyperoid) anatomy. Spikelets loosely spicate to subdigitate. Rachilla persistent, wings absent or nearly so. Scales bicarinate or conduplicate, deciduous. Carpels 2, stigmas 2. Achenes lenticular or biconvex (compressed).

Subgenus *Pycreus* is distinguished by its laterally compressed, lenticular or biconvex achenes (Fig. 23). It comprises about 100 species worldwide; the greatest diversity is in tropical Africa. About 25 species occur in the New World. Only one of the 9 species found in Mexico, *C. aschenbornianus*, is endemic.

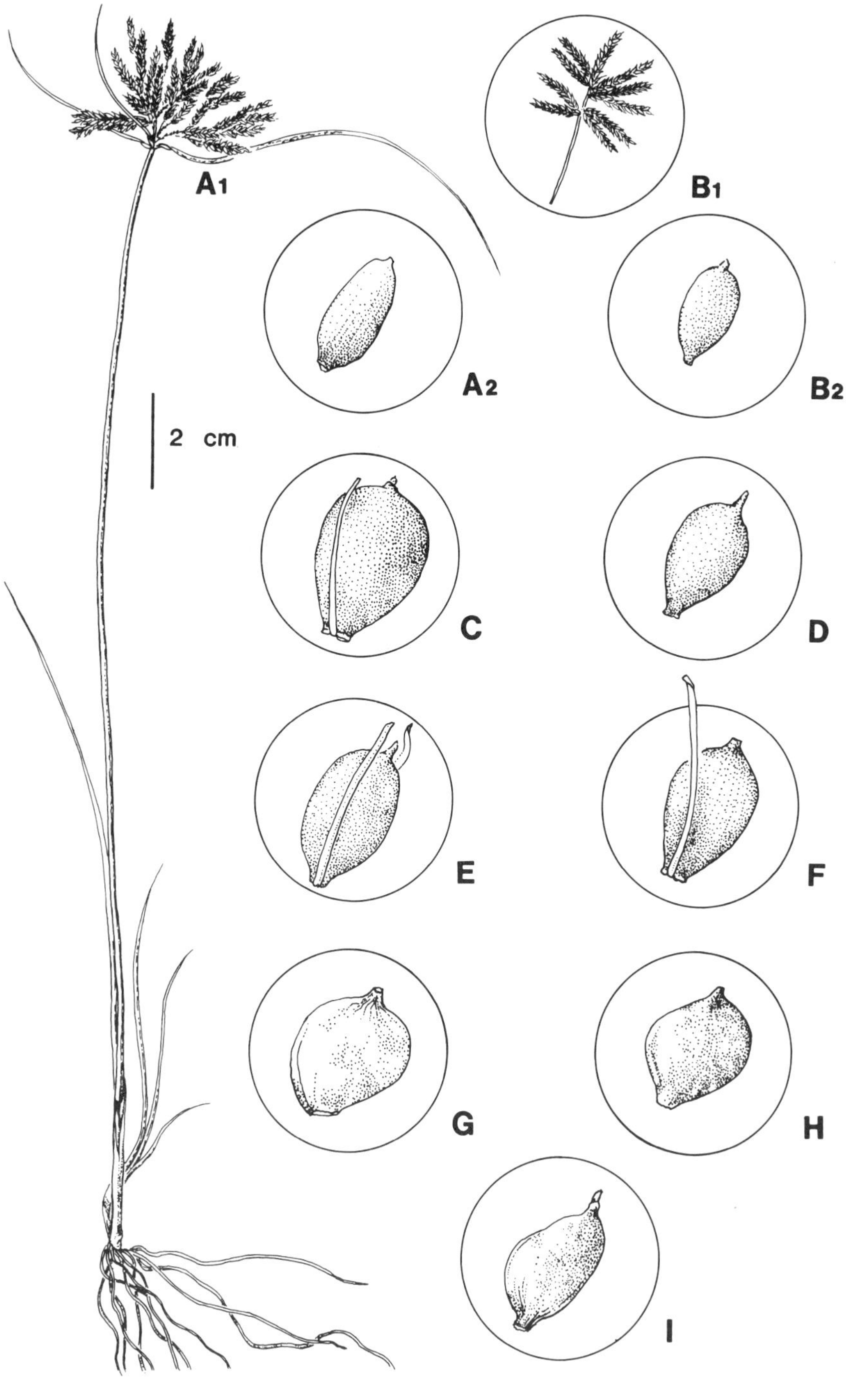

FIG. 23. *Cyperus* subg. *Pycreus* in Mexico. A, *C. polystachyos*. A_1. Habit. A_2. Achene (×2/3). B, *C. fugax*. B_1. Spike (×2/3). B_2. Achene (×12). C–I, Achenes (×12). C. *C. flavicomus*. D. *C. lanceolatus*. E. *C. bipartitus*. F. *C. niger*. G. *C. unioloides*. H. *C. flavescens* var. *piceus*. I. *C. aschenbornianus*. Drawn by Patricia M. Kernan.

30. Cyperus flavicomus Michaux, Fl. bor.-amer. 1: 27. 1803. *Pycreus flavicomus* (Michaux) C. D. Adams, Ann. Missouri Bot. Garden 78: 254. 1991.—TYPE: U.S.A. "Carolina," 1786–96, *Michaux s.n.* (holotype: P; IDC microfiche Herb. Michaux!).

Pycreus albo-marginatus Martius & Schrader ex Nees in Martius, Fl. bras. 2(1): 9. 1842. *Cyperus albomarginatus* (Martius & Schrader ex Nees) Steudel, Syn. pl. glum. 2: 10. 1854.—TYPE: BRAZIL. Piaui, *Martius s.n.* (holotype: M!).

Pycreus sabulosus Martius & Schrader ex Nees in Martius, Fl. bras. 2(1): 10. 1842. *Cyperus sabulosus* (Nees) Steudel, Syn. pl. glum. 2: 10. 1854. *Cyperus albomarginatus* var. *sabulosus* (Martius & Schrader ex Nees) Kükenthal, Pflanzenreich 4(20): 361. 1936.—TYPE: BRAZIL. [Amazonas:] "Prov. Río Nigri, juxta lacum di Coari," *Martius s.n.* (holotype: M!).

Caespitose or single-stemmed annual, 30–75 cm tall. Rhizomes absent. Culms 1–5 mm in diameter, trigonous, smooth, basally soft, easily compressed. Leaves 1–3, usually withered at anthesis, 5–35 cm long, 2–8 mm wide, v-shaped, the margins and keel scabrellate. Inflorescence bracts 3–7, (2–) 8–35 cm long, (1–) 4–10 mm wide, v-shaped, the margins and keel scabrellate, horizontal to ascendent at 30°. Rays (3–) 5–11, 0.4–20 cm long; secondary rays occasionally present, 5–28 mm long. Spikes 5–25 mm long, 8–22 mm wide, loosely ovoid; rachis 4–15 mm long. Spikelets 6–60, 7–30 mm long, 2.0–3.0 mm wide, linear to lanceolate, appearing serrate-margined owing to the spreading scales; rachilla persistent, ca. 0.4 mm wide, 0.2 mm thick, flexuous, quadrate, successive scale scars 1.5–1.7 mm apart, wings 0.1–0.2 mm wide, hyaline. Scales deciduous, 6–24, (1.4–) 1.7–2.3 mm long, 1.4–2.0 mm wide, ovate-obovate, apex acute to obtuse, laterally 1–2-nerved, light brown to deep red, with a conspicuous hyaline border, medially 3-nerved, greenish. Stamens 2–3; filaments 1.5–2.1 mm long; anthers ca. 0.4 mm long, the connective apex up to 0.1 mm, reddish. Style 0.1–0.3 mm long; stigmas 2, 0.7–1.5 mm long. Achenes 1.2–1.6 mm long, 0.6–1.1 mm wide, lenticular, obovoid, apex broadly rounded, base sessile, the faces flattened-depressed, the surface puncticulate, jet black to dark reddish brown. Fig. 23C.

Phenology. Flowering and fruiting from July to October.

Distribution. Pantropic and warm temperate regions; in North America, in the U.S.A. along the Coastal Plain and Piedmont from Florida to New York and westward to Kentucky and New Mexico (Fernald 1950), and along the Pacific Coast of Mexico and Central America (Tucker 1983); in South America from eastern Brazil to Ecuador and Paraguay; also in the Old World tropics (Kükenthal 1935–36). In Mexico (Fig. 24) frequent in moist pastures, along sandy moist roadsides, on sand bars, and at edges of rivers; sea level to 1500 m.

REPRESENTATIVE SPECIMENS. **Mexico.** CHIAPAS: ca. 7 km N of Arriaga, *Tucker 2722* (DUKE).—CHIHUAHUA: near Chihuahua, *Pringle 810* (AC, BM, F, GH, MICH, UC, VT).—GUERRERO: Dtto. Mina, Manchón, *Hinton 9460* (GH, MICH, US); 15 mi SE of Michoacán border on Route 200, *Wilbur 36579* (DUKE, ENCB).—JALISCO: 8 km NW of Ameca, *González & Puga 625* (IBUG); 15 km W of Guadalajara, *Iltis & Lasseigne 629* (IBUG, WIS); ca. 7 mi WNW of Tototlán, *McVaugh 17271* (DUKE, MICH); Unión de Tula, *Wilbur 36761* (DUKE, ENCB, MICH).—MICHOACÁN: 8–10 km SW of Jiquilpan, *Feddema 99* (DUKE, MICH).—NAYARIT: 12 mi SE of Tepic, *Feddema 552* (DUKE,

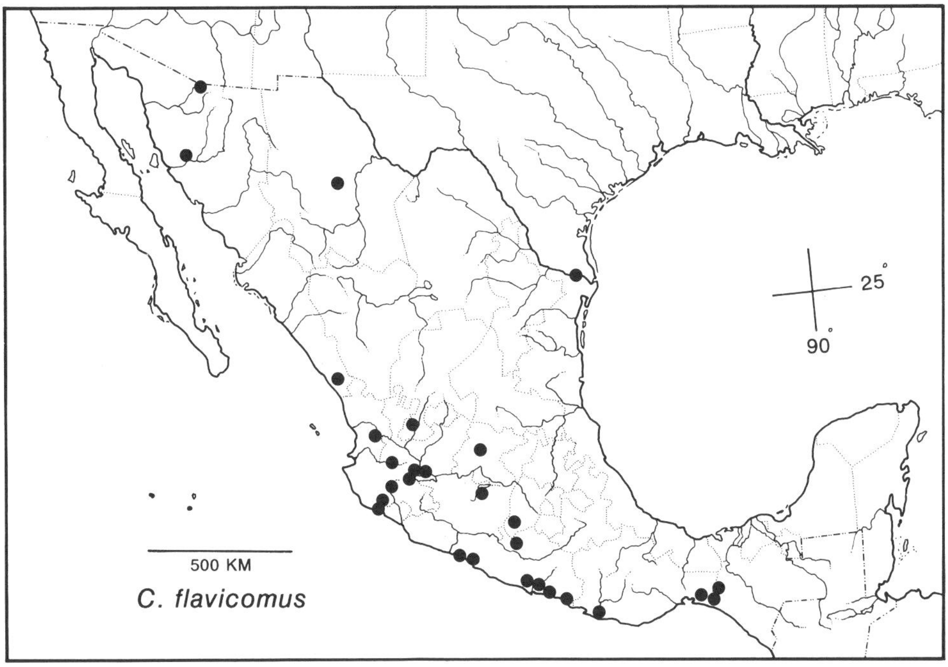

FIG. 24. Distribution of *Cyperus flavicomus* in Mexico.

MICH).—Oaxaca: 20 km E of Pinotepa Nacional, *Tucker 2911* (CAS, DUKE, ENCB, MICH); Tepanatepec, *Wilbur 36074* (DUKE, ENCB).—Sinaloa: Rosario, *Rose 1881* (NY, US).—Sonora: between Hermosillo and Tastiota, *Wiggins & Rollins 275* (DS, GH, NY).—Zacatecas: Teul de González Ortega, *Correa 78* (IBUG).

Cyperus flavicomus is readily identified by the combination of lenticular, obovoid achenes and widely spaced, non-overlapping scales with prominent hyaline borders to which the epithet "*albomarginatus*" refers (Tucker 1983, 1985).

In most recent floristic works, *C. albomarginatus* is attributed to "Mart. & Schrad. ex Nees" or simply "Mart. & Schrad." However, the name published by Nees in *Flora brasiliensis* in 1842 was *Pycreus albo-marginatus* Mart. & Schrad. ex Nees; *C. albomarginatus* was merely a manuscript name listed in synonymy. The combination *Cyperus albomarginatus* was first validly published by Steudel in 1854.

In nineteenth century manuals, e.g., Chapman (1860), and early twentieth century works, e.g., Britton and Brown (1901) and Robinson and Fernald (1908), the name applied to this species is *C. flavicomus* Michx.; however, Fernald (1950), Gleason and Cronquist (1963), and Radford et al. (1967) called this species *C. albomarginatus*. The change seems to have been the result of the publication of Kükenthal's worldwide monograph of *Cyperus* in *Das Pflanzenreich* (1935–36). There, Kükenthal treated this species as *C. albomarginatus* Martius & Schrad. ex Nees and placed the older name *C. flavicomus* Michx. in the synonymy of the common Eastern North American species *C. strigosus* L. Through the courtesy of the Muséum National d'Histoire Naturelle, Paris, I was able to examine a photograph of the type of *C. flavicomus* Michx. (This type can also be seen in the IDC microfiche edition of Michaux's herbarium.) Apparently, Kükenthal never saw the type of *C. flavicomus*, for it is clearly a plant of the species he and recent authors referred to as *C. albomarginatus*. Thus, the oldest name for the species is *C. flavicomus*.

Hooper and Raynal (1969) treated "*C. albomarginatus*" as a synonym of the earlier name *C. macrostachyos* Lam. (Tabl. encycl. 1: 147, 1791). Evidently, they did not see the type of *C. albomarginatus*. Examination of a photograph of the type of *C. macrostachyos* ("ex Africa," Herb. Lamarck, P) shows it to be strikingly different from *C. flavicomus*, notably in its long spikelets with 60 to 70 closely imbricate scales; *C. flavicomus* has only 6 to 24 scales per spikelet. *Cyperus flavicomus* is also distinctive in its spreading, non-overlapping scales (Kükenthal 1935–36; Fernald 1950; Tucker 1983). Thus, the type of *C. macrostachyos* clearly does not belong to the same species as that of *C. flavicomus*. It seems best assigned to the Indian-Asian *C. puncticulatus* var. *quinquagintiflorus* C. B. Clarke, sensu Kükenthal (1935–36). Such a disposition of *C. macrostachyos* was supported by Kunth (1837), who also treated *C. macrostachyos* as a variety of *C. puncticulatus*.

Wiggins (1980) reported *C. flavicomus* as occurring in northeastern Baja California, but I found no specimens from that state.

31. Cyperus fugax Liebmann, Mexic. halvgr. 8. 1850. *Cyperus polystachyos* f. *fugax* (Liebmann) Kükenthal, Repert. Spec. Nov. Regni Veg. 23: 184. 1926. *Pycreus fugax* (Liebmann) C. D. Adams, Ann. Missouri Bot. Garden 78: 254. 1991.—TYPE: MEXICO. Oaxaca: Jocotepec, Jun 1842, *Liebmann 14413* (holotype: C!; isotypes: GH! NY!).

Cyperus tenellus J. Presl & K. Presl in K. Presl, Reliq. Haenk. 1: 176. 1828, non *C. tenellus* L. f., 1781.—TYPE: MEXICO. 1791, *Haenke s.n.* (holotype: PRC!).

Cyperus inconspicuus Liebmann, Mexic. halvgr. 9. 1850. *Cyperus polystachyos* f. *inconspicuus* (Liebmann) Kükenthal, Pflanzenreich 4(20): 372. 1936.—TYPE: COSTA RICA. San José, 1845–48, *Oersted 14392* (holotype: C!).

Caespitose annual, (0.5–) 3–6 (–12) cm tall. Rhizomes absent. Culms (0.2–) 0.4–0.8 mm in diameter, trigonous distally, roundly trigonous proximally, smooth. Leaves 1–4, (1–) 3–5 (–10) cm long, (0.5–) 1.0–2.0 mm wide, v-shaped, scabrellate on the margins and keel. Inflorescence bracts (1–) 2–4 (–5), (1–) 2.5–6.0 (–11) cm long, 0.5–1.8 mm wide, v-shaped, the margins and keel scabrellate, ascendent at 30–45°. Rays (1–) 2–4 (–6), (0.5–) 1–3 (–6) cm long; secondary rays absent. Spikes 8–15 (–25) mm wide, loosely ovoid; rachis 1–3 mm long. Spikelets (2–) 4–6 (–10), (5–) 8–12 (–18) mm long, 1.2–1.5 (–1.8) mm wide, linear-lanceolate, strongly flattened; rachilla persistent, ca. 0.2 mm wide, ca. 0.2 mm thick, geniculate, stramineous to light brown, often with reddish speckles, successive scales (1.0–) 1.2–1.3 mm apart, the wings ca. 0.2 mm wide, hyaline, fragile (or absent). Scales deciduous, (5–) 10–16 (–32), (1.2–) 1.3–1.6 (–2.0) mm long, 1.0–1.2 (–1.4) mm wide, ovate, apex obtuse, mucronate to mucronulate, laterally nerveless, hyaline, stramineous or light reddish brown, medially 3- (–5-) nerved, green. Stamens 2; filaments 1.2–1.4 (–1.6) mm long; anthers 0.1–0.2 mm long, ovoid to oblong-ovoid, the connective apex not prolonged. Style 0.1–0.3 (–0.4) mm long; stigmas 2, 0.3–0.4 mm long. Achenes 0.9–1.1 mm long, 0.4–0.5 mm wide, lenticular, oblong-

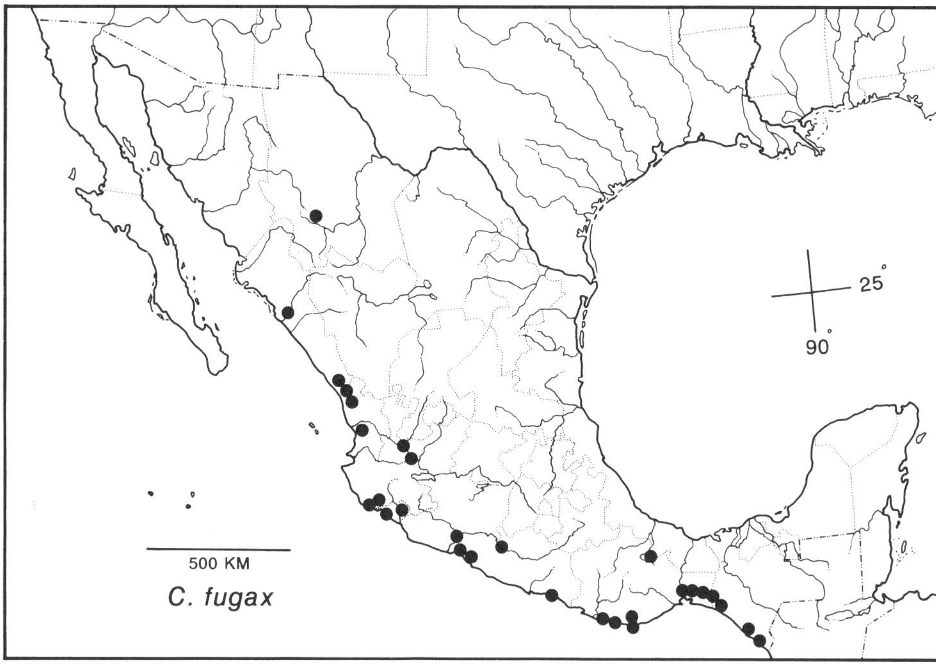

FIG. 25. Distribution of *Cyperus fugax* in Mexico.

obovoid, apex truncate, apiculate, the stipitate base ca. 0.1 mm long but conspicuous, the surface punctate, light to dark brown, glossy or iridescent. Fig. 23B.

Phenology. Flowering and fruiting mostly from July to February, and occasionally throughout the year.

Distribution. Central Texas, U.S.A., south through Mexico and Central America to the Isthmus of Panama, also in Cuba, Isla Margarita (Venezuela), and the Galapagos Islands. In Mexico (Fig. 25) along the coast and western mountains from southern Chihuahua to Chiapas; roadsides, stone fences, sand bars along rivers, disturbed sandy soil, and edges of puddles; sea level to ca. 1000 m.

REPRESENTATIVE SPECIMENS. **Mexico.** CHIAPAS: gravel rd from Escuintla to El Triunfo, *Tucker 2240* (DUKE); along Hwy 200 ca. 5 km N of Tonalá, *Tucker 2706* (DUKE, ENCB).—CHIHUAHUA: SW Chihuahua, *Palmer 63–1/2* in 1885 (BM, GH, MICH, NY, US, VT).—COLIMA: Manzanillo, *Palmer 1079* in 1890 (GH, MICH, NY, US).—GUERRERO: Dtto. Coyuca, Pungarabato, *Hinton 6699* (F, GH, MO, NY, US); ca. 6 km E of Puerto Copala, *Tucker 2928* (DUKE, ENCB, MICH).—JALISCO: Mpio. La Huerta, Est. Biológica Chamela, *Lott 569* (DUKE, MEXU, MICH); 12 km S of La Huerta, S of Autlán, *Tucker 2968* (DUKE, ENCB).—NAYARIT: Matanchen Bay, 3 km SE of San Blas, *Marcks 1174* (WIS).—OAXACA: 30 km E of Niltepec, *Tucker 2741* (DUKE, ENCB); 30 km E of Puerto Escondido, *Tucker 2901* (DUKE, ENCB).—SINALOA: Culiacán, *Gentry 5014* (ARIZ, GH, MICH, NY).—SONORA: 12 mi E of Alamos, *Sanders 9331* (NYS, UCR).—VERACRUZ: Mpio. Cosamaloapan, Playa Vicente, *Martínez Calderón 1983* (F).

Kükenthal (1935–36) and Benedict Ayers (1946) treated *C. fugax* as a form of *C. polystachyos*, but it fully merits specific status. The most conspicuous differences are the annual habit and small size (less than 12 cm tall); *C. polystachyos* is a perennial, although probably often flowering during the first year, of moderate size, mostly taller than 20 cm (Fig. 23A). Other differences are summarized as follows:

1. Perennial with short rhizomes; scales oblong, 1.8–2.4 mm long, closely imbricate; styles 0.6–1.0 mm long; stigmas 1.4–2.0 mm long; filaments 1.8–2.3 mm long; anthers 0.6–0.8 mm long; achenes substipitate to cuneate. *C. polystachyos*.
1. Annual, without rhizomes; scales ovate, 1.3–1.6 mm long, loosely imbricate; styles 0.1–0.2 mm long; stigmas 0.3–0.4 mm long; filaments 1.2–1.4 mm long; anthers 0.1–0.2 mm long; achenes stipitate. *C. fugax*.

I could find no previous record of this species in South America and therefore list the following specimens: ECUADOR: Prov. Guayas, *Asplund 5032* (BR); Galapagos Islands, Chatham Is., *Baur 311* (GH).

32. Cyperus polystachyos Rottbøll, Descr. icon. rar. pl. 39, pl. 11, fig. 1. 1773. *Pycreus polystachyos* (Rottbøll) Palisot de Beauvois, Fl. Oware 2: 48. 1807.—TYPE: probably from India, collector unknown (holotype: Herb. Rottbøll, C!; isotype: Herb. Vahl, C!).

Cyperus olidus Vahl, Enum. pl. 2: 333. 1805.—TYPE: Central America. *Richard s.n.* (holotype: C!).

Cyperus brizaeus J. Presl & K. Presl in K. Presl, Reliq. Haenk. 1: 168. 1828.—TYPE: MEXICO. 1790–91, *Haenke s.n.* (holotype: PRC!).

Cyperus polystachyos var. *leptostachyus* Boeckeler, Linnaea 35: 478. 1868.—TYPE: U.S.A. Texas, *Drummond 454* (lectotype, here designated: GH!).

Caespitose perennial, (3–) 20–45 (–70) cm tall. Rhizomes 1–3 mm in diameter, up to 2 cm long, oblique. Leaves (2–) 6–10 (–16), (2–) 10–30 (–60) cm long, 1–4 mm wide, flat to v-shaped, the margins and keel remotely scabrellate or smooth. Culms (0.8–) 1.2–2 mm in diameter, trigonous or subterete, smooth. Inflorescence bracts (3–) 4–6 (–7), (2–) 5–15 (–23) cm long, 1–4 mm wide, flanged v-shaped, the margins and keel scabrellate, horizontal to ascendent at 45°. Rays 1–8 (–12), 0.5–3 (–6) cm long; secondary rays absent; rays sometimes absent, the inflorescence 12–40 mm in diameter, dense, ± ovoid. Spikes 12–40 mm long, 10–25 mm wide, loosely ovoid to densely turbinate, the spikelets spreading to appressed-ascending; rachis 3–10 mm long. Spikelets (5–) 15–30 (–40), 8–18 (–40) mm long, (1.0–) 1.3–1.6 (–2.0) mm wide, linear-lanceolate to linear, compressed; rachilla persistent, ca. 0.2 mm wide, 0.2 mm thick, geniculate, light brown to stramineous, reddish speckled, successive scales 1.0–1.4 mm apart, the wings ca. 0.2 mm wide, hyaline, fragile. Scales deciduous, (8–) 12–25 (–66), (1.5–) 1.8–2.4 mm long, (1.0–) 1.2–1.4 mm wide, oblong, apex acute to subacute, mucronulate, laterally nerveless, stramineous to yellowish brown, medially 3–5-nerved, green. Stamens 2; filaments 1.8–2.3 mm long; anthers (0.4–) 0.6–0.8 mm long, linear-oblong, the connective apex not prolonged. Style 0.6–1.0 (–1.4) mm long; stigmas 2, (1.0–) 1.4–2.0 mm long. Achenes (0.8–) 0.9–1.1 (–1.2) mm long, 0.4–0.5 (–0.6) mm wide, biconvex to subcylindric, oblong, apex truncate, apiculate, base cuneate to substipitate, the faces convex, the surface punctate to puncticulate with raised cells, light to dark brown. Fig. 23A.

Phenology. Flowering and fruiting as early as May, but primarily from August to January.

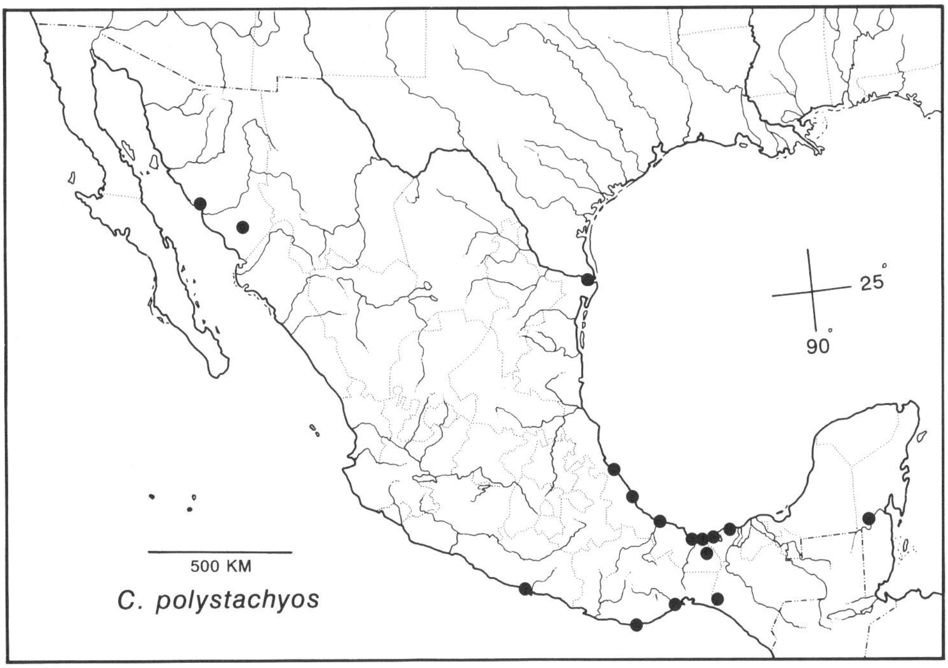

FIG. 26. Distribution of *Cyperus polystachyos* in Mexico.

Distribution. Pantropic regions; in the New World in the U.S.A. (as far north as Maine, West Virginia, and Oklahoma), in Mexico, Central America, and the West Indies, and in South America as far south as Paraguay and Uruguay. Scarce in Mexico, along the Gulf and Pacific Coasts (Fig. 26); wet pastures, stream banks, marshes, edges of mangrove swamps, mineral springs, and disturbed sites with full sun and moist soil; less than 500 m.

REPRESENTATIVE SPECIMENS. **Mexico.** CHIAPAS: 6 km N of Arriaga, *Breedlove 27299* (CAS).—GUERRERO: Acapulco, *Palmer 117* in 1895 (GH, MO, NY, UC, US).—OAXACA: 5 km N of Puerto Angel, *Rzedowski 21168* (DS, ENCB, MICH, MSC).—QUINTANA ROO: 33 km SW of Chetumal, *Chater 79* (BM).—SONORA: Río Mayo, Salitral, swamp in water of mineral springs, *Gentry 1576* (ARIZ, F, GH, MO, NY, UC), *Martin & O'Rourke s.n.* (ARIZ 302751).—TABASCO: Santa Ana, W end of Laguna, 1962, *Barlow s.n.* (GH, UC, WIS).—TAMAULIPAS: valley of the lower Río Grande, 1879–83, *Buckley s.n.* (NY).—VERACRUZ: on Hwy 180 just N of rd to Riachuelos, *Tucker 2480* (DUKE, ENCB); just E of Coatzacoalcos on Hwy 180, *Tucker 2524* (CAS, DUKE, ENCB, F, GH, MICH).

Cyperus polystachyos can be distinguished from other species of *Cyperus* in Mexico by the combination of linear to linear-lanceolate spikelets, and oblong, lenticular achenes with truncate, apiculate apices.

Cyperus polystachyos has several races or varieties, which are distinct in certain areas outside Mexico. Several have been treated as species. Only plants matching the type of *C. polystachyos* were seen from Mexico. Adequate understanding of the infraspecific variations of *C. polystachyos* requires a worldwide examination of the species.

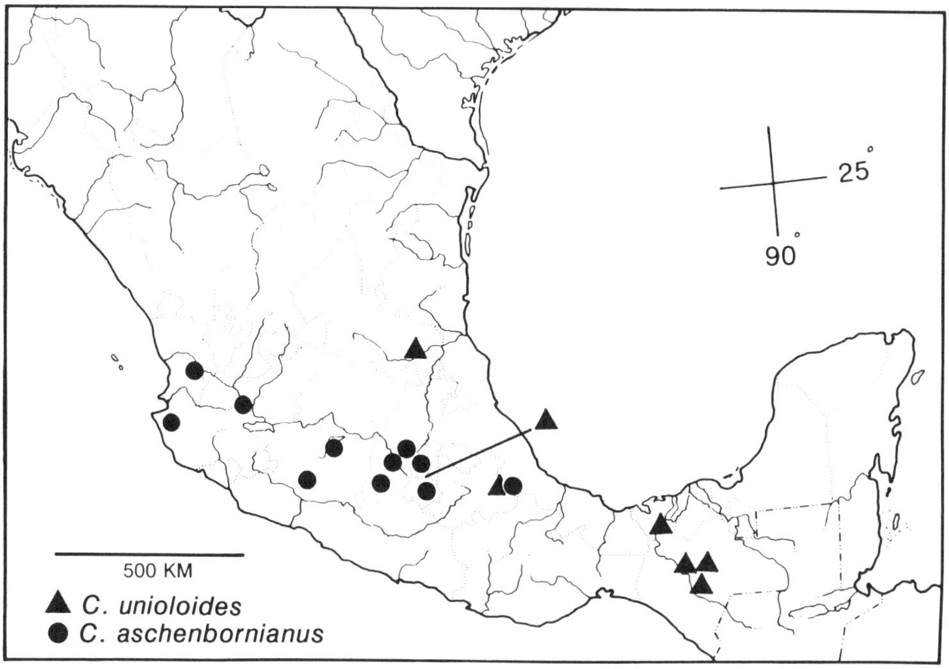

FIG. 27. Distribution of *Cyperus* and *unioloides C. aschenbornianus* in Mexico.

33. Cyperus unioloides R. Brown, Prodr. 216. 1810. *Pycreus unioloides* (R. Brown) C. B. Clarke in Urban, Symb. Antill. 2: 164. 1900.—TYPE: AUSTRALIA. Queensland: Shoalwater Bay, Aug–Sep 1802, *R. Brown 5900* (holotype: BM!; isotype: K!).

Caespitose perennial, (10–) 20–50 (–95) cm tall. Rhizomes 1–3 cm long, approximately horizontal, producing culms 1–4 mm apart. Culms 0.5–2.5 mm in diameter, triquetrous or trigonous, smooth. Leaves 1–5, 7–45 cm long, 1–5 mm wide, v-shaped, margins and keel scabrellate. Inflorescence bracts 2 (–3), 2–40 cm long, 1–3 (–5) mm wide, v-shaped, the margins and keel scabrellate, horizontal or sometimes the longest one erect. Rays 2–4 (–7), 0.3–4.0 (–7) cm long; secondary rays absent. Spikes 10–25 mm long, 10–25 mm wide, broadly ovoid; rachis 1–8 mm long. Spikelets 2–9, (7–) 10–13 (–15) mm long, 3.0–5.0 mm wide, lanceolate to oblong-lanceolate, strongly compressed; rachilla persistent, 0.4–0.6 mm wide, 0.4–0.6 mm thick, flexuous, quadrate, stramineous, successive scale scars ca. 1.2 mm apart, wings absent. Scales deciduous, 10–24, (3.6–) 3.8–4.2 (–4.6) mm long, 2.0–2.4 mm wide, ovate-lanceolate, apex acute, mucronate, laterally nerveless, yellow to yellowish brown, medially bicarinate, strongly 3-nerved, brownish yellow. Stamens 3; filaments 3.0–4.2 mm long; anthers 2.0–2.4 mm long, linear, the connective apex not prolonged. Style 1.0–4.8 mm long; stigmas 2, 1.0–3.3 mm long. Achenes 1.2–1.4 mm long, 0.8–1.0 mm wide, biconvex, obovate to orbiculate, apex apiculate, base sessile, the faces convex, the surface nearly smooth with longitudinal cells, the outer surface weathering to reveal a secondary papillose to punctate inner black layer. Fig. 23G.

Phenology. Flowering and fruiting from May to October.

Distribution. Pantropic regions; in the Old World in Australia and Africa; in the New World in the West Indies and from northern Mexico south through Central America to Argentina. Occasional in Mexico (Fig. 27); marshes, bogs, roadside ditches, and wet seeping slopes in pine woods; 400–2400 m.

REPRESENTATIVE SPECIMENS. **Mexico.** CHIAPAS: Teopisca, *Breedlove 10544* (DS, F, MICH, NY); San Cristóbal de Las Casas, *Breedlove 14701* (DS, F, MICH, NY); near San José del Lago, *Sharp 45458* (GH, TENN).—DISTRITO FEDERAL: Lake Xochimilco on Chinampas, *Harshberger 93* (GH, US).—SAN LUIS POTOSÍ: Las Canoas, *Pringle 3770* (ARIZ, E, F, GH, M, MICH, MO, MTMG, NY, UC, US, VT, Z).—TABASCO: Mpio. Huimanguillo, hacia Fco. Rueda, *Cowan & Magaña 3251* (CHAPA, MO).—VERACRUZ: région d'Orizaba, *Bourgeau 2731* (GH, MO).

Cyperus unioloides is distinguished from other species of *Cyperus* in Mexico and Central America by the combination of very wide spikelets (3–5 mm wide) and biconvex, black achenes. Among Mexican species, only *C. wilburii* has spikelets as wide as those of *C. unioloides*. *Cyperus wilburii* has cuspidate scales and trigonous achenes.

Cyperus unioloides was reported from southern California (Kükenthal 1935–36). I have seen the specimen, *Hasse s.n.* (US), on which this record was based, and it is correctly identified. This site is more than 2000 km from the nearest locality (San Luis Potosí), and I believe that the Hasse collection was perhaps mislabeled.

34. Cyperus aschenbornianus Boeckeler, Linnaea 35: 460. 1868. *Pycreus aschenbornianus* (Boeckeler) C. B. Clarke, Bull. Misc. Inform. Kew, add. ser. 8: 2. 1908.—TYPE: MEXICO. 1842, *Aschenborn 121* (holotype: B!; isotype: Herb. Bothe, B!).

Caespitose perennial, (8–) 15–22 (–40) cm tall. Rhizomes 0.5–3.0 cm long, 2–5 mm in diameter, indurate, horizontal, branching at interval of 5–20 mm. Culms 0.4–0.7 mm in diameter, roundly trigonous, smooth. Leaves 1–3, (8–) 14–20 (–30) cm long, 0.8–1.7 mm wide, v-shaped, sparsely scabrellate distally on the margins and keel. Inflorescence bracts 2 (–3), (1–) 3–10 (–14) cm long, 0.5–1.5 mm wide, v-shaped, the margins and keel smooth or sparsely scabrellate, the longest erect or ascendent, the other(s) ascendent to horizontal. Rays absent or 1–3 (–4), 0.5–2.5 (–3.5) cm long. Spikes 10–20 (–26) mm long, 10–24 (–34) mm wide, loosely hemispherical or ovoid; rachis 2–4 mm long. Spikelets (1–) 3–10, 5–22 mm long, 2.0–3.2 mm wide, linear-lanceolate to oblong-lanceolate, flattened; rachilla persistent, ca. 0.3 mm wide, 0.3–0.4 mm thick, slightly geniculate, greenish brown with reddish splotches, successive scales 0.8–1.2 mm apart, wings absent. Scales deciduous, (12–) 16–30 (–50), 2.4–2.8 (–3.2) mm long, 1.8–2.2 mm wide, oblong-elliptic, apex acute to subacute, laterally nerveless, stramineous to light brown or dark reddish brown, medially bicarinate, 3-nerved, light brown. Stamens 2; filaments 2.2–3.0 mm long; anthers 1.0–1.2 mm long, linear, the connective apex not prolonged. Style 0.8–1.2 mm long; stigmas 2, 0.5–0.7 mm long. Achenes 1.2–1.4 mm long, 0.5 (–0.6) mm wide, lenticular, ellipsoid, apex subacute, conspicuously apiculate, base stipitate to substipitate, the faces convex, the surface puncticulate, dark brown to black. Fig. 23I.

Phenology. Flowering and fruiting from May to October.

Distribution. Endemic to Mexico (Fig. 27), southern Veracruz westward to Nayarit and Jalisco; occasional, bogs, stream banks, marshes, and roadside ditches; 1800–2700 m.

REPRESENTATIVE SPECIMENS. **Mexico.** DISTRITO FEDERAL: Valle de Mexico, *Schaffner 5* (NY).—JALISCO: Km 8 on hwy to P. Vallarta, El Tuito hacia la mina de Zimapan, *Pérez de la Rosa 416* (IBUG); near Guadalajara, *Pringle 2563* (BM, BR, E, F, GH, M, MICH, MO, NY, SMU, US, Z); barranca of Guadalajara, *Pringle 11204* (CU, F, GH, MO, NY, US, Z).—MÉXICO: Dtto. Temascaltepec, Comunidad, *Hinton 1563* (F, GH, K, MO, NY, US); Tenancingo, *Kral 25237* (MO); Mpio. Tepotzotlán, Hac. Lanzarote, *Rzedowski 36032* (ARIZ, ENCB).—MICHOACÁN: vicinité de Morelia, *Arsène 5831* (E, GH, MO, NY, US); ca. 5 km N of Ario de Rosales, *Iltis 401* (MICH, US, WIS).—MORELOS: Lago Zempoala, *Matuda 25607* (NY).—NAYARIT: Tepic, along Río Tepic, *Pennell 19932* (US).—VERACRUZ: Orizaba, *Bourgeau 2731* (BM, C, GH, M, NY).

Cyperus aschenbornianus is morphologically most similar to *C. unioloides*; both have long scales with similar coloration patterns, wide spikelets, and long, linear anthers. *Cyperus aschenbornianus* is distinguished from *C. unioloides* by its shorter scales, narrower spikelets, and ellipsoid, acute rather than obovoid to orbiculate achenes.

35. Cyperus niger Ruiz & Pavón, Fl. peruv. 1: 47. 1798.—TYPE: PERU. Chancay, *Ruiz & Pavón 12/62* (holotype: MA; photos: F! NYS!).

Cyperus melanostachyus H. B. K., Nov. gen. sp. 1 [quarto]: 207. 1816. *Chlorocyperus melanostachyus* (H. B. K.) Rikli, Jahrb. Wiss. Bot. 27: 564. 1895. *Pycreus melanostachyus* (H. B. K.) C. B. Clarke, Bot. Jahrb. Syst. 30, Beiblatt 68: 6. 1901.—TYPE: COLOMBIA. Bogotá, *Humboldt & Bonpland s.n.* (holotype: P-HBK?; isotype: HAL!).

Cyperus variegatus H. B. K., Nov. gen. sp. 1 [quarto]: 208. 1816.—TYPE: MEXICO. San Pedro Jorullo, *Humboldt & Bonpland s.n.* (holotype: P-HBK).

Cyperus polystachyos var. *major* Junghuhn in Chamisso & Schlechtendal, Linnaea 6: 24. 1831.—TYPE: MEXICO. Veracruz: near Jalapa, Sep 1828, *Schiede & Deppe 846* (holotype: HAL!).

Cyperus melanostachyus var. *gracilis* Liebmann, Mexic. halvgr. 14. 1850.—TYPE: MEXICO. Puebla: Chinantla, Jul 1841, *Liebmann 14439* (lectotype, designated by Tucker, 1986: C!; isolectotype: C!).

Cyperus melanostachyus var. *robustus* Liebmann, Mexic. halvgr. 13. 1850. *Cyperus niger* var. *robustus* (Liebmann) Kükenthal, Pflanzenreich 4(20): 344. 1936.—TYPE: MEXICO. Puebla: San Lorenzo in the Tehuacán Valley, Dec 1841, *Liebmann 14441* (lectotype, designated by Tucker, 1986: C!).

Cyperus diandrus var. *castaneus* S. Watson, Bot. Calif. 2: 214. 1880. *Cyperus niger* var. *castaneus* (S. Watson) Kükenthal, Pflanzenreich 4(20): 344. 1936.—TYPE: U.S.A. California: San Francisco, 1866, *Bolander 2330* (lectotype, here designated: GH!).

Cyperus niger var. *pseudo-elegantulus* Kükenthal, Pflanzenreich 4(20): 345. 1936.—TYPE: COSTA RICA. Bords du Río Grande près San Ramón, 18 Jun 1901, *Brenes 14438* (lectotype, here designated: GH!; isolectotype: US!).

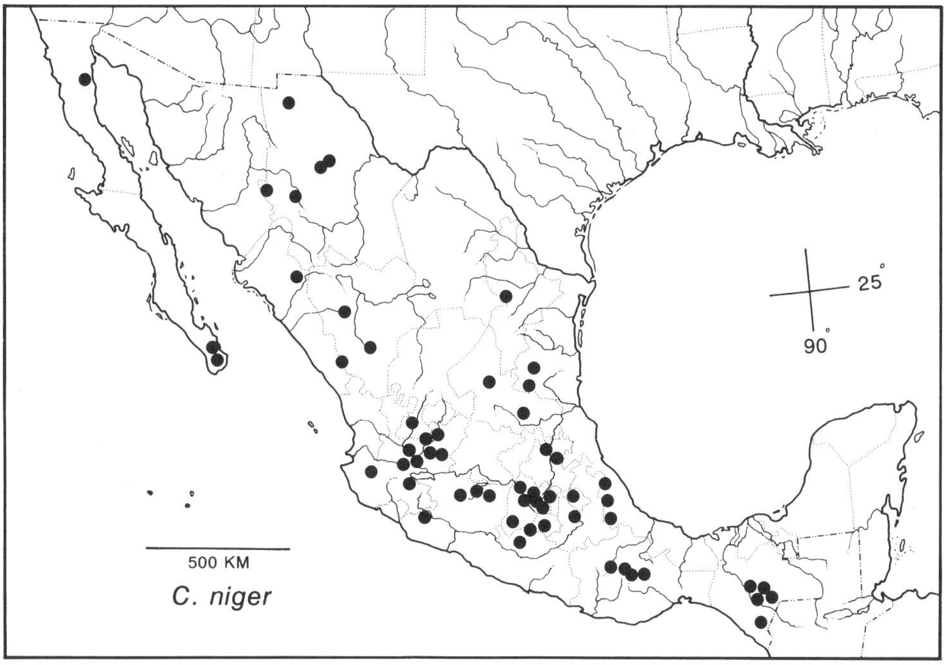

FIG. 28. Distribution of *Cyperus niger* in Mexico.

Cyperus piceus var. *densior* Kükenthal, Pflanzenreich 4(20): 397. 1936.—TYPE: COSTA RICA. Cartago: Turrialba, Nov 1893, *Tonduz 8266* (holotype: B, not located; isotypes: BM! Z!).

Perennial, 5–40 (–95) cm tall. Rhizomes ca. 1 mm in diameter, soft, producing culms 1–30 (–50) mm apart. Culms 0.3–1.4 (–2.0) mm in diameter, trigonous to triquetrous, smooth. Leaves (2–) 3–7, 3–20 (–30) cm long, 1.5–3 mm wide, broadly v-shaped, the margins scabrellate for the entire length, or apically only. Inflorescence bracts 2–3, 1–15 cm long, 1–3 mm wide, broadly v-shaped, the margins and keel scabrellate, horizontal to reflexed downward parallel to the culm. Rays 1–2, 3–45 mm long; rays sometimes absent, the inflorescence 6–20 mm in diameter, densely ovoid. Spikes 7–16 mm in diameter, loosely to densely ovoid to hemispherical; rachis 1–2 mm long, or absent. Spikelets (3–) 5–25 (–60), (3–) 5–9 mm long, 1.8–2.3 mm wide, linear to oblong-linear, flattened; rachilla persistent, 0.4–0.6 mm wide, 0.4–0.5 mm thick, quadrate, stramineous to reddish brown, successive scale scars 0.8–1.1 mm apart, wings absent. Scales deciduous, 4–18, 1.5–2.1 mm long, 1.4–2.2 mm wide, ovate to orbiculate, apex obtuse, laterally nerveless, castaneous, black, or light brown, medially distinctly bicarinate basally, 2 (–3)-nerved, green or greenish brown. Stamens 2; filaments 1.6–2.3 mm long; anthers 0.6–0.8 mm long, narrowly ellipsoid, the connective apex not prolonged. Style 0.4–0.6 mm long; stigmas 2, 0.7–2.1 mm long. Achenes 1.2–1.4 mm long, 0.6–0.8 mm wide, lenticular, ovoid-ellipsoid, apex apiculate, base slightly stipitate, the faces convex, the surface with isodiametric cells forming transverse undulations, puncticulate, brown. Fig. 23F.

Phenology. Flowering and fruiting from July to October and occasionally during the remaining months.

Distribution. Southeastern Oklahoma to western California, U.S.A. (Tucker 1992), south throughout the mountains of Mexico and Central America to northern Argentina. Very common in Mexico (Fig. 28); ditches, wet pastures, seepage, stream banks, and wet meadows; (800–) 1500–2200 (–3200) m.

Representative Specimens. **Mexico.** Baja California: Sierra de la Victoria, SE side of La Laguna Meadow, *Chambers 902* (UC).—Baja California Sur: Sierra de San Francisquito, 22 Mar 1892, *Brandegee s.n.* (UC).—Chiapas: Mpio. San Cristóbal de Las Casas, *Breedlove 37121*, *37132* (DS); 12 km N of Jitotol, *Breedlove & Thorne 21473* (CAS, LL, RSA).—Chihuahua: Guicorichi, Río Mayo, *Gentry 1996* (ARIZ, UC); 50 mi N of Chihuahua, *Gould 8910* (TEX, UC); near Chihuahua, *Pringle 809* (CAS, E, LL, M, MO, NY, UC, US).—Coahuila: vicinity of Saltillo, *Palmer 177* (MO, UC, US); Cañón de San Lorenzo, *Villarreal et al. 1784* (IBUG).—Distrito Federal: Deleg. Cuajimalpa, Puerto de las Cruces, *Rzedowski 34265* (ARIZ, ENCB, SD).—Durango: Mpio. Santiago Papasquiaro, 3.5 km W de La Soledad, *Díaz 81* (DUKE, UTEP).—Guerrero: 20 km WNW of Teloloapan, *Iltis & Doebley 388* (IBUG, WIS).—Hidalgo: Dtto. Ixmiquilpan, by Río Tula, *Moore 3366* (BH).—Jalisco: 20 mi E of Ameca, *McVaugh 18585* (MICH, TEX).—México: Mpio. Tepotzotlán, 3 km W de Magú, *Cruz Cisneros 3772* (MICH, WIS); Mpio. Zumpango, San Juan Citlaltepec, *Cruz del Carmen 170* (ARIZ, ASU, ENCB, IBUG).—Michoacán: vicinité de Morelia, Loma Sta. María, *Arsène 3501* (BM, MO, NY, US).—Morelos: Near Tetela del Volcán, *Flores Crespo 181* (ENCB, LL).—Nuevo León: Cañón de Huestaca, del Río Santa Catarina, *Barkley 14599* (TEX).—Oaxaca: 4 km SW of San Miguel Suchistepec, *Rzedowski 21050* (ENCB, LL); ca. 1 km W of Natividad, *Tucker 2161* (DUKE, ENCB).—Puebla: Puebla, *Barkley et al. 2453* (TEX).—San Luis Potosí: Las Canoas, *Pringle 3789* (ARIZ, E, GH, M, MO, NY, TEX, UC, US, Z).—Sinaloa: Sierra Surotato, Ocurahui, *Gentry 6299* (ARIZ, DS).—Tamaulipas: 18 km NW of Ocampo, *Stanford et al. 1054* (ARIZ).—Tlaxcala: Santa Ana Chiantempan, *Arsène 1726* (BM, C).—Veracruz: Perote, Cruz Blanca, *Balls 5499* (E, UC); Mpio. Banderilla, Rancho la Martinica, *Calzada 7855* (CHAPA).—Zacatecas: Presa de Malpaso, *Puga 2329* (IBUG); 38 km W of Jalpa, *Rzedowski & McVaugh 972* (ASU, ENCB, MICH).

Cyperus niger, one of the most common species of *Cyperus* in the mountains of Mexico, is recognized by its blackish spikelets and lenticular achenes. It is distinguished from *C. bipartitus* by the color and shape of the achenes, by habit, and by the denser clusters of spikelets. It is separated from *C. aschenbornianus* by its narrower spikelets, shorter scales with rounded apices, and less conspicuously apiculate achenes.

36. Cyperus bipartitus Torrey, Ann. Lyceum Nat. Hist. New York 3: 257. 1836. *Pycreus bipartitus* (Torrey) C. B. Clarke, Bull. Misc. Inform. Kew, add. ser. 8: 2. 1908.—Type: U.S.A. New Orleans, *Ingalls s.n.* (holotype: NY!).

Cyperus rivularis Kunth, Enum. pl. 2: 6. 1837. *Chlorocyperus rivularis* (Kunth) Rikli, Jahrb. Wiss. Bot. 27: 564. 1895. *Pycreus rivularis* (Kunth) Palla, Allg. Bot. Zeitschr. Syst. 13: 48. 1907.—Type: U.S.A. Georgia, *Beyrich s.n.* (not located).

Cyperus lagunetto Steudel, Syn. pl. glum. 2: 5. 1854. *Pycreus lagunetto* (Steudel) C. B. Clarke, Bot. Jahrb. Syst. 30: 8. 1901. *Cyperus rivularis* subsp. *lagunetto* (Steudel) Kükenthal, Pflanzenreich 4(20): 383. 1936. *Cyperus rivularis* var. *lagunetto* (Steudel) O'Neill, Rhodora 44: 86. 1942.—Type: Venezuela. Mérida: Bailadores, Jun 1843, *Linden 1396* (holotype: not located; isotype: BR!).

Densely caespitose annual, 3–25 (–30) cm tall. Rhizomes absent. Culms 0.3–1.4 mm in diameter, roundly trigonous, smooth. Leaves 1–3, 1–8 cm long, 1–2 mm wide, v-shaped, the margins scabrellate. Inflorescence bracts 2–3, 1–12 cm long, 0.5–1.5 mm wide, v-shaped to flat, margins and keels scabrellate, horizontal to ascendent at 30° (–45°). Rays 1–4, 0.5–2.0 (–3.0) cm long; secondary rays absent. Spikes 7–14 mm long, 9–14 mm wide, loosely ovoid; rachis 1–2.5 mm long. Spikelets 3–5 (–8), 8–18 mm long, 2.0–2.6 (–3.0) mm wide, oblong to oblong-lanceolate, flattened; rachilla persistent, 0.2–0.3 mm wide, 0.1–0.2 mm thick, quadrate, stramineous to reddish brown, successive scale scars 0.8–1.2 mm apart, wings absent. Scales deciduous, (6–) 10–26 (–32), 1.9–2.7 mm long, 1.8–2.3 mm wide, oblong to ovate, apex obtuse, laterally nerveless, light to dark brown, medially 2-3-nerved, bicarinate, light brown. Stamens 2–3; filaments 2.0–2.5 mm long; anthers 0.4–0.5 mm long, ellipsoid, the connective apex not prolonged. Style 0.6–1.0 mm long; stigmas 2, 1.0–1.5 mm long. Achenes 1.0–1.3 (–1.5) mm long, (0.6–) 0.8 mm wide, lenticular, obovoid to ovoid, apex apiculate, base sessile, the faces convex, the surface with isodiametric cells forming transverse undulations, puncticulate, black. Fig. 23E, 29.

Phenology. Flowering and fruiting from July to January.

Distribution. Maine to Washington, U.S.A., south through Mexico and Central America to Chile. Rare in the mountains of Mexico; ditches, roadsides, trailsides, and pastures; 600–2000 (–3000) m.

ADDITIONAL SPECIMENS EXAMINED. **Mexico.** CHIAPAS: Mpio. La Independencia, 6–10 km NNE of La Soledad, *Breedlove 53204* (CAS); Mpio. San Cristóbal, Rancho Nuevo, *Breedlove & Raven 8318* (DS).—CHIHUAHUA: Sierra Madre, Largo Canyon, 26 Sep 1903, *Jones s.n.* (DS); 41.3 mi SW of El Vergel, *Nesom 5136* (US).—DISTRITO FEDERAL: above Contreras, *Goodman 3472* (F).—MÉXICO: 5 km N de Progreso Industrial, Mpio. Villa Nicolás Romero, *Matamoros 244* (IBUG).—OAXACA: Ixtlán, *Martin 230* (US); on Hwy 175, ca. 55 km N of Pochutla, *Tucker 2888* (DUKE).—PUEBLA: Puebla, *Arsène & Nicolas 5471* (GH, with *Cyperus niger*).—VERACRUZ: Orizaba, *Botteri 30* (US).

Cyperus bipartitus is distinguished from the much more common and morphologically similar *C. niger* by its brownish scales, its obovoid to ovoid black achenes, its densely caespitose habit, and its diffuse spikes of usually 3–5 spikelets.

37. Cyperus lanceolatus Poiret in Lamarck, Encycl. 7: 245. 1806. *Pycreus lanceolatus* (Poir.) C. B. Clarke in Durand & Schinz, Consp. fl. afr. 5: 538. 1894.—TYPE: MADAGASCAR. *Thouars s.n., ex herb. Petit-Thouars s.n.* (holotype: P; probable isotype: B-W no. 1298, IDC microfiche!).

Cyperus densus Link, Jahrb. Gewächsk. 1: 83. 1820. *Cyperus humboldtianus* Schultes, Syst. veg. Mantissa 2: 100. 1824, nom. superfl. *Pycreus densus* (Link) C. B. Clarke in Urban, Symb. Antill. 2: 164. 1900.—TYPE: "America australis." *Humboldt & Bonpland s.n.* (holotype: B-W no. 1311, IDC microfiche!).

Cyperus lanceolatus var. *compositus* J. Presl & K. Presl in K. Presl, Reliq. Haenk. 1: 167. 1828.—TYPE: "Philippine Islands," fide protologue but probably from MEXICO (Kükenthal 1935–36; Kern 1974), *Haenke s.n.* (holotype: not located, not at PR or PRC).

Cyperus olfersianus Kunth, Enum. pl. 2: 10. 1837.—TYPE: URUGUAY. Montevideo, collector unknown (holotype: not located).

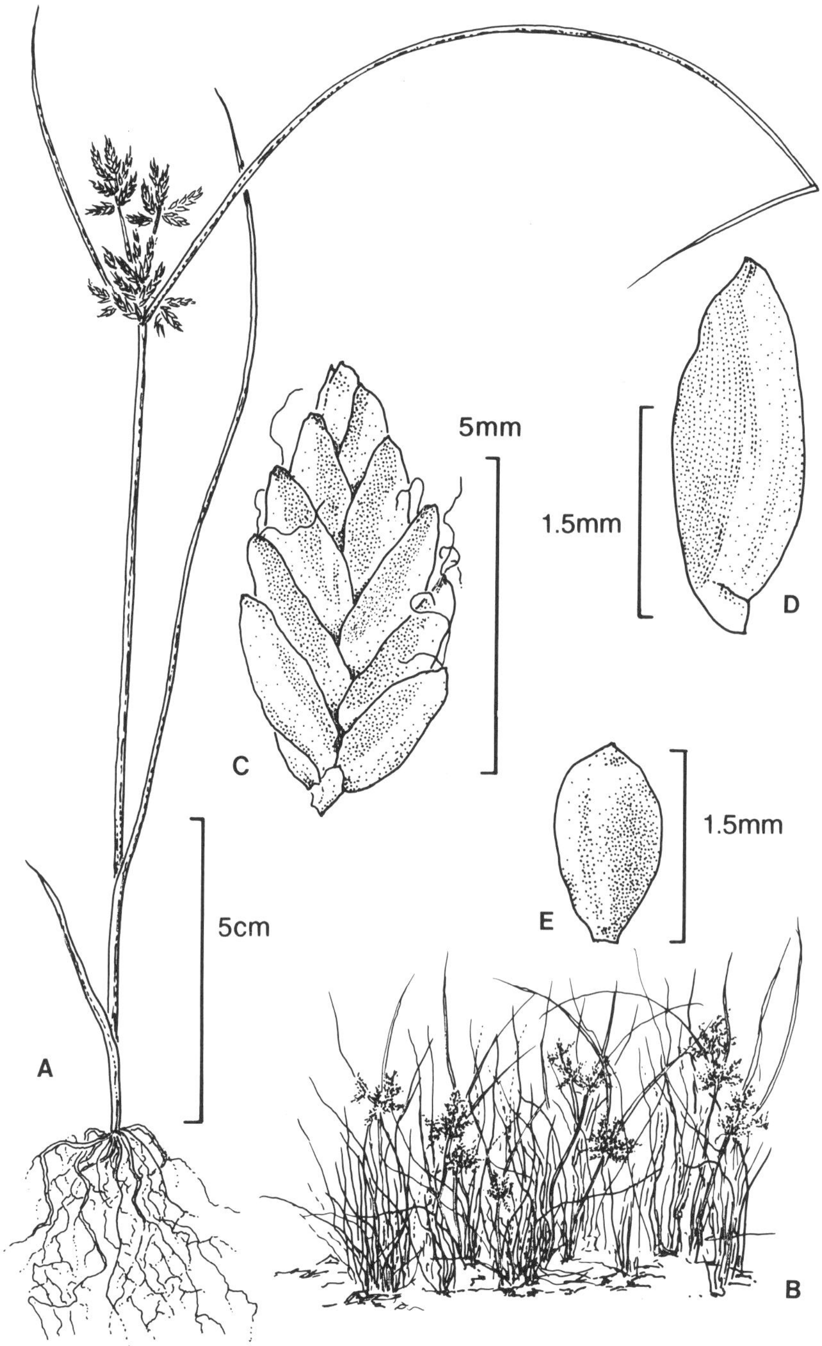

FIG. 29. *Cyperus bipartitus*. A. Habit. B. Aspect of densely grouped plants of a population. C. Spikelet. D. Scale. E. Achene. Drawn by Patricia M. Kernan for the Flora of New York State (Richard S. Mitchell, ed.).

Cyperus ambiguus Liebmann, Mexic. halvgr. 10. 1850.—TYPE: MEXICO. Veracruz: San Antonio Huatusco, Jun 1841, *Liebmann 14382* (holotype: C!; isotypes: F! HAL! K! MICH!).

Caespitose perennial, 15–75 cm tall. Rhizomes ca. 1 mm in diameter, soft, producing caespitose clusters of culms 1–10 mm apart. Culms 0.5–3.0 mm in diameter, compressed-trigonous to triquetrous, smooth. Leaves 1–3, 5–30 cm long, 1–2 mm wide, v-shaped, the margins and keel scabrellate or smooth. Inflorescence bracts 2–3, 2–14 cm long, 0.5–3.5 mm wide, v-shaped, the margins and keel smooth or scabrellate, horizontal to erect. Rays 2–3 (–4), 2–15 mm long, or sometimes absent, the inflorescence (12–) 15–20 mm in diameter, densely ovoid. Spikes 5–23 mm long, 9–21 mm wide, ovoid; rachis 1 (–2) mm long. Spikelets 1–6, 5–26 mm long, 2–3 mm wide, oblong-lanceolate, flattened; rachilla persistent, 0.4 mm wide, 0.2 mm thick, straight, rectangular in cross section, yellowish brown, successive scale scars 0.7–0.9 mm apart, wings absent. Scales deciduous, 12–54, 1.8–2.6 mm long, 1.4–1.7 mm wide, ovate, apex obtuse, laterally 1-nerved, yellowish or yellowish brown, medially 3-nerved, pale brown or occasionally greenish. Stamens 2; filaments 2.0–2.3 mm long; anthers 0.4–0.5 mm long, ellipsoid, the connective apex not prolonged. Style 0.3–1.0 mm long; stigmas 2, 0.6–0.8 (–1.0) mm long. Achenes 1.1–1.3 mm long, (0.5–) 0.6 mm wide, lenticular, obovoid to ellipsoid, apex acute to broadly rounded, base stipitate to substipitate, the faces convex, the surface with isodiametric cells forming transverse undulations, finely papillose or puncticulate, brown. Fig. 23D.

Phenology. Flowering and fruiting from June to January.

Distribution. The West Indies, southern U.S.A., south through Mexico and Central America to Paraguay. Frequent in Mexico (Fig. 30); muddy shores of ponds, stream banks, roadside ditches, springs, seepage, and wet pastures; sea level to ca. 1800 m.

REPRESENTATIVE SPECIMENS. **Mexico.** BAJA CALIFORNIA SUR: San Jacinto, 18 Sep 1893, *Brandegee s.n.* (UC).—CHIAPAS: 13–18 km S of Ocozocoautla, *Breedlove 37791* (DS); 17 km ESE of Palenque, *Chater 138* (BM); Tapachula, *Fisher 35300* (ARIZ, LCU, MO, NY, US); Río Mayo, Guasaremos, *Gentry 2408* (ARIZ, GH, LCU, UC, US).—COLIMA: Colima, *Palmer 7* in 1897 (ARIZ, C, MICH, UC).—GUERRERO: Dtto. Mina, Zihuaqueo, Placeres del Oro, *Hinton 9302* (GH, LCU, MICH, UC); ca. 35 km S of Chilpacingo on Hwy 95, *Tucker 2269* (DUKE).—HIDALGO: Mpio. Molango, boggy margin of Lake Atexca, *Moore 1941* (BH).—JALISCO: Los Colomos, 10 km NW of Guadalajara, 28 Oct 1963, *Bancalari s.n.* (IBUG).—OAXACA: Choapam, *Galleoti 5870* (GH, NY, UC, US).—TAMAULIPAS: Hac. Sta. En Gracia, *Chase 7614* (ARIZ, GH, MICH, MO, NY).—VERACRUZ: région d'Orizaba, *Bourgeau 2733* (C, BM, M); 15 km E of Coatzacoalcos, *Tucker 2529* (DUKE); Cabo Rojo, 10 mi S of Tampico, *Ward 7804* (MICH, SD).

Cyperus lanceolatus is recognizable by its dense clusters of pale yellowish spikelets. Sometimes misidentified as *C. polystachyos*, it differs from that species in having wider spikelets, ovate rather than oblong scales, and acute to broadly rounded rather than truncate achenes.

38. Cyperus flavescens L. var. **piceus** (Liebmann) Fernald, Rhodora 41: 529. 1939.
Cyperus piceus Liebmann, Mexic. halvgr. 12. 1850.—TYPE: MEXICO. Veracruz: San Antonio Huatusco, Sep 1841, *Liebmann 14410* (lectotype, designated by Tucker, 1986: C!; isolectotypes: C!).

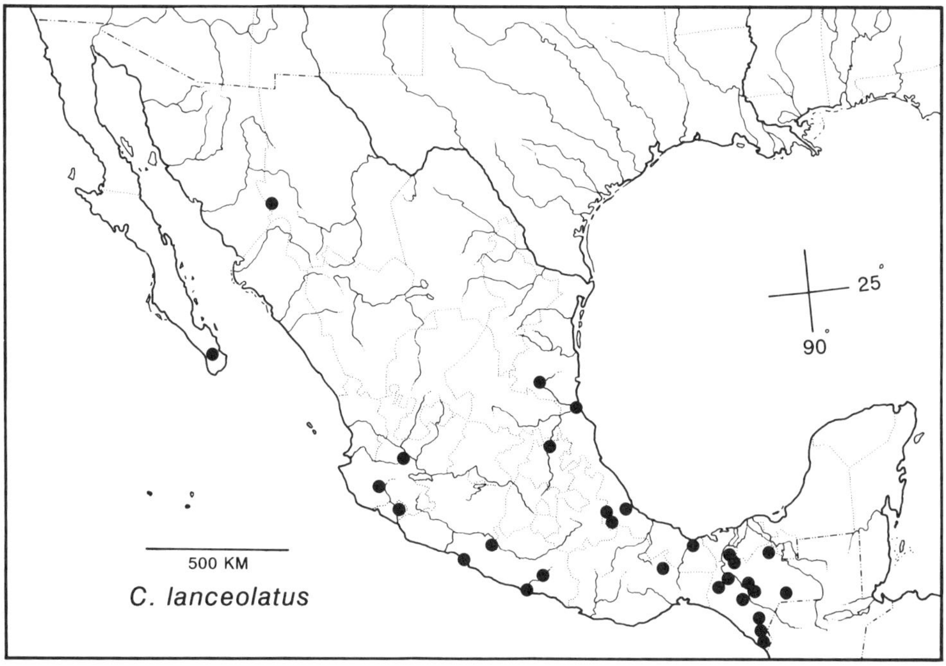

FIG. 30. Distribution of *Cyperus lanceolatus* in Mexico.

Pycreus potosiensis C. B. Clarke, Bull. Misc. Inform. Kew, add. ser. 8: 1. 1908. *Cyperus piceus* f. *potosiensis* (C. B. Clarke) Kükenthal, Pflanzenreich 4(20): 397. 1936.—TYPE: MEXICO. San Luis Potosí, *Schaffner 569* (lectotype, here designated: K!).

Caespitose annual, (2–) 4–30 (–70) cm tall. Rhizomes absent. Culms 0.4–2.0 mm in diameter, trigonous, smooth. Leaves 1–5, (3–) 10–18 cm long, (0.5–) 1.5–2.0 (–2.6) mm wide, v-shaped, the margins and keel scabrellate near the tip; blades often absent, the base of the culm with one reddish sheath bearing a minute blade tooth 1–2 mm long. Inflorescence bracts 2–3, 1–12 cm long, 0.5–2.5 mm wide, the margins and keel scabrellate at least near the apex if not for the entire length, ± horizontal. Rays 1–4 (–6), 0.5–3.0 (–9.0) cm long; secondary rays absent. Spikes 10–30 mm long, 8–26 mm wide, ovoid or subdigitate; rachis 2–6 mm long. Spikelets 1–6, 5–15 mm long, (2.0–) 2.3–2.8 mm wide, oblong-lanceolate, compressed; rachilla persistent, ca. 0.4 mm wide, ca. 0.3 mm thick, flexuous, quadrate, stramineous to reddish brown, successive scale scars 0.8–1.1 mm apart, wings absent. Scales deciduous, (4–) 8–24, (1.5–) 1.8–2.2 mm long, ca. 1.8 mm wide, ovate, apex obtuse, laterally nerveless, castaneous to purplish brown, the edges light brown to hyaline, medially 2–3-nerved, bicarinate basally, green. Stamens 3; filaments 1.2–1.7 mm long; anthers ca. 0.4 mm long, ellipsoid, the connective apex not prolonged. Style 0.5–1.0 mm long; stigmas 2, 0.5–0.8 mm long. Achenes 1.0–1.2 mm long, ca. 0.8 mm wide, lenticular, obovoid, apex apiculate, base slightly stipitate, the faces convex, transversely undulate, the surface with a network of rectangular longitudinally elongate surface cells, jet black to reddish brown. Fig. 23H.

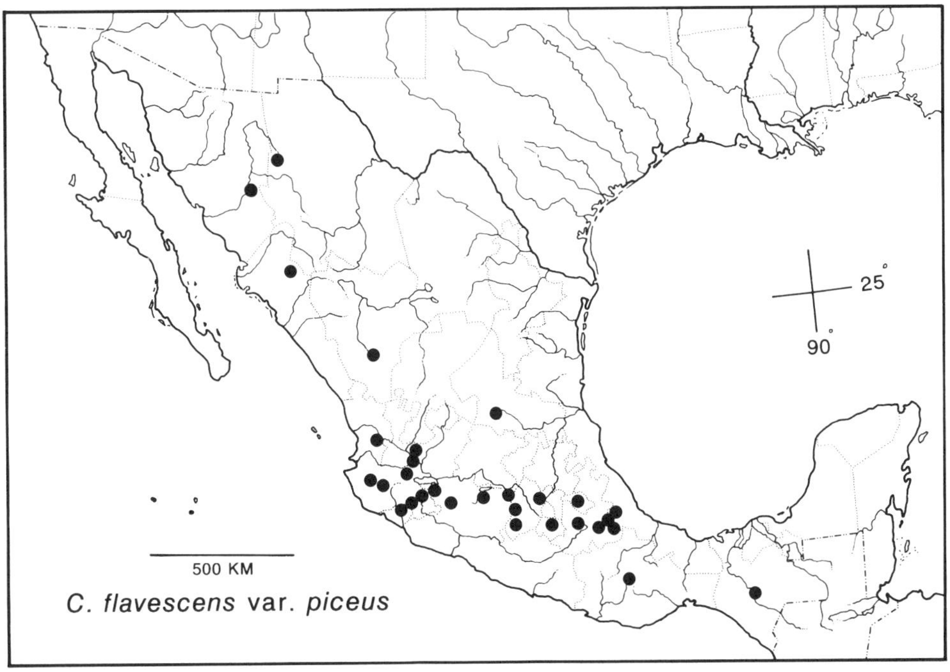

FIG. 31. Distribution of *Cyperus flavescens* var. *piceus* in Mexico.

Phenology. Flowering and fruiting from June to November.

Distribution. Uplands of Mexico, Guatemala, Nicaragua, and Costa Rica, also in Bolivia and the uplands of eastern Brazil. Frequent in Mexico (Fig. 31); marshes, bogs, roadside ditches, lake shores, and stream banks; ca. 1000–2600 m.

REPRESENTATIVE SPECIMENS. **Mexico.** CHIAPAS: Mpio. La Concordia, 96 km S of Hwy 190 on rd to Nueva Concordia, *Breedlove 38504* (DS).—CHIHUAHUA: Mpio. Madera, 20 mi N of Madera, *Muller 3519* (F, LL, MO, UC).—COLIMA: Mpio. Comala, Rancho El Jabali, *Rothschild & Phillips 018, 023* (NYS, UCR).—DURANGO: Durango and vicinity, *Palmer 733* in 1896 (ARIZ, C, MICH, UC, US).—JALISCO: mountains N of Autlán, *McVaugh 19655* (MICH, TEX); 4.5 mi NNE of Talpa de Allende, *McVaugh 20231* (LL, MICH); near Guadalajara, *Pringle 3262* (AC, E, LL, M, MICH, MTMG, UC, Z).—MÉXICO: 8 mi S of Temascaltepec, *Denton 1909* (MICH, UC).—MICHOACÁN: Morelia, *Arsène 5275* (F, GH, M, MO, US); 22 mi S of Jiquilpan, Cerro Potrerillos, *King & Soderstrom 4586* (SMU, TEX, UC); 8 km S of Uruapan, *King & Soderstrom 4714* (SMU, TEX, UC).—NAYARIT: La Labor, 15 mi SE of Tepic, *McVaugh 19431* (MICH, TEX).—OAXACA: on Hwy 175, ca. 15–20 km S of Ixtlán de Juárez, *Tucker 2807* (DUKE).—PUEBLA: Puente del Emperador, *Sharp 44564* (TENN).—SAN LUIS POTOSÍ: San Luis Potosí, *Schaffner 198* (C, MICH).—SINALOA: Sierra Surutato, Mpio. Badiraguato, 0.5 mi N of Los Ornos, *Breedlove & Kawahara 16784* (CAS).—SONORA: Río Mayo region, Mesa La Lagunita, *Martin et al. s.n.* (ARIZ 302745).—VERACRUZ: Huatusco, *Liebmann 14409* (C); inter Huatusco et Orizaba, *Liebmann 14410* (C).

Cyperus flavescens var. *piceus* is distinguished from other Mexican taxa of *Cyperus* by the combination of lenticular obovoid achenes and purplish scales. Frequently, there are frostlike incrustations (thickened, raised anticlinal walls of epidermal cells) on the surface of the achenes, but as is the case for the species worldwide, such incrustations are not consistently present in Mexican collections.

Because *C. flavescens* var. *piceus* has been reported previously only from Mexico and Central America, I list the following South American collections: BOLIVIA: Depto. Santa Cruz, Prov. Velasco, Paraíso, *Bruderreck 292* (NY); BRAZIL: Est. Goiás, 40 km N of Veadeiros, *Irwin et al. 24352* (NY); Est. Pernambuco, Tapera, *Pickel 2449* (NY).

The two varieties of *C. flavescens* are largely allopatric. They have been treated as species (Kükenthal 1935–36), although they differ in only one character, the colors of the scales (Tucker 1983). Thus, I treat them as varieties, following Fernald (1939). The absence of var. *flavescens* in the lowlands of Mexico is curious, since it is common in the southern U.S.A., the West Indies, and Central America.

KEY TO THE VARIETIES OF CYPERUS FLAVESCENS

1. Sides of the scales yellow to yellowish brown, margins hyaline; pantropic and warm-temperate; in the New World, occurring in the southern U.S.A., the West Indies, lowlands of Central America and South America, but absent in Mexico. *C. flavescens* var. *flavescens*.
1. Sides of the scales castaneous to purplish brown, margins hyaline to light brown; uplands of Mexico, Guatemala, Nicaragua, and Costa Rica. *C. flavescens* var. *piceus*.

III. Cyperus subgenus **Juncellus** (Grisebach) C. B. Clarke, J. Linn. Soc., Bot. 21: 33. 1884. *Cyperus* section *Juncellus* Grisebach, Fl. Brit. W. I. 562. 1864. *Juncellus* (Grisebach) C. B. Clarke in Hooker f., Fl. Brit. India 6: 594. 1893.—TYPE: *Cyperus mucronatus* Rottbøll [=*Cyperus laevigatus* L.]; see Goetghebeur (1989).

Acorellus Palla in E. Hallier & A. Brand, Koch's Syn. Deutsch. Schw. Fl., ed. 3, 2: 2557. 1905.—LECTOTYPE, designated by Goetghebeur, 1989: *Acorellus laevigatus* (L.) Palla [=*Cyperus laevigatus* L.].

Culms and leaves with kranz (chlorocyperoid) anatomy. Spikelets spicate. Rachillas persistent, wings absent. Scales rounded, deciduous. Carpels 2, stigmas 2. Achenes lenticular, dorsiventrally flattened, borne with a surface rather than an angle toward the rachilla.

Subgenus *Juncellus* contains only six species. A single species, the pantropical and warm temperate *Cyperus laevigatus*, occurs in Mexico.

39. Cyperus laevigatus L., Mant. pl. 2: 179. 1771. *Pycreus laevigatus* (L.) Nees, Linnaea 10: 130. 1836. *Juncellus laevigatus* (L.) C. B. Clarke in Hooker f., Fl. Brit. India 6: 596. 1893. *Chlorocyperus laevigatus* (L.) Palla, Allg. bot. Zeit. 6: 221. 1900. *Acorellus laevigatus* (L.) Palla in E. Hallier & A. Brand, Koch's Syn. Deutsch. Schw. Fl., ed. 3, 2: 2558. 1905.—TYPE: SOUTH AFRICA. Cape of Good Hope, *Koenig s.n.* (holotype: LINN 70.13, IDC microfiche!).

Cyperus mucronatus Rottbøll, Descr. icon. pl. rar. 19. 1773.—TYPE: "Arabia." *Forsskål s.n.* (holotype:C!).

Cyperus distachyos Allioni, Fl. Pedem. Actuar. 1: 48. 1789. *Cyperus laevigatus* var. *distachyos* (Allioni) Cosson & Durieu, Fl. Algér. 2: 251. 1868. *Acorellus distachyos* (Allioni) Palla, Allg. Bot. Zeitschr. Syst. 9: 68. 1903. *Juncellus distachyos* (Allioni) Turrill, Bull. Misc. Inform. Kew 1926: 375. 1926.—TYPE: ITALY. Inter Nicaeam & flumen le Var, *Allioni s.n.* (holotype: TO). [Application of name based on Allioni's illustration, tab. 2, fig. 5.]

Cyperus laevigatus f. *reptans* C. B. Clarke, J. Linn. Soc., Bot. 21: 78. 1884. *Cyperus laevigatus* var. *reptans* (C. B. Clarke) Kükenthal, Pflanzenreich 4(20): 324. 1936.—Type: Ecuador. *Spruce 5911* (lectotype, here designated: K!).

Perennial, (1–) 8–30 (–60) cm tall. Rhizomes 1.2–2 mm in diameter, horizontal, producing single or clustered culms at intervals of 1–10 (–30) mm. Culms (0.3–) 0.6–1.5 (–2.2) mm in diameter, trigonous to subterete, smooth. Leaf blades usually absent, the base of the culm covered with a reddish sheath 5–25 mm long, bearing at the summit a short tooth 1–2 mm long, or an involute smooth leaf blade 1–7 cm long, 0.3–1 (–1.6) mm wide. Inflorescence bracts (1–) 2 (–3), (0.3–) 2.0–8.0 (–12.0) cm long, 0.2–3.0 mm wide, with a narrow hyaline margin at the base, the longer erect, appearing as a continuation of the culm, the shorter bract(s) horizontal or ascendent. Spikes 5–15 mm long, 5–15 mm wide, loosely subdigitate; rachis 1–4 mm long. Rays absent or sometimes 1, up to 2 mm long. Inflorescence capitate, appearing lateral. Spikelets (1–) 2–7 (–14), 4–7 (–12) mm long, (1.6–) 2.0–3.0 (–3.6) mm wide, elliptic to oblong-lanceolate, compressed-turgid; rachilla persistent, 0.4–0.8 (–1.1) mm wide, 0.2–0.4 mm thick, straight, quadrate, whitish, spongy to firm, successive scale scars 0.4–0.8 mm apart, wings absent. Scales deciduous, (8–) 12–22, 1.5–2.0 mm long, (1.3–) 1.6–2.0 mm wide, oblong-obovate to orbiculate, apex obtuse, laterally nerveless or weakly 2-nerved, off-white, whitish with red speckles, or with a single red spot ca. 0.5 mm wide, medially weakly 1–3-nerved, whitish. Stamens (2–) 3; filaments 1.4–2.0 (–3.0) mm long; anthers (0.5–) 0.7–1.0 mm long, oblong, the connective apex 0.1–0.2 mm long, oblong, reddish. Style 0.8–1.2 mm long; stigmas 2, 0.7–1.0 mm long. Achenes (1.2–) 1.3–1.5 (–1.8) mm long, (0.7–) 0.8–1.0 (–1.2) mm wide, lenticular, dorsiventrally compressed, oblong-ellipsoid to ovoid, apex apiculate, base cuneate to substipitate, the adaxial face flat to concave, the abaxial faces convex, the surface finely reticulate, glossy, dark brown to black. Fig. 1I.

Phenology. Flowering and fruiting throughout the year, but mostly from June to November.

Distribution. Pantropical and warm temperate regions; in the New World from western Texas to southern California, U.S.A., south to southern Mexico, the West Indies, and South America. In Mexico (Fig. 32) from northern Baja California and northern Chihuahua southwards, west of the Sierra Madre Oriental, to central Chiapas, two collections from the lowlands of northern Veracruz; moist grasslands, roadside ditches, stream beds, edges of lakes, in fresh, saline, or alkaline waters; sea level to ca. 2500 m.

Representative Specimens. **Mexico.** Baja California: 23 mi S of Ensenada, *Wiggins & Demaree 4705* (DS); Sierra Juárez, E slope, *Gander 2957* (SD).—Baja California Sur: Arroyo Calamajué, *Moran & Reveal 19614* (ASU, LL, MO, SD, UC, US); San Andrés, NW of Asunción, *Moran & Reveal 19799* (ARIZ, SD, US).—Chihuahua: ca. 60 air mi SW of Cd. Juárez, *Henrickson 14117* (LL); near Lake Santa María, *Nelson 6417* (GH, NY, US).—Distrito Federal: Lago de Texcoco, *Rzedowski 21429* (ENCB, MSC).—Durango: vicinity of Durango, *Palmer 178* in 1896 (F, G, GH, MO, NY, UC, US).—Guanajuato: León, *Hartweg 240* (CGE, G, GH, NY).—Hidalgo: Tula de Allende, *Moore 1370* (BH, GH, MICH, UC).—Jalisco: 7–8 mi S of Acatlán, *McVaugh 19465* (MICH, TEX); 3 km NW of Villa Corona, *McVaugh 19519* (G, MICH, NY, US).—México: 4 km W of Totolcingo, *González 1009* (ARIZ, ASU, ENCB); Dtto. Temascaltepec, Villa Neda, *Hinton 6173* (BM, G, GH, LCU,

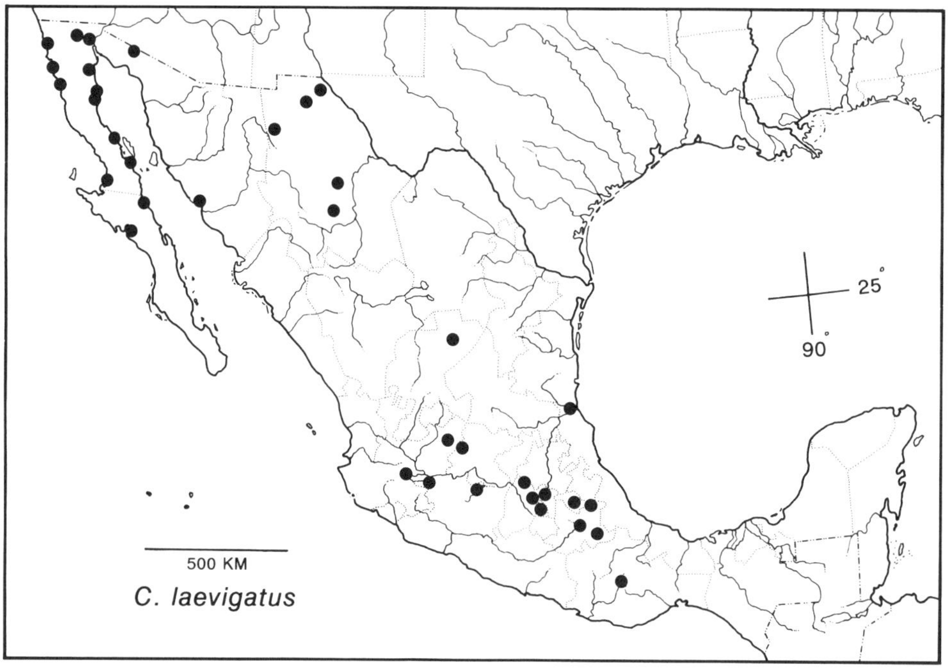

FIG. 32. Distribution of *Cyperus laevigatus* in Mexico.

MO).—MICHOACÁN: Margin of Lake Cuitzeo, *Pringle 4170* (ARIZ, BM, CAS, F, G, GH, MICH, MO, MSC, MTMG, NY, PH, UC, US, WRSL, Z).—OAXACA: Lambityeco, *Carlson 4115* (MICH).—PUEBLA: Shore of Lake Alchichica, *Fassett 29085* (F, MICH, WIS); plains of Tehuacán, *Pringle 6755* (ARIZ, ASU, CAS, GH, MICH, MSC, SD, TX, UC, US, WIS).—SONORA: Quitovac, *Nabhan & Felger 375* (ARIZ).—TLAXCALA/PUEBLA: 3 km al oeste de El Carmen, *Ramos et al. 32* (MICH).—VERACRUZ: between Pánuco and Tampico, *Tucker 2408* (CAS, DUKE, ENCB, F, GH, MICH).—ZACATECAS: ca. 1/2 mi N of San Juan de los Cedros, *Henrickson 13523* (LL).

Cyperus laevigatus is the only species of subg. *Juncellus* in the New World. Thus, it can be distinguished from all other New World species of *Cyperus* by its 2 stigmas and disclike, strongly flattened achene, borne with a face toward the rachilla. Species of subg. *Pycreus*, especially *C. niger*, have been misidentified as *C. laevigatus*, probably because of the dark shiny spikelets. All species of subg. *Pycreus* have lenticular achenes like *C. laevigatus*, but in species of subg. *Pycreus* the achene is attached with an edge toward the rachilla, whereas in *C. laevigatus* a surface of the achene faces the rachilla.

Kükenthal recognized seven different varieties within *C. laevigatus*, all of which occur throughout or nearly throughout the range of the species. He cited specimens for three of these, var. *laevigatus*, var. *reptans*, and var. *distachyos*, from Mexico. Benedict Ayers (1946) did not recognize these three varieties, nor did Scanlon (1942) in her monograph of the Hawaiian species of *Cyperus*. Kükenthal distinguished these varieties on the basis of scale color, spikelet number, height of the culm, and presence of leaf blades. Examination of some 250 herbarium specimens from throughout the range of this pantropic species revealed continuous variation in all these characters. For example, scale coloration shows three patterns: entirely greenish white, white with red speckles laterally, or white-keeled with a dark red spot laterally. In about half of the specimens examined, two or three coloration patterns occurred in specimens of the same collection number, among culms from the same rhizome, or among spikelets or scales on the same culm. These observations support Benedict Ayers's (1946) decision not to recognize varieties in *C. laevigatus*.

IV. Cyperus subgenus **Cyperus.**

Cyperus subgenus *Mariscus* (Vahl) C. B. Clarke, J. Linn. Soc., Bot. Bot. 21: 33. 1884.—TYPE: *Mariscus capillaris* (Swartz) Vahl, typ. cons. [=*Cyperus nanus* Willdenow].

Chlorocyperus Rikli, Jahrb. Wiss. Bot. 27: 563. 1895, nom. superfl. —TYPE: undesignated (see Goetghebeur 1989).

Papyrus Willdenow, Abh. Königl. Akad. Wiss. Berlin 1812–1813: 70. 1816.—TYPE: *Papyrus antiquorum* Willdenow [=*Cyperus papyrus* L.].

Culms and leaves with kranz (chlorocyperoid) anatomy. Spikelets borne in loosely to densely ovoid to cylindric spikes. Rachilla articulate at the base, the spikelets falling as the achenes mature; or the rachilla continuous at the base, persistent even after the achenes mature. Scales persistent or deciduous. Carpels 3, stigmas 3. Achenes trigonous, sometimes dorsiventrally flattened.

Subgenus *Cyperus*, comprising 350–400 species, is the largest subgenus in *Cyperus*. It is diverse in both temperate and tropical regions; 49 species occur in Mexico, of which 9 are endemic.

40. Cyperus squarrosus L., Cent. pl. 2: 6. 1756. *Mariscus squarrosus* (L.) C. B. Clarke in J. D. Hooker, Fl. Brit. India 6: 623. 1893.—TYPE: "India orientalis," *Koenig s.n.* (lectotype: LINN 70.8, IDC microfiche!).

Cyperus inflexus Muhlenberg, Descr. gram. 16. 1817. *Cyperus aristatus* f. *inflexus* (Muhlenberg) C. B. Clarke, J. Linn. Soc., Bot. 21: 92. 1884. *Cyperus aristatus* var. *inflexus* (Muhlenberg) Boeckeler ex Kükenthal, Pflanzenreich 4(20): 504. 1936. *Chlorocyperus inflexus* (Muhlenberg) Palla, Allg. Bot. Zeitschr. Syst. 17: 6. 1911.—TYPE: U.S.A. Pennsylvania, *Muhlenberg herbarium no. 27* (holotype: PH!; isotype: B-W no. 1402, IDC microfiche!).

Isolepis echinulata Kunth, Enum. pl. 2: 205. 1837.—TYPE: VENEZUELA. Cumaná, 1798, *Humboldt & Bonpland s.n.* (holotype: Herb. Kunth, B!).

Cyperus falciculosus Liebmann, Mexic. halvgr. 16. 1850. *Cyperus aristatus* f. *falciculosus* (Liebmann) Kükenthal, Pflanzenreich 4(20): 505. 1936.—TYPE: MEXICO. Veracruz: near Río de Antigua, Feb 1841, *Liebmann 14358* (holotype: C!, photo: BH; isotype: B!).

Cyperus aristatus var. *runyonii* O'Neill, Rhodora 44: 56. 1942.—TYPE: U.S.A. Texas: Kenedy Co., *Runyon 1933* (holotype: LCU!; isotypes: GH! US!).

Caespitose annual, (1–) 2–16 (–20) cm tall. Rhizomes absent. Culms (0.1–) 0.5–0.8 (–1.2) mm in diameter, trigonous, smooth. Leaves 1–3, (1–) 5–10 (–15) mm long, 0.5–2.5 mm wide, flat to v-shaped, the margins smooth. Inflorescence bracts (1–) 2–4, 1–15 cm long, 0.5–3 mm wide, v-shaped, the margins and keel

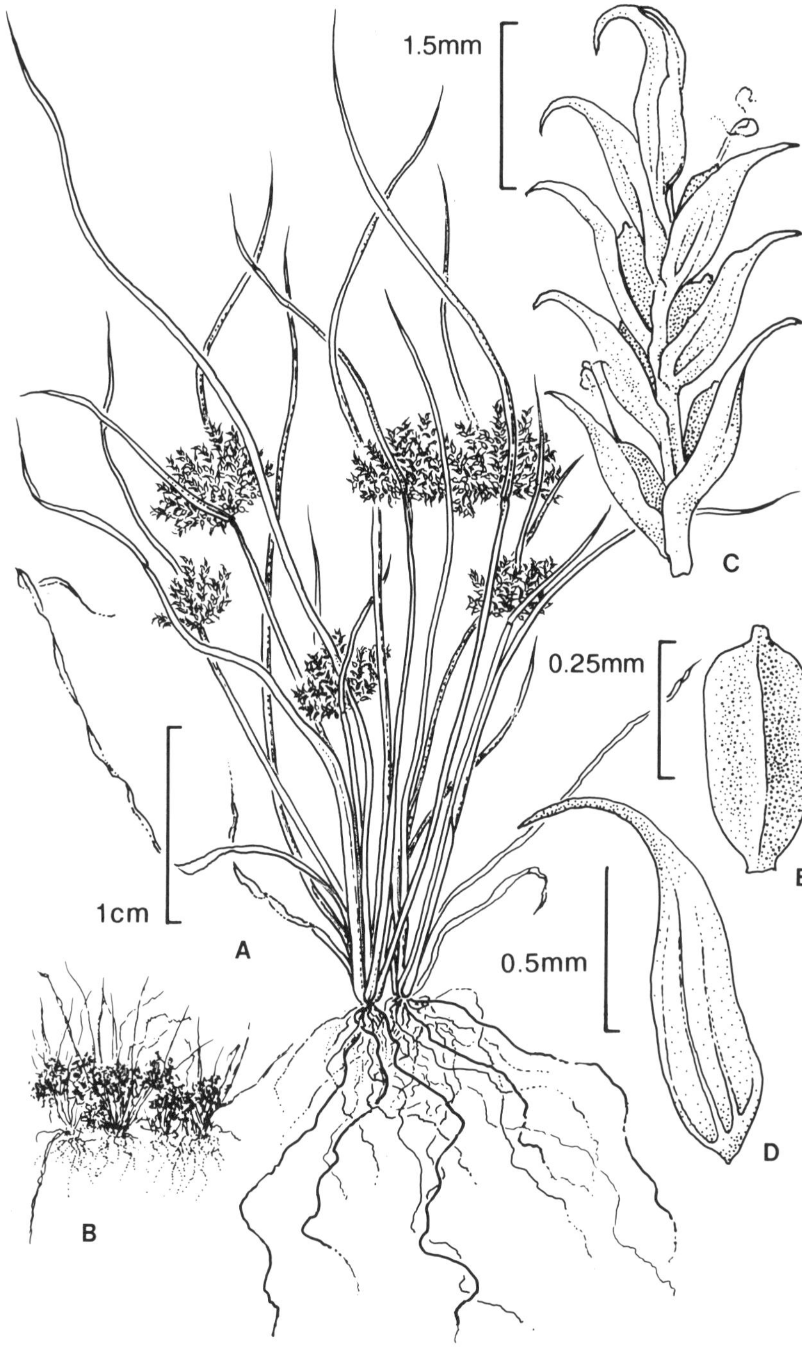

FIG. 33. *Cyperus squarrosus*. A. Habit. B. Aspect of densely grouped plants of a population. C. Spikelet. D. Scale. E. Achene. Drawn by Patricia M. Kernan for the Flora of New York State (Richard S. Mitchell, ed.).

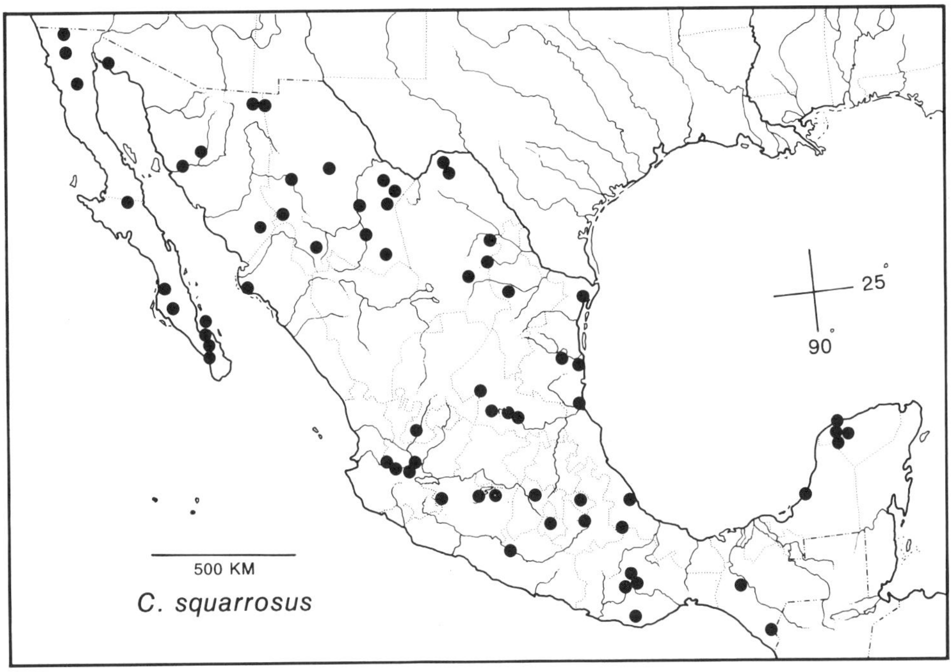

FIG. 34. Distribution of *Cyperus squarrosus* in Mexico.

smooth or sparsely scabrellate, the longest more or less erect, the others oblique. Rays absent or 1–3 (–6), 4–40 mm long. Spikes 6–20 (–40) mm long, 9–15 (–20) mm wide, loosely to densely ovoid to oblong; rachis 5–10 (–15) mm long, smooth. Spikelets (2–) 60–20 (–40), 2.5–10 (–20) mm long, 1.3–2.2 mm wide (excluding awns), ovate-lanceolate to oblong, somewhat compressed; rachilla ± deciduous, ca. 0.3 mm wide, ca. 0.2 mm thick, zigzag, stramineous to reddish brown, successive scales 0.6–0.9 mm apart, wings absent. Scales deciduous, (4–) 10–20 (–34), (1.2–) 1.3–1.8 (–2.2) mm long [the excurved awn an additional 0.5–1.0 (–1.3) mm long], (0.5–) 0.7–0.8 (–1.0) mm wide, oblong-lanceolate, laterally conspicuously (5–) 7–9 (–11) -nerved nearly to the margins, greenish to stramineous or brownish red, medially 1-nerved, greenish. Stamen 1; filament 1–2 mm long; anther 0.2–0.3 mm long, oblong, the connective apex less than 0.1 mm long, reddish. Style 0.3–0.5 mm long; stigmas 3, 0.4–0.7 mm long. Achenes 0.7–0.8 (–1.1) mm long, (0.2–) 0.3–0.4 (–0.5) mm wide, trigonous, obovoid or occasionally linear-spatulate or linear-oblong, infrequently constricted in the middle, apex truncate, apiculate, base stipitate, 0.05–0.1 mm long, ca. 0.1 mm wide, the faces slightly convex to flat, the surface papillose, light brown to nearly black. Fig. 33.

Phenology. Flowering and fruiting mostly from July to November and infrequently during the remaining months.

Distribution. Worldwide, in pantropic and temperate regions; in the New World from New Brunswick, southern Manitoba, and southern British Columbia, Canada, south to central Argentina and Chile. Frequent throughout Mexico (Fig. 34); moist disturbed soils, gravelly roadsides, river flood plains, edges of puddles, and wet muddy places; sea level to ca. 2500 m, most abundant from 1000 to 2000 m.

Representative Specimens. **Mexico.** Baja California: Espíritu Santo Island, *Moran 3674* (SD); N slope of Cerro Blanco, *Moran 16145* (ARIZ, PH, SD); Sierra San Pedro Mártir, *Moran 24471* (SD); 22 mi N of Todos Santos, *Mudie 910* (SD).—Campeche: Champotón, *Steere 1971* (MICH).—Chiapas: 14.7 mi W of Tuxtla Gutiérrez, *Kral 27760* (MO); just S of Motozintla, *Tucker 2229* (DUKE).—Chihuahua: Mpio. Temosachic, 18 mi SE of Madera, *Muller 3497* (LCU, UC); vicinity of Fierro, *Stewart 783* (LL); vicinity of Pirámide, *Johnston 8123* (MO).—Coahuila: Sierra Maderas del Carmen, *Henrickson 15067* (LL); 44.4 mi S of Monclova, *McGill & Keil 7890* (ASU); 24 mi NNE of Monclova, *Palmer 1330* in 1880 (LCU, NY, PH, US); Mpio. Villa Acuña, Cañón de Sentenela, *Wynd & Muller 516* (ARIZ, LCU, MO, NY, US).—Guanajuato: 5 mi S of San Luis de la Paz, *Jones 22575a* (POM, WIS).—Guerrero: Dtto. Coyuca, Querendas, *Hinton 6475* (LCU, MO, NY, US).—Jalisco: Tequila, 1 Oct 1893, *Pringle s.n.* (CAS, MICH, NY, SMU, US); Unión de Tula, *Tucker 2991* (DUKE).—México: 5 km N of Atizapán, *Rzedowski 20888* (ENCB, MICH, MSC).—Michoacán: Cerro Potrerillos, 22 mi S of Jiquilpan, *King & Soderstrom 4644* (MICH, TEX, UC).—Morelos: Cuernavaca, 26 Oct 1962, *Carillo s.n.* (MICH).—Nuevo León: 2 mi NW of Monterrey, *Rollins & Tryon 5828* (LL).—Oaxaca: ca. halfway between Mitla and Oaxaca, *Tucker 2783* (DUKE, ENCB); ca. 15–20 km S of Ixtlán de Juárez, *Tucker 2809* (DUKE).—Puebla: vicinité de Puebla, près Hda. Sta. Bárbara, *Arsène 1380* (MO, US).—San Luis Potosí: Santa Catarina, *Rzedowski 6707* (ENCB, TEX).—Sinaloa: Tapalabampo, *Palmer 247* in 1897 (ARIZ, C, MICH, UC, US).—Sonora: SW part of Pinacate region, *Felger et al. 18788* (ARIZ); near Cornelio, between Hermosillo and Magdalena, *Pennell 20260* (LCU, PH).—Tamaulipas: E of San José de los Leones, *LeSueur 27* (ARIZ, TEX); ca. halfway between Matamoros and San Fernando, *Tucker 1926* (DUKE).—Tlaxcala: Mpio. Santa Cruz Tlaxcala, vic. of San Bernabe Amaxac, *Hernández 367* (MICH).—Veracruz: Antigua, *Purpus 6245* (MO, NY, UC, US).—Yucatán: Izamal, *Gaumer 1023* (CAS, E, GH, MICH, NY, UC, WIS).—Zacatecas: on Hwy 54 ca. 10 km S of Zapoqui, *Tucker 3000* (DUKE).

Cyperus squarrosus can be recognized by its small size and annual habit combined with its oblong-lanceolate scales bearing 5–11 conspicuous nerves and excurved awns. Some collections have been misidentified as *C. cuspidatus* or *C. amabilis*, annual species that have oblong, 3-nerved scales and digitate spikelets.

Cyperus squarrosus exhibits considerable variation in the shape of the mature achenes, more than that encountered in any other species I have studied in Mexico or Central America. In the Mexican material examined, about 70% of the plants produced obovoid, 10% linear-spatulate (more or less constricted medially), 5% linear-oblong, and 15% intermediates between linear-oblong and linear-spatulate achenes. Achene shape showed no obvious correlation with any other morphological features nor did the variants have any discrete geographic range. A collection of plants from southern Texas having linear-spatulate achenes was named as *C. aristatus* var. *runyonii* by O'Neill. This collection simply represents one extreme of the range of variation in the shape of achenes and does not warrant any special taxonomic rank.

A curious feature of *C. squarrosus* is the pleasant odor of the dried plant. This was first recorded by Torrey (1836), who likened it to the smell of dried *Trifolium*. Fernald (1950) compared the fragrance to *Ulmus rubra*, Correll and Johnston (1970) to flowers of the Bombacaceae, and Bruhl (pers. comm.) to fenugreek (*Trigonella foenum-graecum*, Fabaceae). The odor can still be detected on herbarium specimens over a century old, e.g., the type of *C. falciculosus*, collected by Liebmann in Mexico in 1841.

Fernald (1950) called this species *C. inflexus*, although O'Neill (1942) had demonstrated that the New World plants called *C. inflexus* were not distinguishable at any taxonomic level from "*C. aristatus*" [=*C. squarrosus*] of the Old World. O'Neill (1942), Benedict Ayers (1946), and Gleason and Cronquist (1963) applied the name *C. aristatus* Rottb. to this species, which is illegitimate, because Rottbøll cited the earlier name *Scirpus intricatus* L. as a synonym; see also Doubtful and Excluded Names. Kern (1974) showed that the oldest name for the species is *C. squarrosus*, based on a specimen in the Linnaean Herbarium.

41. Cyperus iria L., Sp. pl. 1:45. 1753.—TYPE: "India." *Osbeck s.n.* (lectotype: LINN 70.16, IDC microfiche!).

Cyperus panicoides Lamarck, Tabl. encycl. 1: 145. 1791.—TYPE: Locality unknown, collector unknown (holotype: P-LA, IDC microfiche!).

Caespitose annual, (5–) 20–60 cm tall. Rhizomes absent. Culms (0.2–) 0.6–2.4 mm in diameter, triquetrous, smooth. Leaves (1–) 3–4, 15–40 cm long, 3–6 mm wide, v-shaped, the margins sparsely scabridulous. Inflorescence bracts (4–) 5 (–7), (1.5–) 8–30 (–50) cm long, 1–6 mm wide, v-shaped, the margins and keel scabrellate, ascendent at ca. 45°. Rays 4–8, (1–) 4–11 cm long. Spikes 2–5 (–8), 2–10 cm long, 2–4 cm wide, oblong to ovoid; rachis 2–6 cm long. Spikelets 10–30, 4–20 mm long, 1.5–2.1 mm wide, linear, compressed (appressed to strongly ascending); rachilla persistent, 0.2–0.3 mm wide, ca. 0.2 mm thick, flexuous, light brown, successive scales 0.6–1.2 mm apart, wings absent. Scales deciduous, (2–) 6–26, 1.3–1.8 mm long, 1.2–1.8 mm wide, obovate to nearly orbiculate, apex emarginate, mucronulate, laterally 2-nerved (the outer nerves more noticeable than the inner), brown to golden-brown, becoming hyaline toward the edges, medially 1-nerved, green. Stamens 2 (–3); filaments 1.1–1.4 mm long, their bases coherent as a faint white hypogynous ring; anthers (0.2–) 0.3–0.4 mm long, the connective apex less than 0.1 mm long, reddish. Style less than 0.1 mm long; stigmas 3, 0.2–0.4 mm long. Achenes 1.2–1.4 mm long, 0.5–0.7 mm wide, trigonous, obovoid, apex apiculate, base slightly stipitate, the faces concave, the surface puncticulate, brown.

Phenology. Flowering and fruiting from (May–) July to October.

Distribution. Native in the Old World, introduced in the eastern and southern U.S.A., Mexico, Central and South America. Common in much of Mexico (Fig. 35); weedy roadsides, pastures, puddles and other moist disturbed situations; sea level to 500 (–1300) m. *Cyperus iria* is apparently a recent introduction in Mexico; Benedict Ayers (1946) did not record it. The species was first recorded in the New World from the southeastern U.S.A. in the 1840's.

REPRESENTATIVE SPECIMENS. **Mexico.** CHIAPAS: 15 km NE of Pichucalco, *Tucker 2578* (DUKE, ENCB); 5 km S of Arriaga, *Tucker 2710* (DUKE, ENCB); 3 mi NW of Tonalá, *Wilbur 35975* (DUKE).—COLIMA: [Cerro de] Ortega, *Tucker 2958* (DUKE, ENCB); Las Brisas across harbor from Manzanillo, *Wilbur 36741* (DUKE, ENCB).—GUERRERO: Petatlán, *Tucker 2937* (DUKE, ENCB); along Hwy 200, 50 km W of Pinotepa Nacional, *Tucker 2920* (DUKE, ENCB); along Hwy 95, between Acapulco and Tierra Colorada, *Tucker 2266* (DUKE).—JALISCO: 2 km N of La Huerta, *Tucker 2967* (DUKE, ENCB); Villa Purificación, *Puga & Guzmán 1205* (IBUG).—MICHOACÁN: 22 km S of Uruapan, *King & Soderstrom 4829* (TEX, UC); El Tícuiz, *Tucker 2955* (DUKE, ENCB).—NAYARIT: Mpio. Ruiz, orillas Río San Pedro, 1 May 1978, *Flores s.n.* (IBUG).—OAXACA: 8 km SW of Juchitán, *Tucker 2772* (DUKE, ENCB); 8 km SW of Tehuantepec, *Tucker 2768* (DUKE, ENCB); Tepanatepec, *Tucker 2736* (DUKE, ENCB); 15 km W of Pochutla, *Tucker 2893* (DUKE, ENCB).—SINALOA: 5 mi S of Rosario, *Stevens & Fairhurst 2037* (CAS, DS, DUKE, MSC, TEX).—SONORA: 11.5 mi by rd E of Alamos, *Sanders et al. 8812* (NYS, UCR).—TABASCO: on Hwy 180, 30 km E of Río Tonalá, *Tucker 2550*

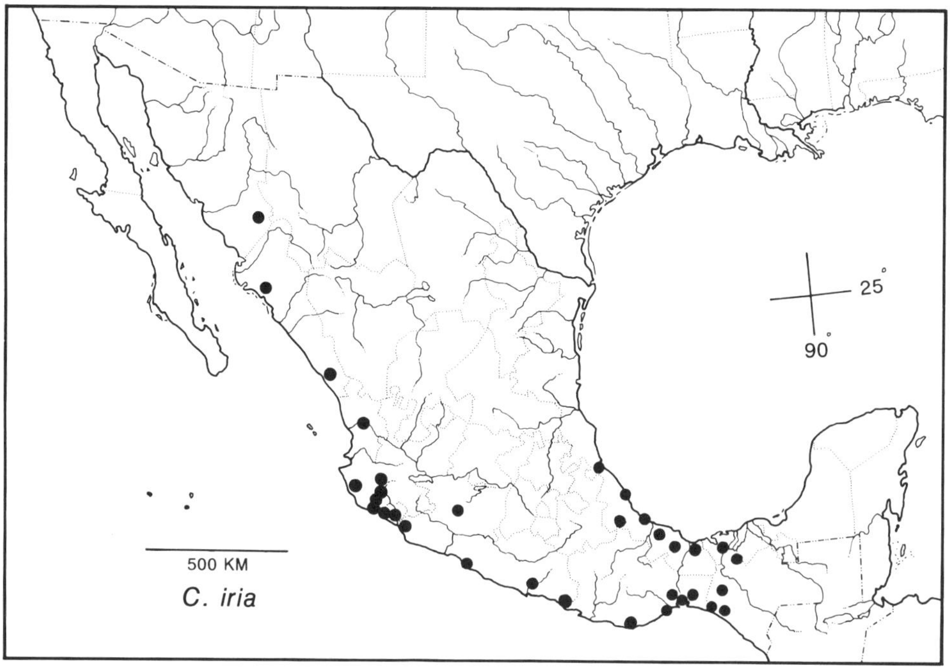

FIG. 35. Distribution of *Cyperus iria* in Mexico.

(DUKE, ENCB).—VERACRUZ: 50 km S of Casitas, *Tucker 2484* (DUKE); roadside of Hwy 180, 30 km S of Tuxpan, *Tucker 2462* (DUKE); 16 km S of Lerdo de Tejada, *Tucker 2512* (DUKE, ENCB).

Cyperus iria can be separated from all other species in Mexico by its ascending, appressed spikelets and its emarginate, strongly 5-nerved scales. The scales are widely spaced on the rachilla, with the apex of one scale barely reaching the base of the next on the same side of the rachilla. *Cyperus laxus* might be mistaken for this species in that it has similar ovoid achenes and sessile stigmas; however, *C. laxus* can be distinguished by its cuspidate scales and digitate spikelets.

42. Cyperus prolixus H. B. K., Nov. gen. sp. 1 [quarto]: 206. 1816.—TYPE: COLOMBIA. Near Bogotá, 3000 m, *Humboldt & Bonpland s.n.* (holotype: P-HBK, fide Raynal, 1967; isotypes: B-W no. 1397, IDC microfiche!, P!).

Cyperus amplissimus Steudel, Syn. pl. glum. 2: 31. 1854.—TYPE: "America." Collector unknown (holotype: P, fide Raynal, 1967).

Large perennial, 0.5–3 m tall. Rhizomes 3–20 cm long, 1–2 cm in diameter, indurate. Culms 0.5–3 cm in diameter, stoutly trigonous, smooth, basally indurate, thickened. Leaves 3–10, 40–130 cm long, 10–15 mm wide, v-shaped, with cross-veins, the margins and keel scabrous. Inflorescence bracts (6–) 9–10, (5–) 15–90 cm long, 0.5–20 mm wide, v-shaped or inversely w-shaped, the margins scabrous, ascendent at 45–75°. Rays (5–) 9–12, 2–22 (–30) cm long; secondary rays 1–10 cm long; tertiary rays 0.5–3 cm long. Spikes (2–) 3–45 mm long, (4–) 7–10 mm wide, narrowly cylindric (appressed-ascendent); rachis 2.5–4.0 cm long. Spikelets (4–)

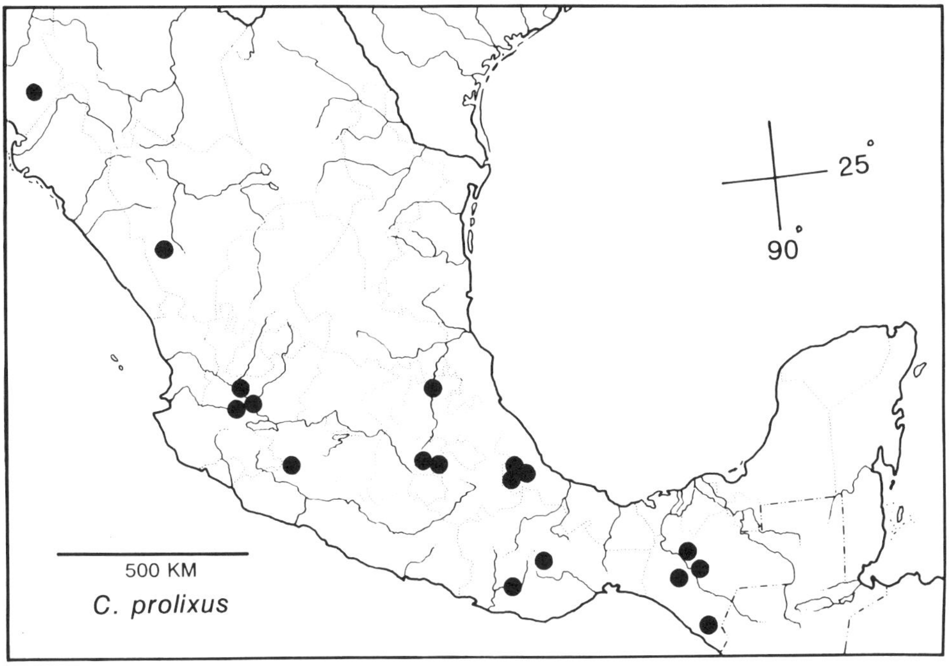

FIG. 36. Distribution of *Cyperus prolixus* in Mexico.

10–35, 7–12 (–15) mm long, 0.6–1.2 mm wide, linear, compressed (appressed-ascendent); rachilla persistent, 0.3–0.4 mm wide, ca. 0.2 mm thick, flexuous, light brown, successive scales 1.5–2.2 mm apart, wings 0.2–0.4 mm wide, hyaline. Scales deciduous, 6–20, (3.2–) 3.8–4.3 mm long, 1.2–1.7 mm wide, oblong-ovate, apex acute to obtuse, mucronulate, spreading, marginally hyaline, laterally 2–3-nerved, brown, medially 1-nerved, green. Stamens 3; filaments 2–4 mm long; anthers 0.7–0.9 mm long, the connective apex up to 0.1 mm long, oblong, bright red. Style 0.8–1.3 mm long; stigmas 3, 2.0–3.3 mm long. Achenes 1.8–2.5 mm long, 0.8–1.3 mm wide, trigonous, narrowly ellipsoid, apex subacute but not apiculate, base sessile, the faces flat or slightly concave, the surface puncticulate, brown.

Phenology. Flowering and fruiting from July to November.

Distribution. Mostly in neotropical regions; Louisiana, U.S.A., northwestern Mexico south through Central America and South America to northern Argentina. Occasional in the uplands of Mexico (Fig. 36); along streams, rivers, and roadside ditches; (500–) 1000–2000 m.

REPRESENTATIVE SPECIMENS. **Mexico.** CHIAPAS: Mpio. Chenalho, Colonia Choro, *Ton 1569* (DS, F, MICH, NY).—DISTRITO FEDERAL: Ixtapalapa, *Matuda 25654* (NY).—DURANGO: Sianori, *Ortega 5255* (US).—HIDALGO: Dtto. Molango, Mpio. Laguna de Atexca, *Moore 2016* (BH).—JALISCO: Río Blanco, *Palmer 213* in 1886 (G, MICH, MO, NY, US); W of Guadalajara on Hwy 15, *Tucker 2277* (DUKE, ENCB, MICH).—MÉXICO: near Villa Guerrero, Rancho Tobías, *Alexander & Hernández 1965* (NY).—MICHOACÁN: 22 km S of Uruapan, *King & Soderstrom 4879* (MICH, TEX, UC, US).—OAXACA: Trapiche de la Concepción, *Liebmann 14472* (C); rd from Mitla to Ayutla, ca. 35 km from Mitla, *Tucker 2198* (CAS, DUKE, ENCB).—SONORA: 7 km SW of Alamos, *Van Devender et al. 92-764* (ARIZ, NYS).—VERACRUZ: Ruisseaux de l'Antigua, *Galeotti 5792* (BR, G); Orizaba, *Mohr & Botteri 12* (NY).

Both *Cyperus prolixus* and *C. giganteus* reach 3 m in height and are the tallest species of the genus in Mexico and Central America. They are readily separable with the following key.

1. Leaf blades absent; inflorescence hemispherical, 20–100 cm wide; spikelets borne at approximately right angles to the rachis. *C. giganteus.*
1. Leaf blades present; inflorescence narrow, 15–30 cm wide; spikelets appressed-ascendent. *C. prolixus.*

43. Cyperus distans L. f., Suppl. pl. 103. 1781. *Cyperus elatus* Rottbøll, Descr. icon. rar. pl. 37. 1773, non *Cyperus elatus* L., 1756.—TYPE: "Asia." Collector unknown (holotype: Herb. Rottbøll-C!; isotype: Herb. Vahl-C!).
Cyperus purpureo-vaginatus Boeckeler, Beitr. Cyper. 2: 8. 1890.—TYPE: BRAZIL. *Warming s.n.* (holotype: C!).

Perennial, (10–) 40–80 (–120) cm tall. Rhizomes 0.5–3 cm long, 3–10 mm in diameter. Culms 1.2–2.5 (–5) mm in diameter, trigonous, smooth. Leaves 2–5, (10–) 20–50 (–90) cm long, 2–5 (–10) mm wide, inversely w-shaped, the margins and keel scabrellate distally, smooth proximally. Inflorescence bracts 4–7, (5–) 10–40 (–60) cm long, (2–) 3–8 (–11) mm wide, inversely w-shaped, the margins and keel scabrellate, ascendent at 30–45°; secondary bracts 1–4 per ray, 5–10 (–30) mm long, 0.5–1 mm wide, scabrellate. Rays (4–) 5–6 (–8), smooth, (1–) 3–8 (–16) cm long; secondary rays absent or 1–5, 5–15 (–50) mm long. Spikes 1–4 (–7), (20–) 25–35 (–42) mm long and wide, loose, broadly ovoid to ovoid-pyramidal; rachis (10–) 12–18 (–23) mm long. Spikelets (10–) 20–35 (–45), (10–) 13–18 (–25) mm long, 1.2–1.5 (–1.6) mm wide, linear, seemingly flattened proximally from the spreading scales, becoming terete apically; rachilla persistent, ca. 0.2 mm wide, 0.1 mm thick, the internodes slightly curved, greenish with red speckles, or reddish brown, successive scales 1.8–2.4 mm apart. Scales deciduous, (10–) 12–20 (–25), 1.7–2.0 (–2.4) mm long, 0.9–1.2 mm wide, oblong-ovate, apex obtuse and barely reaching or as much as 0.4 mm distant from the base of the next scale, laterally light reddish brown to dark red, nerveless (the margins clasping the achene), medially 3–5-nerved, greenish. Stamens 3; filaments 1.5–2.5 mm long; anthers 0.4–0.5 mm long, oblong, the connective apex at most 0.1 mm long, reddish. Style 0.3–0.6 mm long; stigmas 3, 1.0–1.8 (–2.0) mm long, reddish punctate. Achenes (1.3–) 1.4–1.6 (–1.7) mm long, (0.3–) 0.4–0.5 mm wide, trigonous, narrowly oblong, apex acute but not apiculate, base cuneate to substipitate, the faces flat, the surface papillose, dark brown.

Phenology. The only known Mexican collection was made in October. Central American collections examined were gathered in January, June, and July.

Distribution. Widespread in the Old World tropics, from West Africa to New Guinea; in the New World, from the Dominican Republic south through the Lesser Antilles to eastern Brazil and northern Argentina, but in Central America known only from the mountains of Nicaragua. In Mexico known only from a single collection.

ADDITIONAL SPECIMENS EXAMINED. **Mexico**. GUERRERO: Dtto. Mina, Manchón, barranca at 1200 m, 11 Sep 1936, *Hinton 9496* (DS, K, MICH, US).

Cyperus distans is easily distinguished from other species in Mexico by its slender spikelets (1.2–1.6 mm wide) and its divaricate, non-overlapping scales, which are laterally nerveless and red to reddish brown.

44. Cyperus articulatus L., Sp. pl. 1: 44. 1753.—LECTOTYPE: Sloane, Voy. Jamaica Nat. Hist. 1: 121, t. 81, fig. 1. 1707.

Cyperus corymbosus Rottbøll, Descr. pl. rar. 19. 1772.—TYPE: SURINAME. *Rolander s.n.* (holotype: Herb. Rottbøll-C!).

Cyperus diphyllus Retzius, Observ. bot. 5: 11. 1788.—TYPE: "India orientalis." *Koenig s.n.* (holotype: C!).

Cyperus nodosus Willdenow, Enum. hort. berol. 72. 1809. *Cyperus articulatus* var. *nodosus* (Willdenow) Kükenthal, Pflanzenreich 4(20): 79. 1935.—TYPE: "America meridionalis," *Humboldt & Bonpland s.n.* (holotype: B-W no. 1282, IDC microfiche!).

Cyperus subnodosus Nees & Meyen, Nov. Actorum Acad. Caes. Leop-Carol. German. Nat. Cur. 19: 59. 1843. *Cyperus nodosus* var. *subnodosus* (Nees & Meyen) Boeckeler, Linnaea 36: 276. 1870. *Cyperus corymbosus* var. *subnodosus* (Nees & Meyen) Kükenthal, Repert. Spec. Nov. Regni Veg. 23: 185. 1926.—TYPE: PERU. Tacna, *Meyen s.n.* (not located).

Cyperus articulatus var. *multiflorus* Kükenthal, Pflanzenreich 4(20): 79. 1935.—TYPE: U.S.A. Florida: Lake Co., vic. of Eustis, Lake Norris, 16–25 Aug 1894, *Nash 1701* (holotype: B, not located; isotypes: CONN! MICH! NY!).

Perennial, 40–140 (–200) cm tall. Rhizomes 2–125 cm long, 2–3 mm in diameter, indurate, with prominent septa 5–15 mm apart, clothed with brownish, lanceolate scales 2–4 cm long. Culms 4–12 mm in diameter at the base, (1.2–) 1.5–2.5 mm in diameter at the apex, terete (occasionally trigonous for the apical third, scarcely compressed in drying) with transverse septa (conspicuous when dried) ca. 3 cm apart basally, decreasing in separation to 3–5 mm apart apically, smooth or infrequently scabrous on the angles apically. Leaves usually bladeless, the base of the culm with 2–3 loose papery sheaths (5–) 10–25 cm long; blades occasionally present, 1 (–4), (1–) 10–20 (–40) cm long, 4–6 (–9) mm wide, with prominent cross veins especially abaxially, the margins and keel scabrous. Inflorescence bracts 2 (–4), 0.2–2.0 (–9) cm long, 1.5–4.5 mm wide, deltate to lance-linear, v-shaped, the margins and keel scabrous, the apex sharp, thornlike, erect, the longest appearing to be a continuation of the culm. Rays 5–7 (–10), 0.3–8 (–12) cm long, slender, wirelike, erect to drooping; secondary rays occasionally present, 5–20 mm long, borne spicately on the distal portion of the primary rays. Spikes 15–45 mm long, 10–30 (–40) mm wide, broadly ovoid to subumbellate; rachis 5–15 mm long. Spikelets (1–) 5–10, 10–35 (–45) mm long, 1.1–2.0 mm wide, linear, compressed, flatly elliptic in cross section; rachilla persistent, ca. 0.4 mm wide, ca. 0.2 mm thick, straight to slightly undulate, whitish to hyaline, successive scales 1.4–2.0 (–2.4) mm apart, wings ca. 0.4 mm wide. Scales deciduous, 15–45, 2.9–3.7 mm long, 1.1–1.6 (–1.8) mm wide, oblong-elliptic to ovate, apex acute, laterally 1–2-nerved, light brown, medially 3-nerved, green to brown. Stamens 3; filaments 1.3–1.7 mm long, light reddish brown; anthers linear-oblong, 1.7–2.4 mm long, the connective reddish, prolonged at most 0.1 mm. Style 1.2–3.6 mm long; stigmas 3, 1.7–6.0 mm long. Achenes 1.2–1.6 mm long, 0.4–0.6 mm wide, trigonous, slightly dorsiventrally flattened, obovoid-ellipsoid, apex markedly apiculate, base substipitate to stipitate and ca. 0.1 mm long, the faces concave, the surface puncticulate, brown.

Phenology. Flowering and fruiting throughout the year, most frequently in July and August.

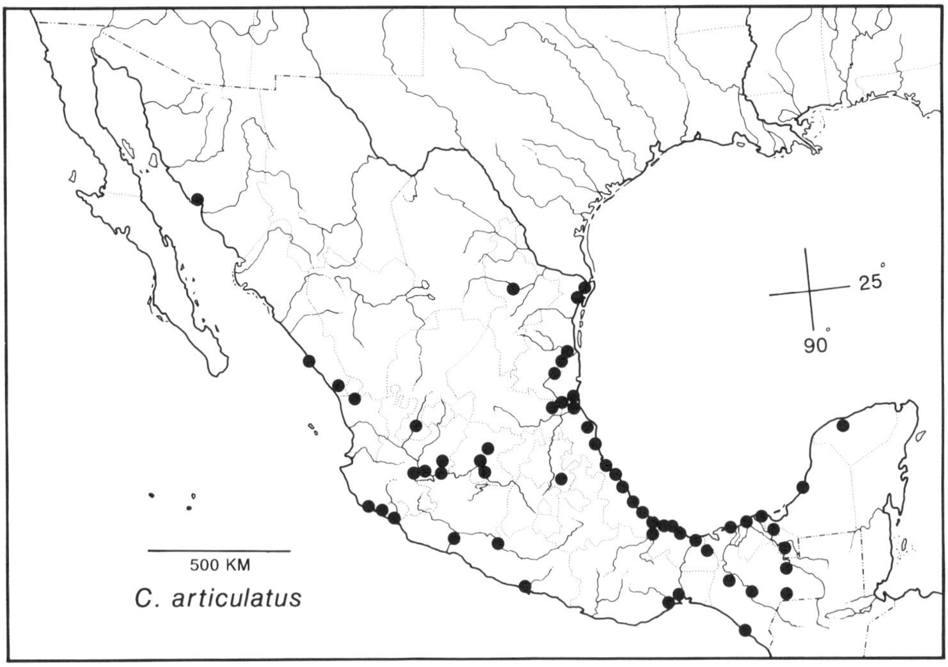

FIG. 37. Distribution of *Cyperus articulatus* in Mexico.

Distribution. Pantropic and warm-temperate regions; in the New World from South Carolina to Texas, U.S.A., and western Mexico south to northern Argentina. Common along the Gulf Coast of Mexico, occasional along the Pacific Coast, and around Lago de Chapala, Michoacán, and Guadalajara, Jalisco (Fig. 37); marshes, waterholes, lake shores, and ditches; sea level to 400 (–1700) m.

REPRESENTATIVE SPECIMENS. **Mexico.** CAMPECHE: Palizada, *Matuda 3868* (F, GH).—CHIAPAS: Mpio. Las Margaritas, W side of Laguna Miramar, *Breedlove 33314* (DS).—COLIMA: 15 mi SE of Manzanillo, *McVaugh 15676* (MICH, US).—GUANAJUATO: Laguna de Yuriria, *Rzedowski 27589* (ARIZ, ENCB, SD).—GUERERRO: Pie de la Cuesta, NW of Acapulco, *Barkley 14167* (GH, MO, TEX, US); Dtto. Coyuca, Santa Barbara in Río Balsas, *Hinton 5831* (F, GH, MO, US).—HIDALGO: Chapulhuacán, *Kenoyer 1013* (F).—JALISCO: 1 mi W of Ayo el Chico, *McVaugh 17213* (MICH, TEX); Ameca, *McVaugh 18578* (DUKE, MICH, TEX); on Hwy 200, at intersection near turnoff to Barra de Navidad, *Tucker 2965* (DUKE, ENCB).—MICHOACÁN: halfway between Jacona and Zamora along Hwy 15, *Iltis et al. 650* (IBUG, WIS).—NAYARIT: along Hwy 15, near Sinaloa border, *Gibson 2158* (CAS).—NUEVO LEÓN: E of Monterrey, *Rasmussen 37* (TEX).—OAXACA: 4 km NNE of Tehuantepec, *King 1332* (MICH, TEX, UC, US); ca. 8 km SW of Juchitán, *Tucker 2774* (DUKE, ENCB).—SAN LUIS POTOSÍ: 10 mi W of Ebano, on Hwy 70, *Tucker 2005* (DUKE, ENCB).—SINALOA: N of Mazatlán, *Mexia 86* (E, US).—SONORA: Guaymas, *Palmer 636* in 1887 (US).—TABASCO: Mecacán, *Barlow 28/24 B* (MEXU).—TAMAULIPAS: 10 mi N of San Fernando, *Tucker 1937* (CAS, DUKE, ENCB); 37.5 mi S of Soto la Marina, *Tucker 1977* (DUKE, ENCB); ca. 21.5 mi N of Manuel, *Tucker 1983* (DUKE, ENCB).—VERACRUZ: 5 mi E of Coatzacoalcos along coast, *Davidse 9342* (MO, SD); Pital, *Liebmann 14419* (C); 15 km NW of Alvarado, *Roe et al. 1292* (UC, US, WIS); on Hwy 70, E of turnoff to Pánuco, *Tucker 2416* (DUKE, ENCB); 3 km N of Lerdo de Tejada, *Tucker 2506* (DUKE, ENCB).—YUCATÁN: Mérida, *Souza-Novelo 237* (US).—ZACATECAS: 3 mi from Santa Rosa along E side of Río Juchilipa, *Taylor 6058* (US).

Both Kükenthal (1935–36) and Benedict Ayers (1946) cited several collections of *C. corymbosus* from Mexico. Kükenthal distinguished *C. corymbosus* by its having 1–2 leaf blades per culms, subtrigonous, non-septate culms, and inflorescence bracts about as long as the rays; yet, no representatives of this taxon from either Mexico or Central America were encountered in this study. According to Kükenthal, *C. articulatus* is always leafless; however, approximately 35 leafy-stemmed collections of *C. articulatus* from Mexico and Central America were examined. The leaf blades of these specimens varied from 1–40 cm in length; thus, there was no clear division between leafy and leafless individuals. In eight of these collections, the culms were subtrigonous to nearly trigonous in the distal 10 cm, but terete and septate proximally. Among these leafy plants, the inflorescence bracts ranged in length from 1–9 cm, from much shorter than to slightly exceeding the length of the rays. Thus, among Mexican and Central American collections, I am unable to separate *C. corymbosus* from *C. articulatus* on the basis of any of the characters used by Kükenthal or Benedict Ayers.

To investigate further the status of *C. corymbosus*, some 400 Old World and South American collections labelled as "*C. articulatus*" or "*C. corymbosus*" were studied with regard to the characters used by Kükenthal and Benedict Ayers. This extensive survey confirmed the results obtained from the examination of Mexican and Central America collections. The character differences proposed by Kükenthal could not be substantiated with the large number of collections examined. Thus, *C. corymbosus* is treated here as a synonym of *C. articulatus*.

45. Cyperus esculentus L., Sp. pl. 1: 45. 1753.—TYPE: The plate cited by Linnaeus, "*Cyperus rotundus esculentus angustifolius* . . ." in Morison, Pl. hist. univ. 3: 236, sect. 8, t. 11, no. 8 ["10"]. 1699 (lectotype, here designated).

Cyperus phymatodes Muhlenberg, Descr. gram. 23. 1817. *Cyperus esculentus* var. *leptostachyus* Boeckeler, Linnaea 36: 290. 1870. *Chlorocyperus phymatodes* (Muhlenberg) Palla, Allg. Bot. Zeitschr. Syst. 9: 69. 1903. *Cyperus esculentus* var. *phymatodes* (Muhlenberg) Kükenthal, Repert. Spec. Nov. Regni Veg. 23: 185. 1926. —TYPE: U.S.A. Pennsylvania: Muhlenberg Herb. no. 28 (holotype: PH!; isotype: B-W no. 1377, IDC microfiche!).

Cyperus fulvescens Liebmann, Mexic. halvgr. 30. 1850. *Cyperus lutescens* var. *fulvescens* (Liebmann) C. B. Clarke, J. Linn. Soc., Bot. 21: 182. 1884.—TYPE: MEXICO. Oaxaca: Trapiche de la Concepción, Jul 1842, *Liebmann 14386* (lectotype, designated by Tucker, 1986: C!).

Cyperus esculentus var. *sprucei* C. B. Clarke, J. Linn. Soc., Bot. 21: 181. 1884.—TYPE: ECUADOR. Chanduy, *Spruce 6414* (holotype: K).

Perennial, (20–) 30–40 (–65) cm tall. Stolons soft, spongy, flexible when dried, 1–8 (–15) cm long, bearing tubers (3–) 6–11 mm in diameter. Culms 0.6–3.4 mm in diameter, trigonous, smooth. Leaves 3–7, (3–) 10–25 (–35) cm long, 2–6.5 mm wide, v-shaped to flanged v-shaped, the margins scabrellate in the distal half. Inflorescence bracts (3–) 4–5 (–7), (1.5–) 5–30 cm long, 2–8 mm wide, v-shaped to flanged v-shaped, the margins scabrellate in the distal half or throughout, borne horizontal to ascendent at 45°. Rays 5–10, (0.3–) 2–12 (–20) cm long; secondary

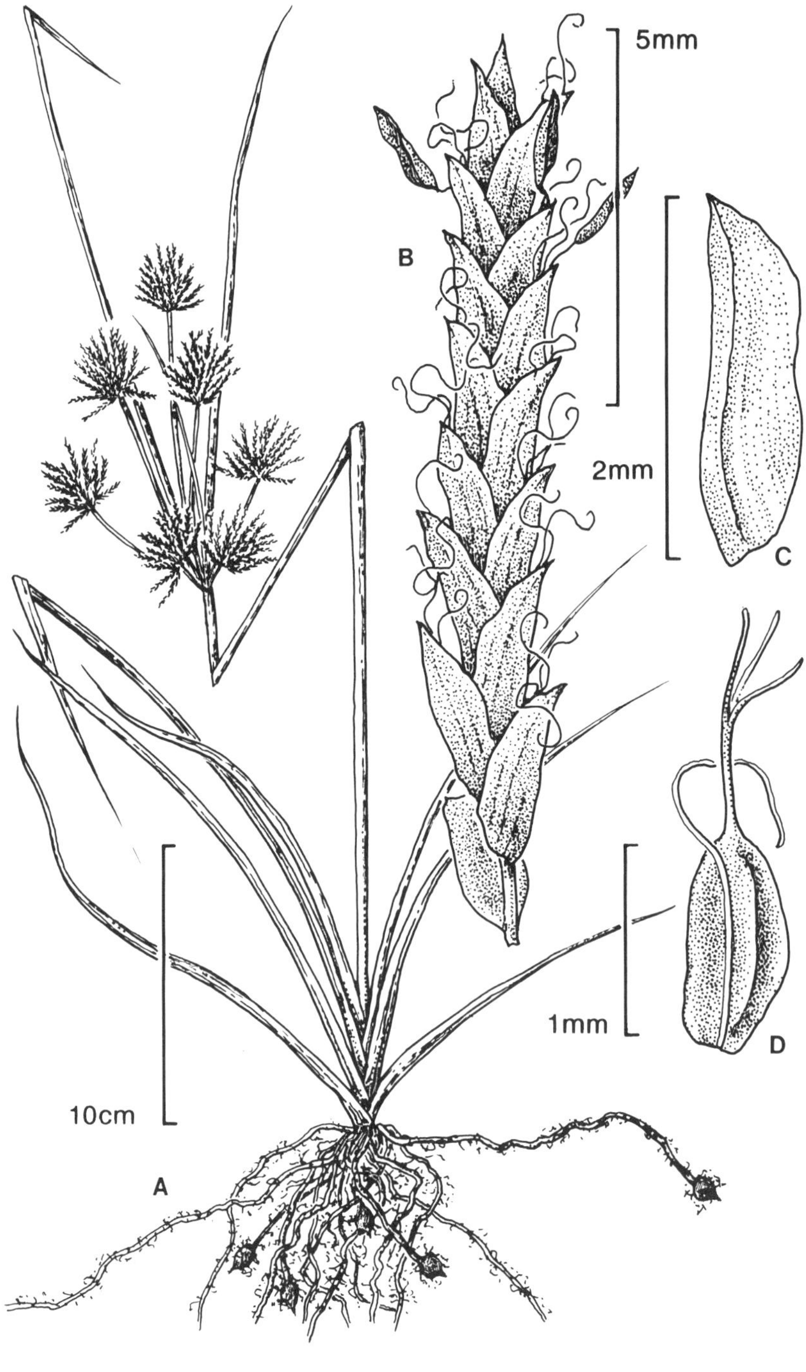

FIG. 38. *Cyperus esculentus*. A. Habit. B. Spikelet. C. Scale. E. Achene. Drawn by Patricia M. Kernan for the Flora of New York State (Richard S. Mitchell, ed.).

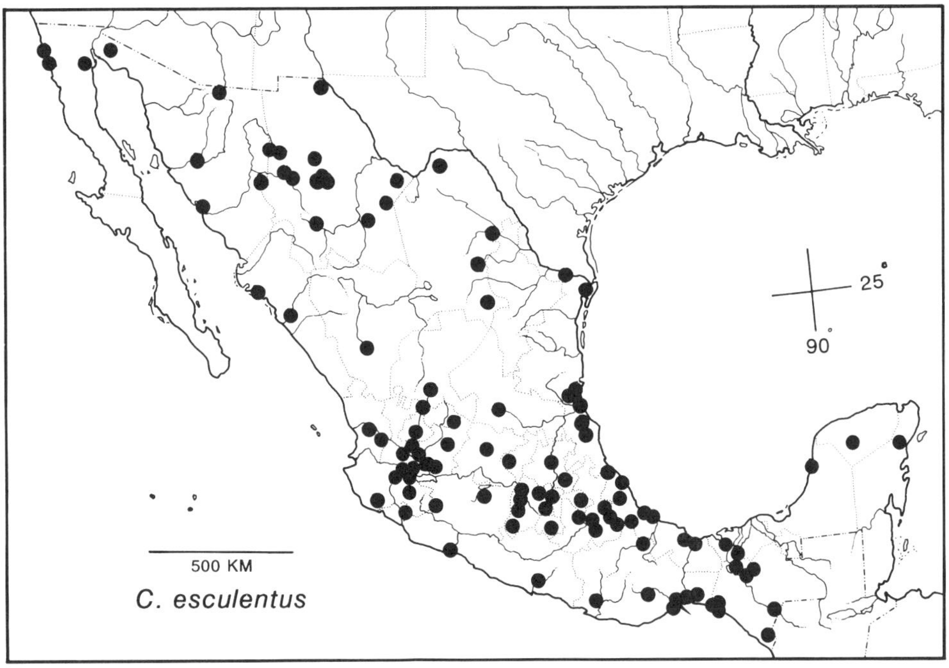

FIG. 39. Distribution of *Cyperus esculentus* in Mexico.

rays 2–5 mm long, frequently present on the longer rays. Spikes 1 (–3), 18–30 mm long, 18–35 mm wide, broadly ellipsoid; rachis 5–18 mm long. Spikelets (3–) 10–20 (–25), (5–) 10–25 (–34) mm long, (1.2–) 1.5–1.6 (–2.0) mm wide, linear, subquadrate (i.e., flattened-rhombic in cross section); rachilla persistent, 0.2 (–0.3) mm wide, 0.1 (–0.2) mm thick, straight, light brown, successive scales 1.4–1.8 mm apart, wings (0.2–) 0.5 mm wide, hyaline. Scales persistent, 9–42, (1.9–) 2.7–3.1 mm long, 1.4–2.1 mm wide, ovate to elliptic, apex obtuse to subacute, laterally 2–3-nerved, yellowish brown to dark brown, medially 3-nerved, brownish, reddish, or greenish. Stamens 3; filaments 1.7–2.4 mm long; anthers 1.3–1.8 mm long, narrowly ellipsoid, the connective apex ca. 0.1 mm long, ovate, reddish. Style 0.6–2.0 mm long; stigmas 3, 1.8–3.0 mm long. Achenes (1.1–) 1.3–1.5 (–1.6) mm long, 0.3–0.6 (–0.8) mm wide, trigonous, ellipsoid, apex obtuse, base sessile, the adaxial face slightly concave, the abaxial faces convex, the surface puncticulate, brown. Fig. 38.

Phenology. Flowering and fruiting from June to October.

Distribution. Cosmopolitan, in tropical, warm, and cool-temperate regions; in the New World from New Brunswick to southern British Columbia, Canada (Mulligan & Junkins 1976), south through Mexico and Central America to Argentina, also in the West Indies. Common in Mexico (Fig. 39); beaches, roadsides, ditches, croplands, pastures, and disturbed situations; (sea level–) 1000–2000 (–3100) m.

Representative Specimens. **Mexico.** Aguascalientes: Aguascalientes, *Hartweg 254* (CGE, E, G, NY).—Baja California: S edge of Ensenada, *Moran 29313* (SD); Mpio. Constitución, *Rodríguez 1713* (ARIZ).—Campeche: between Playa Bonita and Punta Morros, *Sauer & Gade 3189* (WIS).—Chiapas:

Tuxtla Gutiérrez, *Breedlove 10623* (DS, F, NY); 5 km N of Tonalá, *Tucker 2702* (DUKE, ENCB).—CHIHUAHUA: 8 mi W of Chihuahua, *Jones 22478* (UC, WIS); Chuhuichupa, *LeSueur 1107* (ARIZ, NYS, TEX).—COAHUILA: Del Carmen Mts, *Marsh 748* (F, TEX).—COLIMA: Río Tuxpan below Hwy 110, *Stevens & Fairhurst 1837* (MO, MSC).—DISTRITO FEDERAL: San Angel, *Fisher 128* (F, MO, NY, US, WIS); Cerro Xochitepec, *Rzedowski 24270* (ARIZ, ENCB, MO, SD).—DURANGO: W of Durango, *Bell & Duke 16588* (UC); Otinapa, *Palmer 334* in 1906 (F, MO, NY, UC, US).—GUANAJUATO: Mpio. Yuriria, *Zizumbo V. Z-697* (CHAPA).—GUERRERO: El Cuarenta near Tierra Colorada, *Tucker 2261* (DUKE, ENCB).—HIDALGO: 4 km W of Tolcayuca, *Hernández 9* (TEX, WIS).—JALISCO: Cd. Guzmán, 29 Oct 1978, *Hernández s.n.* (IBUG); Mpio. Zapopan, 20 Sep 1978, *Juárez s.n.* (IBUG); Río Blanco, *Palmer 194* in 1886 (F, G, MO, UC, US); 54 km N of Guadalajara on Hwy 54, *Tucker 2998* (DUKE, ENCB).—MÉXICO: Ixtapán, *Hinton 1153* (F, NY, US); La Gavia, *Sharp 44332* (TENN).—MICHOACÁN: vic. Morelia, *Arsène 3050* (BH, MO, NY, US); 1 km E of Tuxpan, *Iltis & Doebley 41* (WIS); Playa Azul, *Tucker 2950* (DUKE, ENCB).—MORELOS: autopista México-Cuernavaca, *Puga 1796* (IBUG).—NAYARIT: Lower slopes of Volcán Ceboruco, *Tucker 2274* (DUKE, ENCB).—OAXACA: beach of Laguna Superior, *King 1572* (MICH, TEX, UC, US); ca. 100 km W of Tehuantepec on Hwy 190, *Tucker 2777* (DUKE); Zaachila, *Tucker 2829* (DUKE, ENCB).—PUEBLA: Mpio. Felipe los Angeles, La Candelaria, *Ventura 2158* (ENCB, MO, SD).—QUERÉTARO: Querétaro, *Agniel 10516* (F, MO, US).—QUINTANA ROO: Coba, *Lundell 7836* (F, TEX, US).—SINALOA: Mpio. Guasave, *Rodríguez 1569* (IBUG, MO).—SONORA: Guaymas, *Palmer 203* in 1887 (NY, UC, US).—TABASCO: 30 km W of Cárdenas on Hwy 180, *Tucker 2560* (DUKE, ENCB).—TAMAULIPAS: 25 km N of Tampico, *Tucker 2376* (DUKE, ENCB); 5 km S of Aldama, *Tucker 2389* (DUKE, ENCB).—TLAXCALA: Amaxac de Guerrero, *Hernández X. 384* (LL).—VERACRUZ: Perote, *Balls 5457* (E, UC, US); région d'Orizaba, *Bourgeau 2386* (C, G, NY, US); 10 mi SE of Jalapa, *Tucker 2137* (DUKE, ENCB); 30 km S of Pánuco on Hwy 127, *Tucker 2420* (DUKE, ENCB); La Piedra SE of Veracruz, *Tucker 2488* (DUKE, ENCB).—YUCATÁN: Chichankanab, *Gaumer 1511* (F, G, NY, UC).—ZACATECAS: Hwy 54, 10 km S of Zapoqui, *Tucker 3003* (DUKE, ENCB), *Wilbur 36781* (CAS, DUKE, ENCB, F, MICH); 40 km N of Villanueva on Hwy 54, *Tucker 3005* (DUKE).

Cyperus esculentus is distinguished from other species of *Cyperus* in the New World by its linear brownish spikelets with closely appressed, overlapping scales. *Cyperus esculentus* and *C. rotundus* are the only two species of subg. *Cyperus* in the New World that produce tuberiferous stolons. Also, the two species have persistent scales and persistent spikelets, a combination of characters not found in any other New World species of *Cyperus*.

Mulligan and Junkins (1976), emphasizing weed control and management, provided a thorough summary of the biology of *C. esculentus*. Horak and Holt (1986) analyzed isozymes in 10 widely separated populations of *C. esculentus* in California; genetic variation served to determine the relative importance of sexual and asexual reproduction. Their results indicate that reproduction by achenes is unimportant in maintenance of populations in croplands, and that stolons and tubers are the primary means of reproduction. Variation in germinability of achenes from northeastern populations ranged from 7% to 95%; such variation was believed to be genetic. Achenes from a fifty-year-old herbarium specimen had a 5% germination rate. *Cyperus esculentus* is self-incompatible (Holt & Horak 1986).

46. Cyperus rotundus L., Sp. pl. 1: 45. 1753.—TYPE: "India." Hermann Herbarium (holotype: BM); see Trimen (1888).

Perennial, 10–35 (–40) cm tall. Stolons indurate, wiry, springy when dried, (2–) 5–12 cm long, 1–2 mm in diameter, bearing tubers 3–8 (–12) mm in diameter. Culms 0.7–3.4 mm in diameter, trigonous, smooth, basally indurate. Leaves (4–) 8–10 (–20), 5–30 cm long, 2–6 mm wide, v-shaped to flanged v-shaped, the margins

FIG. 40. *Cyperus rotundus*. a. Ray showing linear spikelets and prophyll at base (×0.9). b. Portion of spikelet, with one scale and wing at left removed to show wing on rachilla partially covering the achene on the side opposite the scale (×13.5). c. Distal portion of scale, abaxial view, flattened (×9). d. Achene with style and stigmas (×13.5). e. Habit (×0.45). f. Spikelet (×3.6). g. Base of plant with stolon and tubers (×0.9). Drawn by Mary Barnas Pomeroy. Published with permission of Herbert L. Mason and The University of California Press, taken from Fig. 132, *Flora of the Marshes of California*, by H. L. Mason, 1957.

scabrellate in the distal half. Inflorescence bracts (2–) 3–5, 0.5–10 cm long, 0.5–4 mm wide, v-shaped to flanged v-shaped, the margins and keel scabrellate, borne horizontal to ascendent to 45°. Rays (3–) 4–6 (–7), 0.2–10 cm long; secondary rays absent. Spikes 1 (–3), (12–) 15–25 (–30) mm long, (12–) 20–30 (–50) mm wide, broadly ellipsoid; rachis 3–10 (–14) mm long. Spikelets (2–) 3–7 (–12), 4–40 mm long, 1.3–1.8 mm wide, linear, compressed; rachilla persistent, 0.2–0.3 mm wide, 0.1–0.2 mm thick, straight, successive scales 1.5–1.8 mm apart, wings 0.5–1.0 mm wide, hyaline, often red-speckled. Scales persistent, 6–36 (–42), (1.8–) 2.6–3.4 mm long, 2.2–3.0 mm wide, ovate, apex obtuse, laterally 2–3-nerved, purple to reddish brown, with a narrow hyaline border, medially 3-nerved, green. Stamens 3; filaments 2.4–3.5 mm long; anthers 1.0–2.5 mm long, the connective apex 0.1–0.2 mm, ovate, red. Style 1.3–3.5 mm long; stigmas 3, (1.8–) 2.0–3.3 mm long. Achenes 1.4–1.7 (–1.9) mm long, 0.8–1.0 mm wide, trigonous, ellipsoid, apex obtuse, base sessile, the adaxial face concave, the abaxial faces convex, the surface puncticulate, black. Fig. 40.

Phenology. Flowering and fruiting from July to November, with a few collections in the remaining months.

Distribution. Pantropic and warm temperate regions; in the New World from eastern Virginia, Missouri, and California, U.S.A., south through Mexico and Central America to northern Argentina, also in the West Indies. Frequent in the uplands of Mexico, very common in the lowlands (Fig. 41); sea level to 500 (–2200) m.

Representative Specimens. **Mexico.** Baja California Sur: Mulege, Club Aeres, *Wiggins 18062* (DS).—Chiapas: Tuxtla Gutiérrez, *Breedlove 10622* (DS, F, LL, NY); Huixtla, *Wilbur 35873* (DUKE, ENCB).—Chihuahua: Cd. Juárez to Chihuahua at Potrero del Bajio, *Martínez 105* (TEX).—Guerrero: Dtto. Adama, Santo Tomás, *Mexia 8917* (ARIZ, BH, G, LL, MO, NY, UC, US).—Jalisco: Mpio. Zapopan, Las Agujas, 28 Sep 1979, *Flores Ruben s.n.* (IBUG); Mpio. La Huerta, La Manzanilla, *González 7545* (IBUG); outskirts of Cocula, *Wilbur 36750* (DUKE, ENCB).—Michoacán: Dtto. Huetamo, Tacupa, *Hinton 5815* (LCU, LL, NY, UC, US).—Nuevo León: near Monterey, *Kenoyer 323* (F).—Oaxaca: 16 km NNE of Tehuantepec, *King 1418* (MICH, NY, TEX, UC, US); between Xoxocotlán and Zaachila, *Tucker 2825* (DUKE, ENCB).—San Luis Potosí: Mpio. El Pujal, bank of Río Tampaón, *Chase 7489* (ARIZ, F, M, NY).—Sinaloa: Mazatlán, *Dunn et al. 21841* (LL, MO, NY).—Sonora: Mpio. Hermosillo, carr. a la costa, *Rodríguez 1663* (IBUG, MO); 1.8 km NNE of Alamos, *Van Devender et al. 93-212* (NYS).—Tabasco: 30 km E of Río Tonalá, *Tucker 2549* (DUKE).—Tamaulipas: vic. of Tampico, *Palmer 150* (F, G, MO, NY, US); 7 mi N of Victoria, *Tucker 1942A* (CAS, DUKE, ENCB, F, MICH).—Veracruz: Mpio. Santiago Tuxtla, *Martínez Calderón 1477* (F, LL, MEXU, MO, SD, TENN, UC); Orizaba, *Seaton 52* (F, NY, US); between Pánuco and Tampico, *Tucker 2407* (DUKE, ENCB); 10 km N of Riachuelos, *Tucker 2466* (DUKE, ENCB).—Yucatán: Chichankanab, *Gaumer 2173* (C, F, G, NY, UC); Playa Progreso, *Puga 3491* (CHAPA, IBUG).—Zacatecas: Mpio. Tlaltenango, a 5 km de Los Ramos, 15 Sep 1979, *Filiberto s.n.* (IBUG).

Cyperus rotundus is distinguished from other species of the genus in the New World by its open spikes composed of linear reddish spikelets borne on a conspicuous slender rachis. *Cyperus rotundus* is generally acknowledged to be among the world's worst weeds. It occurs nearly throughout Mexico, but is less common in the mountains. In the U.S.A., it does not grow north of the mean 1°C January isotherm (Stoller 1973). *Cyperus esculentus* is also a serious weed in much of the world, especially in cooler regions where the more tropical *C. rotundus* does not grow. *Cyperus esculentus* is able to tolerate colder air temperatures (as low as –18°C; Stoller 1973). The two species apparently also differ in their thermal optima for

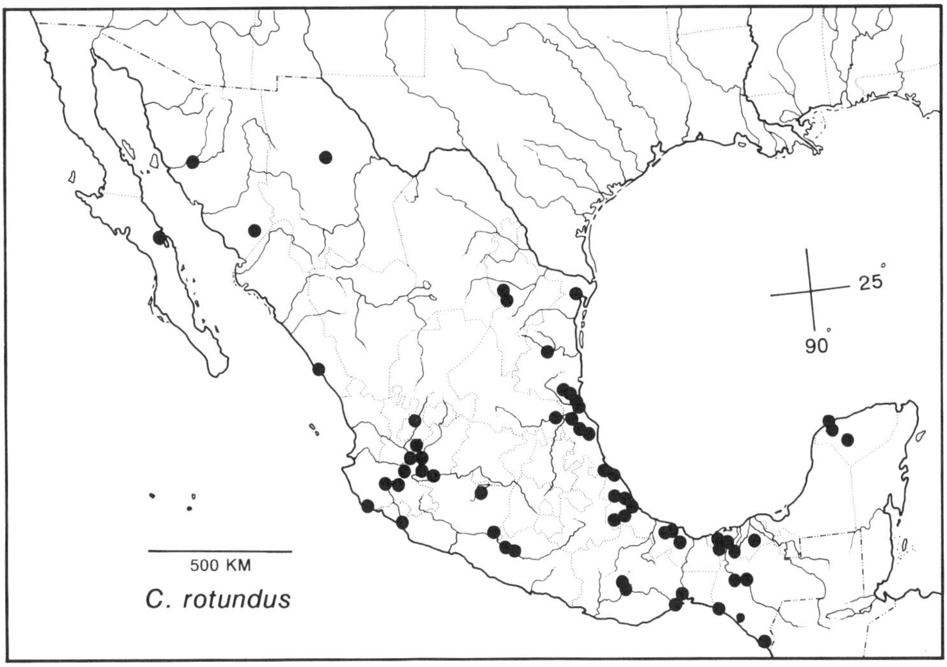

FIG. 41. Distribution of *Cyperus rotundus* in Mexico.

growth. In Mexico, *C. esculentus* is found from sea level to ca. 2600 m, whereas *C. rotundus* occurs from sea level to ca. 1500 m. The stoloniferous nature of *C. rotundus* and *C. esculentus* underlies their success as weeds. A single tuber can produce an infestation covering 2–4 m^2 in two months (Horowitz 1972). The sharp-pointed stolons cause puncture wounds in the hands of farm workers and curious taxonomists, and can grow through root crops such as potatoes and yams.

It is unclear whether *C. rotundus* and *C. esculentus* are native in the New World. *Cyperus esculentus* was first collected in Mexico by Sessé and Mociño (*no. 5302*, F), ca. 1790. *Cyperus rotundus*, however, was not collected in Mexico until 1860, and interestingly, never collected by C. G. Pringle. Both species were present in the southeastern U.S.A. by the early nineteenth century, since Elliot (1816–1824) noted that *C. rotundus* was a great trouble to farmers in Georgia and South Carolina.

47. Cyperus compressus L., Sp. pl. 1: 46. 1753. *Chlorocyperus compressus* (L.) Palla, Denkschr. Akad. Wiss. Wien Math.-Nat. Kl. 84: 451. 1909.—TYPE: U.S.A. Virginia, *Clayton Herbarium 598* (holotype: BM).

Cyperus pectiniformis Roemer & Schultes, Syst. Veg. Mantissa 2: 128. 1824. *Cyperus pectinatus* Roxburgh, Fl. Ind. 1: 195. 1817, non *Cyperus pectinatus* Vahl, 1805. *Cyperus compressus* var. *pectiniformis* (Roemer & Schultes) C. B. Clarke, J. Linn. Soc., Bot. 21: 97. 1884.—TYPE: INDIA. Coromandel, *Roxburgh s.n.* (not located).

Caespitose annual, (2–) 6–15 (–35) cm tall. Rhizomes absent. Culms (0.3–) 1.0–1.8 mm in diameter, trigonous, smooth. Leaves 1–5, (1–) 5–15 (–22) cm long,

(0.5–) 1.5–3.0 mm wide, flat to v-shaped, the margins sparsely scabrellate. Inflorescence bracts (1–) 4–5 (–6), 1–15 (–20) cm long, 0.5–3.5 mm wide, v-shaped, the margins and keel scabrellate, ascendent at 45–60°; the longest bract in some plants erect, looking like a continuation of the culm. Rays 1–6, 2–12 cm long, or rays absent and the inflorescence a sessile cluster of (1–) 3–10 spikelets. Spike 1 per ray, 7–40 mm long, 10–20 mm wide, subdigitate to subglobose; rachis 1–3 (–7) mm long. Spikelets (1–) 3–12 (–16), 6–34 (–40) mm long, 2.0–3.0 (–3.6) mm wide, linear to linear-lanceolate, compressed-quadrate; rachilla persistent, 0.5–0.7 mm wide, ca. 0.2 mm thick, straight, greenish, successive scales 1.3–1.6 mm apart, wings absent. Scales deciduous, (8–) 16–36 (–42), (2.4–) 2.6–3.0 mm long, (1.9–) 2.0–2.4 (–2.5) mm wide, ovate, apex acute to subacute, the cuspidate apex (0.3–) 0.5–0.7 (–0.8) mm long, laterally weakly 1–2-nerved, whitish, greenish, or pale brownish, medially weakly 5-nerved, green. Stamens 3; filaments (1.8–) 2.5–3.0 mm long; anthers 0.6–0.7 mm long, oblong, the connective apex ca. 0.1 mm, ovate, reddish. Style 0.9–1.3 mm long; stigmas 3, 0.5–0.8 mm long. Achenes (1.0–) 1.2–1.4 (–1.5) mm long, 0.9–1.0 (–1.1) mm wide, trigonous, obovoid, apex obtuse to emarginate, base substipitate to sessile, the faces concave, the surface very finely puncticulate, brown.

Phenology. Flowering and fruiting from July to October, with a few collections during the remaining months.

Distribution. Pantropic and warm-temperate regions; in the New World from Delaware, Missouri, and New Mexico, U.S.A., south through Mexico to Argentina, also in the West Indies. Frequent to common in Mexico (Fig. 42); open disturbed areas; sea level to 1200 m.

Representative Specimens. **Mexico.** Chiapas: Mpio. Arriaga, 15 km NW of Arriaga, *Breedlove 36781* (DS); Huixtla, *Wilbur 35871* (DUKE, ENCB).—Colima: Colima, *Palmer 4* in 1897 (ARIZ, C, DS, MICH, TEX, US).—Guerrero: El Tibor, *Langlassé 293* (G, US); Acapulco, *MacDaniels 146* (BH, F).—Jalisco: Barra de Navidad, *McVaugh 19732* (G, MICH, US); Tequila, *Pringle 4617* (AC, CAS, E, F, G, MICH, MO, MSC, MTMG, UC, US, Z).—Michoacán: Presa Zicuirán, *González & Puga 777* (IBUG).—Nayarit: vic. San Blas, *Ferris 5405* (DS, F, US), *Wright 1332* (DS, F, G, MO, MSC, US).—Oaxaca: 3 km SW of Puerto Angel, *Rzedowski 21153* (BH, DS, MICH, MSC, TEX).—Sinaloa: Mazatlán, *Ortega 5763* (C, CAS, DS, GH, MO, POM, US, Z).—Sonora: 1–2 mi SW of Alamos, *Sanders et al. 1866* (UCR).—Tabasco: on Hwy 180, 30 km E of Río Tonalá, *Tucker 2558* (DUKE, ENCB).—Tamaulipas: 25 km N of Tampico, *Tucker 2380, 2381* (DUKE).—Veracruz: 2 km NE of Catemaco, *Rzedowski 26395* (DS, ENCB, MICH).

The low annual habit of the plants, combined with their greenish spikelets and obovoid achenes, distinguish *C. compressus* from other species of *Cyperus* in the New World. *Cyperus compressus* is separated from *C. wilburii* by its smaller achenes and scales (see further comments under *C. wilburii*, no. 48).

In the protologue Linnaeus gave the type locality as "in Americae septentrionalis pratis arenosis" with the characters taken almost verbatim from Gronovius *Flora Virginica* ("Gron. virg. 131"). I erred in suggesting that Sloane's plate (Voy. Jamaica 1: 117, t. 76, fig. 1. 1707) was the type (Tucker 1983) and thank Rogers McVaugh for bringing this lapse to my attention.

48. Cyperus wilburii G. C. Tucker, Rhodora 88: 508. 1986.—Type: Mexico. Oaxaca: 8.5 km W of Niltepec on Hwy 200 at the turnoff to Cerro Iguana,

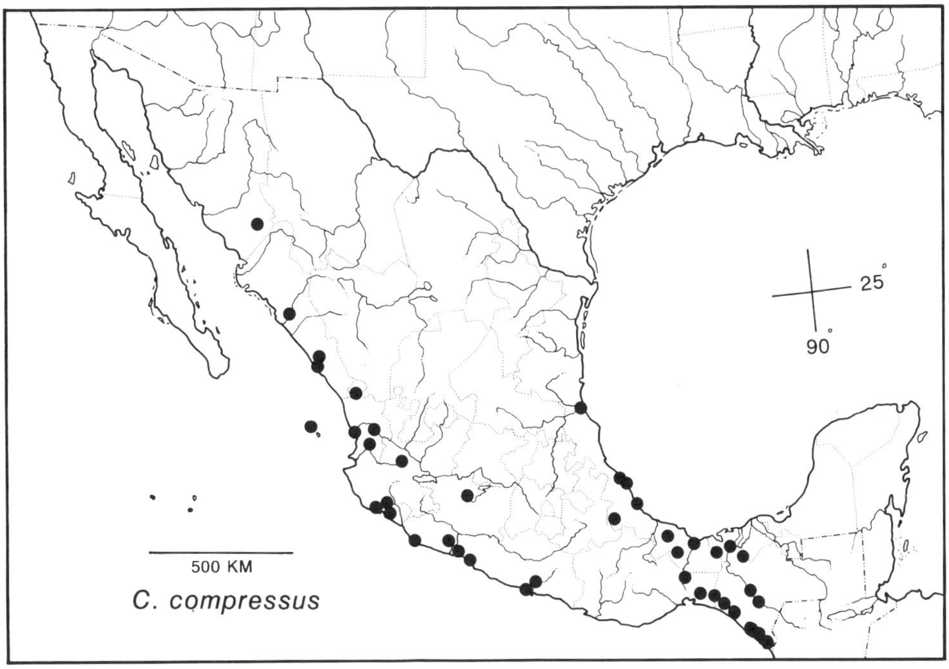

FIG. 42. Distribution of *Cyperus compressus* in Mexico.

31 Jul 1984, *Tucker 2756* (holotype: DUKE!; isotypes: BH! C! CAS! ENCB! F! GH! K! MICH! MO! NY! US!).

Caespitose annual, (5–) 10–25 (–30) cm tall. Rhizomes absent. Culms 0.5–1.5 mm in diameter, trigonous, smooth. Leaves (1–) 3–4, (1–) 12–20 (–30) cm long, 0.5–3 mm wide, conduplicate, the margins and keels scabrellate distally. Inflorescence bracts (2–) 3–5 (–6), (2–) 8–15 (–25) cm long, 0.5–4 mm wide, v-shaped to flat, the margins and keels scabrellate, ascendent at 45–60°; the longest bract in some plants erect, looking like a continuation of the culm. Rays 1–3 (–5), 1.5–5 cm long; rays occasionally absent and the inflorescence a loose cluster of 3–6 spikelets. Spike 1 per ray, 20–35 mm long, 20–50 mm wide, loosely ovoid; rachis 1.5–3.5 (–6) mm long. Spikelets (1–) 3–7 (–10), (10–) 14–20 (–32) mm long, (3.5–) 5.0–6.5 (–7.0) mm wide, oblong to oblong-lanceolate, strongly compressed; rachilla persistent, 0.5–0.8 mm wide, 0.2–0.3 mm thick, slightly geniculate, brown to reddish brown, successive scales 2–2.4 (–2.6) mm apart, wings absent. Scales deciduous, 10–16 (–30), (3.8–) 4.0–4.6 mm long, (2.4–) 2.6–3.2 (–3.6) mm wide, ovate-oblong, the cuspidate apex (1.1–) 1.3–1.5 (–1.6) mm long, stout, straight to slightly excurved, laterally conspicuously 2-nerved midway between the center and margins, stramineous to reddish brown, medially weakly 3–5-nerved, greenish. Stamens 3; filaments 4–5 mm long; anthers 0.8–0.9 mm long, oblong, the connective apex minute. Style 2.0–2.5 mm long; stigmas 3, ca. 0.6 mm long. Achenes 1.7–2.2 mm long, 1.0–1.4 mm wide, trigonous, obovoid, apex blunt, base stipitate, 0.3–0.4 mm long and about as wide, the faces concave, the surface puncticulate, brownish. Fig. 2A–C.

Phenology. A single flowering collection was made in June; collections made from July through October had mature achenes.

Distribution. Endemic to Mexico in the Pacific Coastal lowlands of the Isthmus of Tehuantepec in Oaxaca; apparently uncommon, disturbed open soil among thorny shrubs and cacti and in pastured thorn savannas; sea level to ca. 100 m.

ADDITIONAL SPECIMENS EXAMINED. **Mexico**. OAXACA: Ixtepec, 17 Aug 1935, *Fisher 35258* (ARIZ, F, MO, NY, SMU, TEX, US, VT); along the hwy to Tehuantepec, 6 km N of Salina Cruz, 30 Jun 1958, *King 917* (MICH, TEX, US); San Gerónimo, 5 Oct 1933, *Mell 2115* (NY, US); Route 200, ca. 5.3 mi W of Niltepec near turnoff to Cerro Iguana, *Wilbur 35922* (DUKE, ENCB).

Cyperus wilburii is similar to *C. compressus* in its annual habit, loosely clustered spikelets, cuspidate scales, and trigonous obovoid achenes. To the unaided eye, *C. wilburii* is easily distinguished from *C. compressus* by its wide spikelets.

49. Cyperus erythrorhizos Muhlenberg, Descr. gram. 20. 1817. *Chlorocyperus erythrorhizos* (Muhlenberg) Palla, Allg. Bot. Zeitschr. Syst. 9: 69. 1903.—TYPE: U.S.A. Pennsylvania, North Carolina, and Georgia, *Muhlenberg's herbarium, sheet 27.2* (lectotype, here designated: PH!).

Cyperus cupreus J. Presl & K. Presl in K. Presl, Reliq. Haenk. 1: 172. 1828. *Cyperus erythrorhizos* var. *cupreus* (J. Presl & K. Presl) Kükenthal, Pflanzenreich 4(20): 59. 1936.—TYPE: U.S.A. California: Monterey, 1791, *Haenke s.n.* (holotype: PR!; isotype: HAL!).

Cyperus occidentalis Torrey, Ann. Lyceum Nat. Hist. New York 3: 259. 1836.—TYPE: U.S.A. Northwest coast, "mouth of the Oregon [Columbia] River, from Dr. Hooker," collector not indicated but probably David Douglas (holotype: NY!).

Cyperus washingtonensis Gandoger, Bull. Soc. Bot. France 66: 296. 1919.—TYPE: U.S.A. Washington: Bingen, *Suksdorf 587* (holotype: LY; isotype: NY!).

Caespitose annual, (0.5–) 5–25 (–105 in many specimens from the U.S.A.) cm tall. Rhizomes absent; roots bright red. Culms 1.0–2.5 (–7.5) mm in diameter, trigonous to roundly trigonous, smooth. Leaves 5–25 (–90) cm long, 2–5 (–11) mm wide, flat to inversely w-shaped, the margins and keel scabrellate. Inflorescence bracts (3–) 5–7 (–11), (3–) 10–20 (–70) cm long, 1–3 (–12) mm wide, inversely w-shaped, the margins and keel scabrellate, borne horizontal to ascendent at 30°. Rays 2–6 (–12), 1–8 (–28) cm long; secondary rays absent. Spikes 1–3 (–6), (4–) 10–30 (–45) mm long, (6–) 10–16 (–23) mm wide, cylindric-ovoid; rachis (4–) 10–25 mm long. Spikelets (20–) 40–80, 3–8 (–18) mm long, 1.0–1.5 mm wide, linear, quadrate; rachilla persistent, ca. 0.2–0.3 mm wide, ca. 0.1 mm thick, greenish, successive scales 0.5–0.7 mm apart, wings 0.3 (–0.4) mm wide, pale cream to bronze, at maturity becoming laterally free but remaining firmly attached basally. Scales deciduous, 6–16 (–30), 1.3–1.5 mm long, 0.8–1.2 mm wide, oblong-ovate to obovate, apex obtuse, mucronulate, laterally nerveless, light brown with reddish speckles, medially 3-nerved, greenish. Stamens 3; filaments 1.2–1.6 mm long; anthers 0.2–0.3 mm long, oblong, the connective apex minute. Style 0.7–0.8 mm long; stigmas 3, ca. 0.4 mm long. Achenes 0.7–1.0 mm long, 0.4–0.6 mm wide, trigonous, dorsiventrally flattened, ovoid, apex rounded, apiculate, base sessile, the adaxial face flat, the abaxial faces flat to slightly concave, the surface smooth, light grayish to brown.

Phenology. Scarcely known in Mexico, but judging from specimens from the southwestern U.S.A., probably flowering and fruiting from April to September.

Distribution. Endemic to North America, from southern New Hampshire, southern Ontario, Minnesota, and Washington south to Florida, and Texas and northern Baja California; sandy edges of lakes, streams, and rivers. Rare in Mexico; probably mostly near sea level.

Additional Specimen Examined. Baja California: Gardner's Laguna, 21 Apr 1894, *Schoenfeldt 2897* (US).

The annual habit and light grayish to brown, dorsiventrally flattened achenes separate *C. erythrorhizos* from all other species of the genus in Mexico. Also, the orange-red to blood-red roots are a distinctive feature not found in any other North American species of *Cyperus*.

According to Benedict Ayers (1946), the Muhlenberg Herbarium included three sheets under the name *Cyperus erythrorhizos* in folio 27. Because the specimens in Muhlenberg's herbarium were mounted after his death (Merrill & Hu 1950), it is unclear whether the three sheets represent one collection (later mounted on three sheets) or two or three separate collections. Since all these specimens are clearly *C. erythrorhizos*, the application of Muhlenberg's name is not in doubt. One sheet, 27.2, has been stamped by a curator as "holotype." It is the best of these three sheets and the only sheet that shows the roots, a feature that forms the name of the species in Muhlenberg's protologue. Thus, I designate it as lectotype.

Shreve and Wiggins (1964) in their *Vegetation and Flora of the Sonoran Desert* reported this species as occurring in southern California and southern Arizona. They stated that *C. erythrorhizos* "almost certainly occurs in northeastern Baja California." Benedict Ayers (1946) did not include this species in his treatment of the genus in Mexico. Thus, the nineteenth century specimen cited above, which lay unidentified at US, represents the first record of this species in Mexico.

50. Cyperus digitatus Roxburgh, Fl. Ind. 1: 209. 1820.—Type: India. *Roxburgh s.n.* (not located).

Cyperus mexicanus Liebmann, Mexic. halvgr. 39. 1850. *Cyperus digitatus* var. *mexicanus* (Liebmann) Kükenthal, Pflanzenreich 4(20): 57. 1935.—Type: Mexico. Veracruz: Santa María Tlapacoyo [Tlapacoyan], May 1841, *Liebmann 14369* (holotype: C!).

Coarse perennial, 50–150 cm tall. Rhizomes 1–4 cm long, 1–2 cm in diameter, indurate. Culms 2–15 mm in diameter, trigonous, smooth or rarely sparsely scabrellate on the angles just below the bracts. Leaves 1–3, 40–100 cm long, 5–15 mm wide, inversely w-shaped or v-shaped, the margins and keel scabrous with brownish prickles. Inflorescence bracts 8–12, (5–) 20–80 cm long, 3–15 mm wide, the margins and keel scabrous with dark brown prickles, borne ascendent at 45–60°; secondary bracts 3–7 cm long, 2–4 mm wide. Rays 8–10, (1–) 15–35 cm long; secondary rays 1–3 cm long. Spikes (2.5–) 3.5–5 cm long, 1.2–1.5 cm wide, cylin-

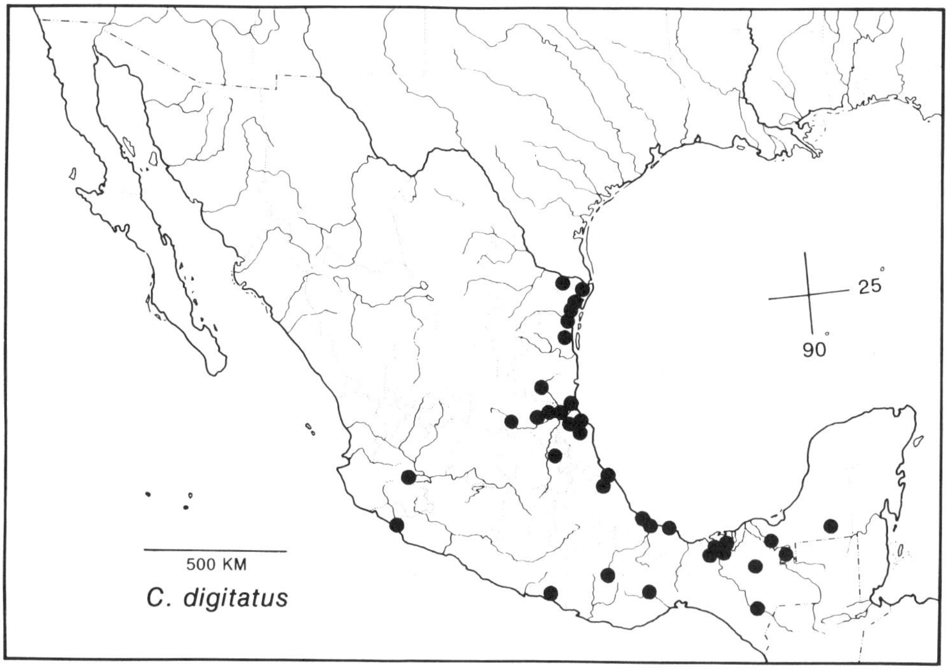

FIG. 43. Distribution of *Cyperus digitatus* in Mexico.

dric, often with 1–3 sessile smaller spikes at the base; rachis 15–40 mm long. Spikelets 35–65, 5–8 mm long, 0.8–1.1 mm wide, linear, more or less quadrangular but slightly compressed; rachilla persistent, ca. 0.2–0.3 mm wide, ca. 0.1 mm thick, straight, successive scales 0.9–1.2 mm apart, wings 0.2–0.3 mm wide, golden-brown, during achene maturation the adaxial edge of the wing detaching from the rachilla but the base remaining firmly attached. Scales deciduous, 12–16, 1.6–1.8 mm long, 1.1–1.3 mm wide, ovate, apex mucronulate, laterally weakly 1–2-nerved, reddish near the midvein, marginally hyaline, medially strongly 5-nerved, green. Stamens 3; filaments 1.6–1.9 mm long; anthers 0.4–0.5 mm long, ellipsoid, the connective apex less than 0.1 mm long. Style 0.8–1.0 mm long; stigmas 3, ca. 0.4 mm long. Achenes ca. 0.9 mm long, ca. 0.4 mm wide, trigonous, dorsiventrally compressed, ellipsoid, slightly wider at the base, apex apiculate, base sessile, the surface finely puncticulate, brown or whitish.

Phenology. Flowering and fruiting throughout the year.

Distribution. Pantropic regions; in the New World from Texas, U.S.A., south through Central America to northern Argentina, also in the West Indies. Common in the Gulf lowlands of eastern Mexico, scarce in the Pacific lowlands (Fig. 43); ditches, moist fallow fields, roadsides, and marshy pastures; sea level to 700 (–1700) m.

REPRESENTATIVE SPECIMENS. **Mexico.** CAMPECHE: Palizada, *Matuda 3851* (F, GH, LCU, MICH).—CHIAPAS: Mpio. Venustiano Carranza, banks of Río Blanco, *Breedlove 41164* (CAS, DS).—COLIMA: 2.5 km W of Ortega heading towards Tecomán, *Tucker 2961* (CAS, DUKE, ENCB).—GUERRERO: Marquelia, 140 km E of Acapulco, *Tucker 2925* (DUKE, ENCB).—HIDALGO: Mpio. Metztitlán, Vegas de Metztitlán, *Moore 2117* (BH).—JALISCO: Lago de Chapala, *Barkley et al. 7608* (TEX).—OAXACA: Etla, *Conzatti 501* (GH).—SAN LUIS POTOSÍ: Cárdenas, *Fisher 37240* (ARIZ, F, MO, NY, UC).—TABASCO: 18 km E of Río Tonalá, *Tucker 2548* (DUKE, ENCB).—TAMAULIPAS: Hwy 101, just N of intersection to Valle Hermoso, *Tucker 1934* (DUKE, ENCB).—VERACRUZ: Hwy 127, 30 km S of Pánuco, *Tucker 2428* (DUKE, ENCB); 3 km N of Lerdo de Tejada, *Tucker 2509* (DUKE, ENCB).

Cyperus digitatus is recognizable by its large coarse habit (50–150 cm tall), cylindric spikes, and ellipsoid, slightly compressed achenes. It is distinguished from *C. semiochraceus* by its shorter anthers and pale greenish white rather than yellowish spikelets. Incomplete specimens of *C. giganteus* have been misidentified as *C. digitatus*, but can be easily distinguished by their conspicuous subulate anther connectives, 0.2–0.5 mm long.

51. Cyperus semiochraceus Boeckeler, Flora 61: 29. 1878.—TYPE: MEXICO. Valle de Mexico, 1850–55, *Schaffner s.n.* (B, not located). [Boeckeler gave no collection number; Kükenthal (1935–36) cited only *Schaffner 11* from Mexico. I have found no Mexican collections of this species made by Schaffner in any herbaria. My application of the name is based on the other specimens cited by Kükenthal.]

Cyperus bourgeaui C. B. Clarke ex Lundell, Carnegie Inst. Wash. Publ. 436: 287. 1934.—TYPE: MEXICO. Distrito Federal, *Bourgeau 215* (lectotype, here designated: K!; isolectotypes: F! GH!).

Coarse perennial, (40–) 60–100 (–150) cm tall. Rhizomes 1–25 cm long, 3–10 mm in diameter, indurate, clothed with stiff, orbiculate, cucullate scales 7–12 mm long, producing culms 1–3 cm apart and horizontal stolons 2–5 mm in diameter. Culms 6–10 mm in diameter, trigonous, smooth, light green to yellowish brown. Leaves 5–12, 30–110 cm long, 10–20 mm wide, v-shaped basally, becoming flat distally, the margins and keel harshly scabrous with brownish prickles. Inflorescence bracts 7–10, 15–60 (–90) cm long, (2–) 10–20 mm wide, v-shaped to inversely w-shaped, the margins and keel scabrous, borne ascendent at (45–) 60–75°, the inflorescence open, 12–30 (–42) cm high, (4–) 10–25 cm wide; secondary bracts 4–7, 2–15 cm long, 2–8 mm wide, scabrellate-margined. Rays 6–12, 2–18 (–28) cm long; secondary rays sometimes present, 1.0–4.0 (–7.5) cm long. Spikes (4–) 10–16, (10–) 25–50 mm long, (5–) 12–16 mm wide, cylindric, flattened, ellipsoid in cross section, rachis (8–) 12–50 mm long. Spikelets (4–) 30–80 (–120), 4–10 mm long, 1.1–1.3 mm wide, oblong to linear-oblong, subcylindric, slightly compressed; rachilla persistent, ca. 0.5 mm wide, less than 0.1 mm thick, straight, reddish to greenish brown, with a stramineous stripe down the center, successive scales ca. 0.8 mm apart, wings 0.2–0.4 mm wide, golden to reddish brown, at maturity the lateral edges detaching but the base remaining firmly attached. Scales deciduous, 6–12, 1.5–2.0 (–2.8) mm long, 1.2–1.5 mm wide, elliptic to ovate, widest at the middle, rounded to subacute, mucronulate, laterally nerveless, golden-brown to glossy reddish brown, reddish striate, medially 3-nerved, greenish. Stamens 3; filaments 1.8–2.2 mm long; anthers 1.1–1.3 (–1.5) mm long, linear, the connective apex ca. 0.1 mm long, blunt, reddish. Style 1.5–2.2 mm long; stigmas 3, 0.8–1.4 mm long. Achenes 1.0–1.1 (–1.2) mm long, (0.4–) 0.6 mm wide, plano-convex (the flat face toward the rachilla), ovoid-ellipsoid, widest at the middle or toward the base, apex

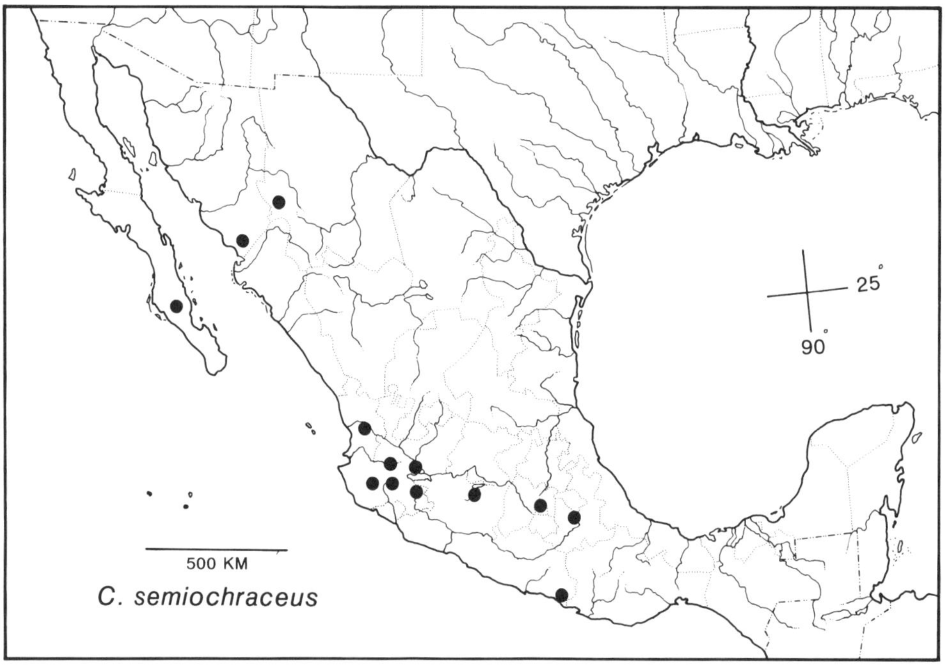

FIG. 44. Distribution of *Cyperus semiochraceus*.

rounded or sometimes slightly apiculate, base sessile, the surface faintly reticulate, whitish at first, grayish brown.

Phenology. Flowering and fruiting from May to January.

Distribution. Endemic to Mexico (Fig. 44), from southern Baja California and southwestern Chihuahua to southern Guerrero, and in mountain valleys in Jalisco, México, the Distrito Federal, and Puebla; as scattered plants in marshy pastures, ditches, irrigation canals, along lake shores, and muddy pools; (sea level–) 1000–2300 m. The specimens from the Yucatan Peninsula cited as *C. semiochraceus* by O'Neill (1940) are *C. digitatus*.

REPRESENTATIVE SPECIMENS. **Mexico.** BAJA CALIFORNIA SUR: Sierra de la Giganta, W side of Mesa de San Alejo, *Carter 4309* (GH).—CHIHUAHUA: Río Mayo, Guasaremos, *Gentry 1567* (ARIZ, GH, LCU, MO, UC, US).—DISTRITO FEDERAL: in lacu prope Mexico, Apr 1827, *Karwinsky s.n.* (M, MO); valley of Mexico, *Pringle 3895* (ARIZ, CAS, E, G, K, LCU, M, MICH, MO, MSC, MTMG, NY, UC, VT, Z).—GUERRERO: along Hwy 200, 50 km W of Pinotepa Nacional, *Tucker 2923* (CAS, DUKE, ENCB, F, GH, MICH, MO).—JALISCO: 1 mi W of Ayo el Chico, *McVaugh 17215* (MICH); on Hwy 80, 8 km N of Unión de Tula, *Tucker 2997* (DUKE, ENCB).—MÉXICO: Mpio. Zumpango, SW of Lago de Zumpango, *Arredondo 28* (ARIZ, ENCB, G, IBUG, SD, UC).—MICHOACÁN: vic. Morelia, Rincón, *Arsène 5471* (MO, US); lakeside marsh, Lake Pátzcuaro, 21 May 1948, *Martin s.n.* (CU).—NAYARIT: 12 mi SE of Tepic, *Feddema 543* (DUKE, MICH, TX).—PUEBLA: près du cimetière à Puebla, *Arsène 2239* (BH, BM, C, CAS, K, MICH, MO, Z).—SONORA: Mpio. Huatabampo, Hwy 15, 5 mi N of Estación Don, *Breedlove & Thorne 18654* (CAS); Dtto. Guaymas, valle del Yaqui, *Pacheco M. 155* (CHAPA).

In the field, *C. semiochraceus* can be rapidly distinguished from *C. digitatus* by the golden color of the upper culm and the spikelets. Also, the scales of *C. semiochraceus* are not so closely appressed as those of *C. digitatus*; thus, the spikelets appear slightly compressed in cross section. The achenes of *C. semiochraceus* are ovoid-ellipsoid and the anthers are 1.1–1.3 (–1.5) mm long, whereas the achenes of *C. digitatus* are ellipsoid and its anthers are only 0.4–0.5 mm long.

52. Cyperus imbricatus Retzius, Observ. bot. 5: 12. 1789.—TYPE: INDIA. Tranquebar, *Koenig s.n.* (holotype: LD).

Cyperus radiatus Vahl, Enum. pl. 2: 369. 1805.—TYPE: GUINEA. 1784, *Isert s.n.* (lectotype, designated by Tucker & McVaugh, 1993: C!).

Cyperus confertiflorus Schultes in Roemer & Schultes, Syst. Veg. Mantissa 2: 131. 1824. *Cyperus densiflorus* Link, Jahrb. Gewächsk. 3: 88. 1818, non *Cyperus densiflorus* G. Meyer, 1818.—TYPE: South America. *Humboldt & Bonpland s.n.* (holotype: B-W no. 1400, IDC microfiche!).

Cyperus spicatus J. Presl & K. Presl in K. Presl, Reliq. Haenk. 1: 173. 1828.—TYPE: MEXICO. 1790–91, *Haenke s.n.* (holotype: PR!).

Caespitose perennial, (15–) 30–100 cm tall. Rhizomes 1–3 cm long, 0.5–1.0 cm in diameter, indurate. Culms (3–) 5–15 mm in diameter, trigonous, smooth. Leaves 2–7, 10–80 cm long, 2–15 (–18) mm wide, v-shaped, the margins and keel scabrous. Inflorescence bracts 3–5, 5–50 cm long, 2–10 mm wide, v-shaped, margins and keel scabrellate, borne horizontally to ascendent at 30–60°; secondary bracts 2–4, 1–5 cm long, 0.4–1.0 mm wide. Rays 4–10, 1–15 (–20) cm long; secondary rays 0.5–5.0 cm long. Spikes (1–) 3–10, 2–5 (–8) cm long, (4–) 10–18 mm wide, dense, narrowly cylindric. Spikelets 50–100 (–150), 4–11 mm long, 1.2–1.5 mm wide, linear-lanceolate, compressed, ascendent, their apices oriented toward the apex of the spike before anthesis, becoming somewhat more divergent in maturity; rachilla persistent, ca. 0.5 mm wide, less than 0.1 mm thick, straight, yellowish green, successive scales 0.7–0.9 mm apart, wings absent. Scales deciduous, 10–30, 1.1–1.7 mm long, 0.7–0.9 mm wide, ovate, cuspidate [the cusps 0.1 mm long in proximal scales, increasing to 0.3 (–0.4) in distal ones], laterally 1–2-nerved, stramineous to reddish brown, medially 3-nerved, greenish. Stamens 3; filaments 0.9–1.3 mm long; anthers (0.2–) 0.3–0.5 mm long, the connective apex not prolonged. Style 0.5–0.8 mm long; stigmas 3, 0.4–0.5 mm long. Achenes ca. 0.5 mm long, ca. 0.4 mm wide, trigonous, dorsiventrally compressed, ovoid, apex broadly rounded, apiculate, base sessile, the adaxial face flat, the abaxial faces broadly rounded, the surface finely puncticulate, dull whitish to pale brown.

Phenology. Flowering and fruiting from May to December.

Distribution. Pantropical regions, in the New World from western Mexico and Cuba south through the West Indies and Central America to northern Argentina. Occasional in Mexico (Fig. 45), from Nayarit to Chiapas and Tabasco; ditches, marshes, stream banks, and sandbars; sea level to 500 m.

REPRESENTATIVE SPECIMENS. **Mexico.** CHIAPAS (or TABASCO): margines inundadas del Río Grijalva, *Rovirosa 258* (K, NY).—COLIMA: on Hwy 200, 2.5 km W of Ortega, *Tucker 2960* (DUKE, ENCB).—GUERRERO: La Cofradía, *Langlassé 135* (G, GH); Dtto. Adama, N of Río Balsas, Santo Tomás, *Mexia 8915a* (UC).—JALISCO: Laguna de Zapotlán, 3 mi N of Cd. Guzmán, *McVaugh & Koelz 1113* (MICH, US); 15 km NE of Autlán, *Pérez Gómez s.n.* (IBUG).—MICHOACÁN: Huetamo Dtto., Tacupa, *Hinton 5818* (GH, K, NY, US).—NAYARIT: Acaponeta, *Rose et al. 14236* (NY; US).—TABASCO: ad ripas fluminis González inter Ocuilzapotlán et Santa Anita, *Rovirosa 700* (US).—VERACRUZ: Mpio. Minatitlán, Bodega on Río Uxpanapa, *Cházaro & Gallegos 2788* (WIS).

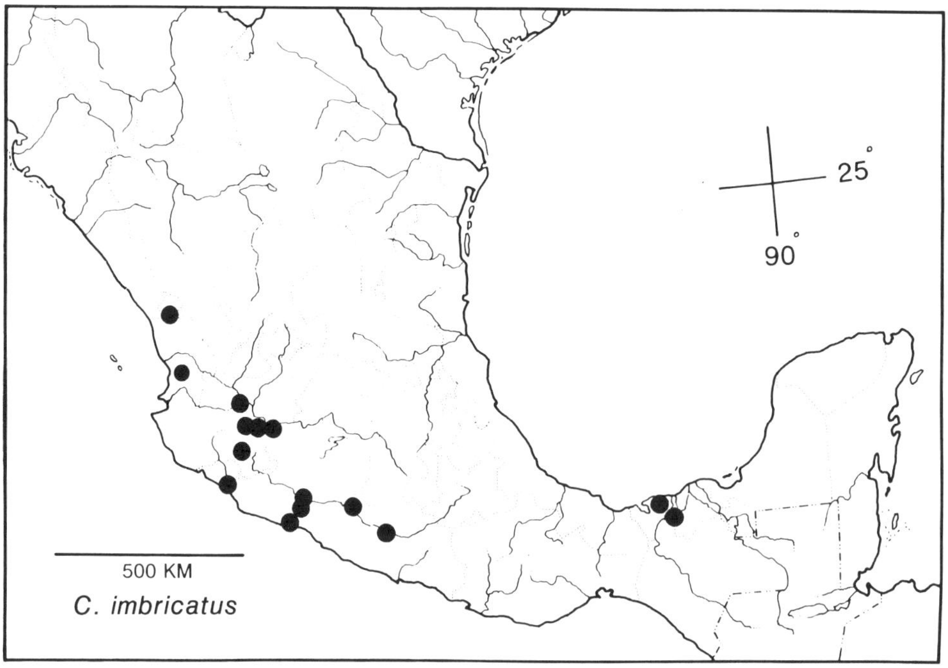

FIG. 45. Distribution of *Cyperus imbricatus* in Mexico.

Cyperus imbricatus is distinguished from *C. digitatus* and *C. semiochraceus*, which it resembles in habit, by its cuspidate scales and wingless rachilla. The excurved scale apices and ascendent-appressed spikelets of *C. imbricatus* are characters apparent before anthesis, and thus are useful for determining immature collections.

53. Cyperus giganteus Vahl, Enum. pl. 2: 364. 1805.—TYPE: PUERTO RICO. *West s.n.* (holotype: C!, photo: F!; isotypes: B! C!).

Papyrus comosa Willdenow, Abhandl. Königl. Akad. Berlin 1812–1813: 74. 1816, non *Papyrus comosa* H. B. K., 1816.—TYPE: ECUADOR. Guayaquil, *Humboldt & Bonpland 3803* (holotype: B-W no. 1416, IDC microfiche!).

Papyrus comosa H. B. K., Nov. gen. sp. 1 [quarto]: 16. 1816, non *Papyrus comosa* Willdenow, 1816.—TYPE: ECUADOR. Guayaquil, *Humboldt & Bonpland 3803* (P-HBK, IDC microfiche!).

Cyperus comosus Poiret in Lamarck, Encycl. Suppl. 5: 185. 1817. *Cyperus giganteus* var. *comosus* (Poiret) Kükenthal, Repert. Spec. Nov. Regni Veg. 32: 74. 1933.—TYPE: ECUADOR. Guayaquil, *Humboldt & Bonpland 3803* (P-HBK, IDC microfiche!).

Cyperus princeps Kunth, Enum. pl. 2: 66. 1837. *Papyrus elegans* Schrader ex Nees in Martius, Fl. bras. 2(1): 59. 1842, non *Papyrus elegans* (L.) G. Don, 1830, nec *Cyperus elegans* L., 1753.—TYPE: BRAZIL. Río de Janeiro, Serra do Mar, *Sellow s.n.* (holotype: B, not located; isotype: M!).

Cyperus trinidadensis Boeckeler, Beitr. Cyper. 2: 7. 1890.—TYPE: TRINIDAD. *Hart 2096* (holotype: B!).

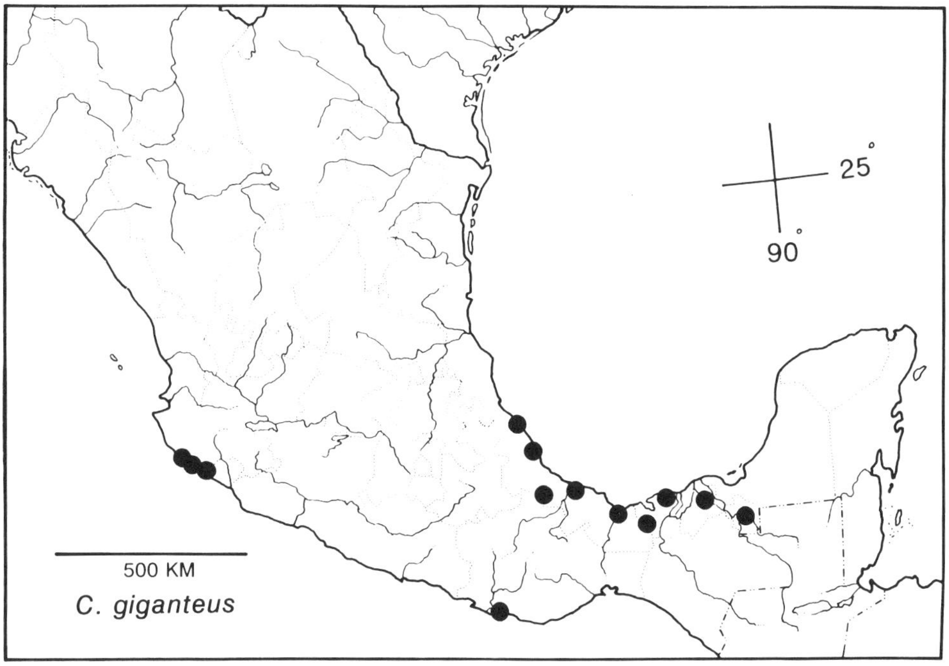

FIG. 46. Distribution of *Cyperus giganteus* in Mexico.

Perennial, 2.0–3.5 m tall. Rhizomes 5–30 cm long, 1–3 cm in diameter, indurate. Culms (0.5–) 1.0–5.0 cm in diameter, roundly trigonous, smooth, basally indurate. Leaves lacking blades but with 1–4 blunt sheaths along the lower culm. Inflorescence bracts 10–12, (8–) 20–45 cm long, 4–10 mm wide, inversely w-shaped, the margins scabrous with colorless prickles, borne approximately horizontally to ascendent at 45°; secondary bracts (5–) 15–25 cm long, 1.5–5.0 mm wide. Rays 5–8, 20–25 cm long; secondary rays 5–15 cm long; tertiary rays 0.5–5.0 cm long. Spikes (1.5–) 2.0–4.0 cm long, (0.8–) 1.2–1.6 cm wide, cylindric; rachis (7–) 20–30 mm long. Spikelets (15–) 50–80, 5–12 mm long, 0.6–1.4 mm wide, linear, more or less quadrate but slightly compressed; rachilla persistent, ca. 0.2–0.3 mm wide, ca. 0.1 mm thick, straight, greenish white, successive scales 0.7–0.9 mm apart, wings 0.3–0.4 mm wide, white, separating from the rachilla laterally but remaining firmly attached basally. Scales deciduous, 8–20, 1.8–2.2 mm long, 1.2–1.5 mm wide, ovate, apex acute to obtuse, laterally nerveless, reddish white grading to hyaline near the margins, medially 5-nerved, green. Stamens 3; filaments 1.4–1.8 mm long; anthers 0.7–1.0 mm long, ellipsoid, the connective apex 0.2–0.5 mm long, subulate, red, sometimes setose. Style 0.2–0.5 mm long; stigmas 3, (0.8–) 1.2–1.8 mm long. Achenes 0.9–1.0 mm long, 0.4–0.5 mm wide, trigonous, dorsiventrally flattened, oblong, apex scarcely apiculate, base sessile, the adaxial face flat, the abaxial faces flat to slightly concave, the surface puncticulate, pale brown.

Phenology. Flowering and fruiting from June to October, and rarely during remaining months.

Distribution. Neotropical regions, West Indies, western Mexico south through Central America to northern Argentina. Occasional in the lowlands of Mexico (Fig.

46), southward from Río Tecolutla, Veracruz, on the Gulf Coast and southwestern Jalisco on the Pacific Coast; in pure stands or in mixed populations with *Typha*, *Pontederia*, and other emergent aquatics, in marshes and edges of rivers; sea level to 100 (–500) m.

REPRESENTATIVE SPECIMENS. **Mexico.** COLIMA: Laguna de Cuyutlán, *Díaz Luna 4185* (ENCB, MICH).—JALISCO: just N of Barra de Navidad, *McVaugh 19742* (G, MICH, NY, TX, US).—OAXACA: 40 km W of Puerto Escondido on Hwy 200, *Tucker 2910* (CAS, DUKE, ENCB, MICH).—TABASCO: Balancan, S. Isidro, *Matuda 3350* (F, GH, K, LCU, MICH, MO, NY).—VERACRUZ: River of Antigua, *Galeotti 5787* (BR, G, K); on Hwy 180 just N of turnout to Riachuelos, *Tucker 2473* (CAS, DUKE, ENCB, MICH); 3 km N of Lerdo de Tejada, *Tucker 2507* (CAS, DUKE, ENCB).

Cyperus giganteus is conspicuous in the field by its tremendous size, 2–3.5 m, leafless culm, and open hemispherical inflorescence. Incomplete specimens can be reliably identified by their conspicuous secondary inflorescence bracts, narrow subcylindric spikelets, and anthers with prominent connective apices 0.2–0.5 mm long.

54. Cyperus ischnos Schlechtendal, Bot. Zeit. 7: 99. 1849. *Mariscus ischnos* (Schlechtendal) C. B. Clarke in J. D. Hooker, Fl. Brit. India 6: 623. 1893.—TYPE: MEXICO. Veracruz: Jalapa, Sep 1828, *Schiede 876* (lectotype, designated by Tucker & McVaugh, 1993: HAL!).

Chlorocyperus arsenei Palla, Österr. Bot. Zeitschr. 63: 403. 1913. *Cyperus ischnos* var. *arsenei* (Palla) Kükenthal, Pflanzenreich 4(20): 462. 1936.—TYPE: MEXICO. Michoacán: Morelia, *Arsène 5318* (holotype: W; isotype: BM!).

Cyperus botterii Boeckeler, Allg. Bot. Zeit. Syst. 2: 2. 1896.—TYPE: MEXICO. Veracruz: Orizaba, *Botteri s.n.* (holotype: B!; isotype: B!).

Perennial, (20–) 40–70 (–85) cm tall. Rhizomes 1–3 cm long, 0.5–1.8 mm in diameter, indurate, with barely overlapping, closely appressed, ovate, purplish brown scales 1–2 mm long. Culms (0.5–) 0.8–2.0 mm in diameter, trigonous, smooth or barely scabrid on the angles apically to strongly scabrous on the angles, or infrequently scabrous on the surface as well as the angles, basally usually clothed with the fibrous remains of leaf sheaths. Leaves 1–4 (–5), 5–30 cm long, 1.0–3.5 mm wide, v-shaped, the margins and keel scabrellate, the adaxial surface occasionally scabrous as well. Inflorescence bracts 3–5, 1–20 cm long, 1–3.5 mm wide, v-shaped, the margins and keel scabrellate, the adaxial surface occasionally scabrellate as well, horizontal to reflexed. Rays absent or 1–4, 0.5–2.5 cm long. Spikes (5–) 8–15 mm long, (5–) 8–12 mm wide, dense, subglobose to oblong-ovoid; rachis 3–8 mm long. Spikelets (20–) 40–80, 5–9 mm long, (1.5–) 2.0–3.0 (–4.0) mm wide, lanceolate to oblong-ovate, appearing compressed owing to the scales spreading at ca. 30–45°; rachilla ± deciduous, (0.3–) 0.4–0.7 mm wide, ca. 0.1 mm thick, straight, whitish to faintly greenish white, sometimes reddish spotted, successive scales (1.3–) 1.4–2.0 (–2.4) mm apart, wings 0.1–0.3 mm wide, hyaline to firm, nearly transparent to nearly opaque, whitish or occasionally bronze near the edges, usually clasping the achene, or frequently absent entirely. Scales deciduous, 4–8 (–12), 2.2–3.0 mm long, 1.0–1.6 mm wide, lanceolate to ovate-lanceolate, apex excurrent as mucro 0.1–0.2 mm long, tipped with a tuft of crystalline prickles, laterally 2–4-nerved, whitish but often reddish lineolate as

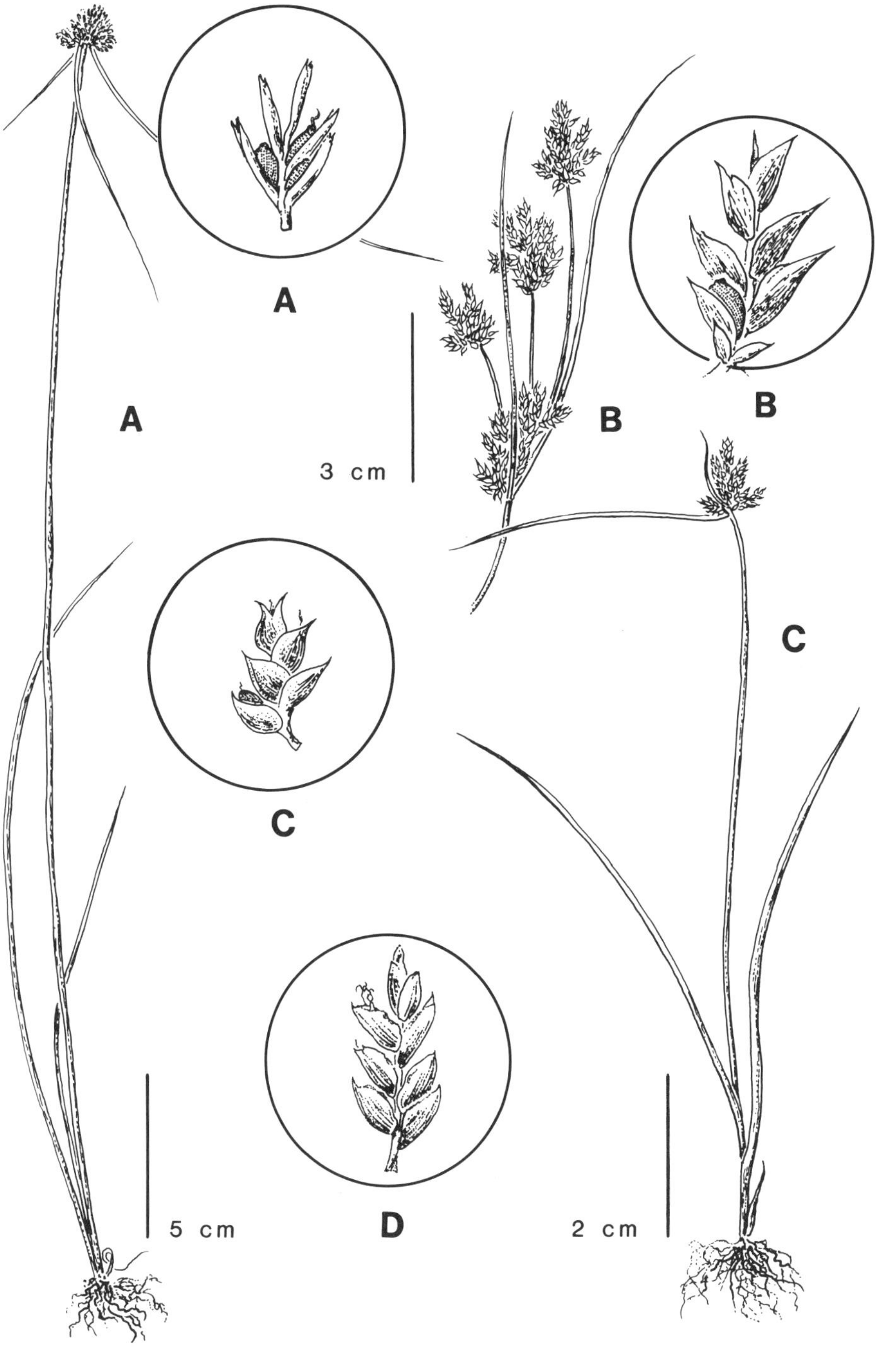

FIG. 47. Diagnostic features of *Cyperus ischnos*, *C. fendlerianus*, *C. schweinitzii*, and *C. spectabilis*. A. *Cyperus ischnos*, habit and spikelet. B. *C. schweinitzii*, inflorescence and spikelet. C. *C. fendlerianus*, habit and spikelet. D. *C. spectabilis*, spikelet. (Spikelets ×5.) Drawn by Patricia M. Kernan.

well but greenish white when immature, medially 1–3-nerved, whitish or occasionally greenish, smooth or rarely with a few prickles distally. Stamens 3; filaments 2–3 mm long; anthers (0.3–) 0.4–0.6 mm long, ellipsoid to oblong-ellipsoid, the connective apex not prolonged. Style (0.8–) 1.0–2.2 mm long; stigmas 3, 0.5–1.5 mm long. Achenes 1.3–1.6 (–1.8) mm long, (0.5–) 0.6–0.9 (–1.2) mm wide, trigonous, obovoid to ellipsoid or rarely ovoid, apex obtuse to truncate, apiculate, base cuneate, the faces concave, the surface distinctly papillose, reddish brown to dark brown or nearly black. Fig. 47A.

Phenology. Flowering and fruiting from June through October.

Distribution. From southwestern Sonora and northern Sinaloa through the mountains of western and southern Mexico (Fig. 48) to Belize and Guatemala, south to Volcán Chiriquí in western Panama, also in the mountains of Colombia and Venezuela; pine forests, oak woods, edges of fields, roadsides, and grasslands; (500–) 1000–2000 m.

Representative Specimens. **Mexico.** Chiapas: Teopisca, *Breedlove 15054* (DS, F); between Oxchuc and Ocosingo, *Tucker 2223* (DUKE).—Chihuahua: Mojarachic, *Knobloch 5529* (MSC).—Durango: Mpio. El Salto, 8–11 km NE of El Palmito, *Breedlove 43926* (CAS).—Guerrero: Dtto. Mina, Manchón, *Hinton 9465* (K, LCU, MICH, US).—Jalisco: Mpio. Tecalitlán, Llano Verde, *Rzedowski 17463* (DS, ENCB, F, MICH, MSC).—México: Dtto. Temascaltepec, Chorrera, *Hinton 1826* (ARIZ, F, K, LCU, MO).—Michoacán: 17 mi N of Aguililla, *Denton 2029* (MICH).—Nayarit: NE of Santa María del Oro, *Feddema 613* (DS, DUKE, MICH, TEX).—Oaxaca: 5 km N of Matías Romero, *King 828* (MICH, US).—Sinaloa: Ocurahui, Sierra Surotato, *Gentry 6367* (ARIZ, DS, LCU, MICH, MO, NY).—Sonora: Río Mayo, Sierra Charuco, *Gentry 1736* (LCU).—Veracruz: Orizaba, Serra del Borrego, *Muller 1994* (BR, NY); Consoquitla, *Liebmann 14432* (C).

Cyperus ischnos is distinguished by the usually whitish color of the scales, the dense spikes, and the single culms borne on slender rhizomes. The tridentate scale apices may aid in the identification as well, although this character is not readily apparent or easily interpreted.

Intraspecific variation in *C. ischnos* led O'Neill and Benedict Ayers (1944) and Benedict Ayers (1946) to maintain the segregate species *C. botterii*. According to these authors, *C. botterii* has a conspicuously winged rachilla and scabrid culms, whereas *C. ischnos* is wingless and smooth. Examination of 65 collections from Mexico and Central America shows that these two characters are not correlated (Fig. 48). The distribution of characters in the 65 collections examined from Mexico and Central America follows: wings absent, culms smooth: 26; wings absent, culms rough: 5; wings present, culms smooth: 22; wings present, culms rough: 12. In compiling this information, the arbitrary distinction between "wingless" and "winged" plants was made at 0.1 mm wide, since a rachilla wing less than 0.1 mm wide has been described by most authors as "wingless." A full range of widths from approximately 0 to nearly 0.3 mm wide was observed. Roughness of the culm also showed variation. Two collections, *Arsène 5580* and *McVaugh 20365*, have culms scabrid on the angles and faces, whereas the other collections with rough culms are scabrid on the angles only. The roughness varies from just a few prickles below the inflorescence bracts to many prickles on the upper 10–20 cm of the culm. Mapping the distribution of these two characters indicates both occur throughout the range of the species. Neither character is allopatric with

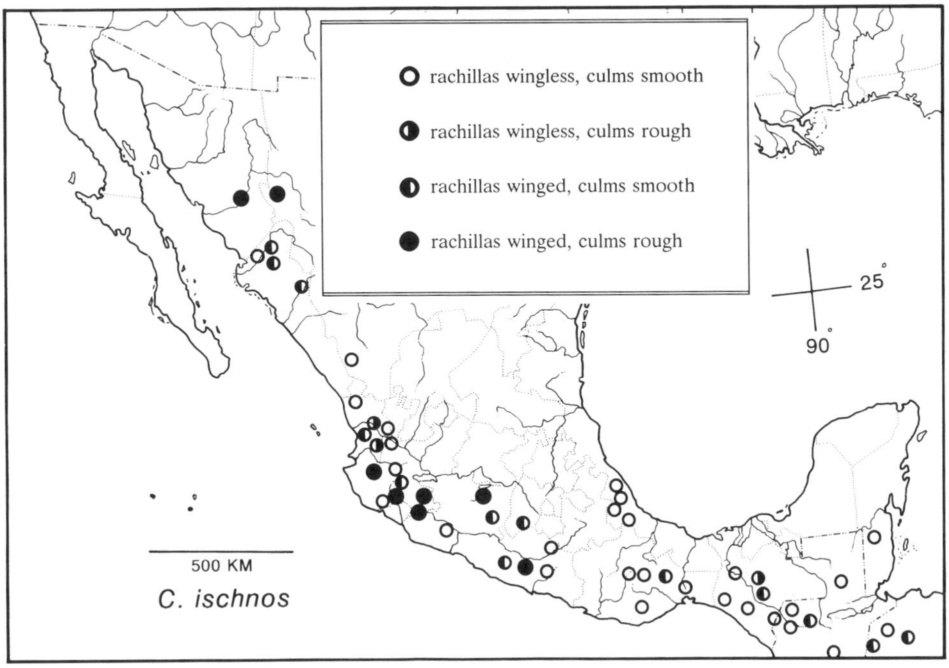

FIG. 48. Distribution, showing infraspecific variation, of *Cyperus ischnos* in Mexico and northern Central America.

respect to the other. There is no single character that will consistently separate *C. botterii* from *C. ischnos*. Kükenthal (1935–36) treated *C. botterii* as a synonym of *C. ischnos* without comment. The holotype of *C. botterii* has smooth culms and hyaline rachilla wings at most 0.1 mm wide. Thus, it is a typical plant of *C. ischnos*, lacking both the conspicuously winged rachilla and rough culms ascribed to *C. botterii* by O'Neill and Ayers.

A few collections of *C. ischnos* have distinctly ovoid achenes and ovate-lanceolate scales. The most extreme example is *McVaugh 18658* (MICH, US) from western Jalisco, in which the achenes are 1.7 mm long and 1.2–1.3 mm wide. It seemed that this collection might represent an undescribed taxon. However, two other collections with ovoid-ellipsoid achenes were also seen: *Vera Santos 3193* from Oaxaca and *Norris & Taranto s.n.*, 30 Jul 1970 (MICH) from Nayarit. These two collections have ovate-lanceolate scales with achenes 0.9–1.1 mm wide, thus bridging the morphological gap between typical representatives of *C. ischnos* and *McVaugh 18658* from Jalisco.

In the protologue, Schlechtendal stated that he had two specimens, *Schiede 876*, from Jalapa. In the herbarium at Halle are two sheets of this species examined by Schlechtendal, one labeled "876 Prope Jalapam" in Schiede's hand, the other bearing a label "Hac. de la Laguna," a locality close to Jalapa. Thus these two sheets should be regarded as syntypes since Schlechtendal stated that he had two sheets from Jalapa. The sheet with two culms [HAL 53192] was designated as lectotype.

55. Cyperus sanguineoater Boeckeler, Bot. Jahrb. Syst. 1: 363. 1881. *Mariscus atrosanguineus* Liebmann, Mexic. halvgr. 40. 1850. *Cyperus atrosanguineus* (Liebmann) Hemsley, Biol. Centr.-Am. Bot. 3: 445. 1885, non *C. atrosanguineus* Hochstetter ex Steudel, 1854.—TYPE: MEXICO. Puebla: Chinantla, "in montibus graminosis inter Pinus," Jul 1841, *Liebmann 14378* (lectotype: C!, photos: LCU! MICH!; isolectotypes: C! HAL!).

Chlorocyperus michoacanensis Palla, Österr. Bot. Zeitschr. 63: 403. 1913.—TYPE: MEXICO. Michoacán: vic. of Morelia, 1 Sep 1910, *Arsène 5316* (holotype: W, photo: LCU! isotypes: MO! NY!).

Cyperus sanguineo-ater var. *floribundus* Kükenthal, Pflanzenreich 4(20): 464. 1936.—TYPE: MEXICO. Morelos: Cuernavaca, 20 Jul 1904, *Pringle 13231* (holotype: B, destroyed; isotypes: CU! K! MICH! US! VT!).

Perennial, (5–) 25–65 cm tall. Rhizomes 2–3 cm long, 1–2 mm in diameter, indurate, closely clothed with brownish scales or their fibrous remnants. Culms (0.6–) 0.9–1.8 in diameter, trigonous, smooth or scabrellate apically. Leaves 2–4, 10–25 cm long, 1.7–2.8 mm wide, flat to v-shaped, the margins and keel smooth or scabrellate in the distal portion. Inflorescence bracts (2–) 3–5 (–8), 3–15 (–24) cm long, 0.8–2.8 mm wide, the margins and keel scabrellate, horizontal to reflexed downward parallel to the culm. Rays absent, the inflorescence a single sessile spike, very rarely with a single smaller spike produced at the base. Spike (9–) 15–20 (–25) mm long, (6–) 9–14 (–20) mm wide, densely ovoid to oblong-cylindric; rachis 5–20 mm long. Spikelets (10–) 30–75, 5–11 mm long, 2–4 mm wide, oblong-lanceolate, compressed-quadrate; rachilla ± deciduous, ca. 0.6 mm wide, 0.1–0.2 mm thick, geniculate, whitish or stramineous, successive scales 1.2–1.4 (–1.8) mm apart, wings 0.2–0.4 mm wide, white or red. Scales deciduous, 3–12, 2.4–2.8 (–3.2) mm long, (1.0–) 1.4–1.6 (–2.0) mm wide, ovate-lanceolate to oblong-ovate, the mucro 0.4–0.8 (–1.0) mm long (occasionally the proximal scales or rarely all the scales merely mucronulate), laterally 2–4-nerved, light to deep red, medially stramineous to red or green basally, green distally. Stamens 3; filaments ca. 2 mm long; anthers 0.6–1.3 (–1.4) mm long, slenderly ellipsoid to oblong, the connective apex minute, reddish. Style 0.8–2.0 mm long; stigmas 3, 0.6–1.4 mm long. Achenes (1.2–) 1.4–2.0 mm long, (0.6–) 0.8–1.3 mm wide, trigonous, oblong-ellipsoid to oblong-ovoid, apex truncate, apiculate, base sessile, the faces concave, flat, or subtly convex, the surface coarsely papillose, brown.

Phenology. Flowering and fruiting in July, August, and September.

Distribution. Mexico, and in the mountains of Guatemala and Peru. In Mexico (Fig. 49) in the Sierra Madre Occidental of Sinaloa and Durango, the Trans-Mexican Volcanic Belt from western Jalisco to the Distrito Federal, and in the Sierra Madre del Sur of Oaxaca; in pine and oak forests, along trails, in clearings and moist or disturbed situations; 1600 to ca. 3000 m.

REPRESENTATIVE SPECIMENS. **Mexico.** CHIHUAHUA: Río Mayo region, Talayotes Camp, *Martin et al. s.n.* (ARIZ 302763).—DISTRITO FEDERAL: Rancho El Conejo, deleg. Xochimilco, *Ventura 2897* (ARIZ, DS, ENCB); 1 mi SE of San Andreas, *Barkley & Webster 2171* (TEX).—DURANGO: 20 km NE of El Salto on Mex. 40, *Marcks 1265* (MICH, WIS); Otinapa, *Palmer 560* in 1906 (US); 5 1/2 mi E of El Salto, *Waterfall & Wallis 13591* (TEX).—GUERERRO: Dtto. Mina, Manchón, *Hinton 9465* (UC).—JALISCO: mountains N of Autlán, 3–5 mi above Mina San Francisco, *McVaugh 19663* (MICH); Sierra de la Venta, 20 km W of Guadalajara, *Puga 1509* (IBUG); Guadalajara, *Pringle 3844* (VT).—MÉXICO: Dtto. Te-

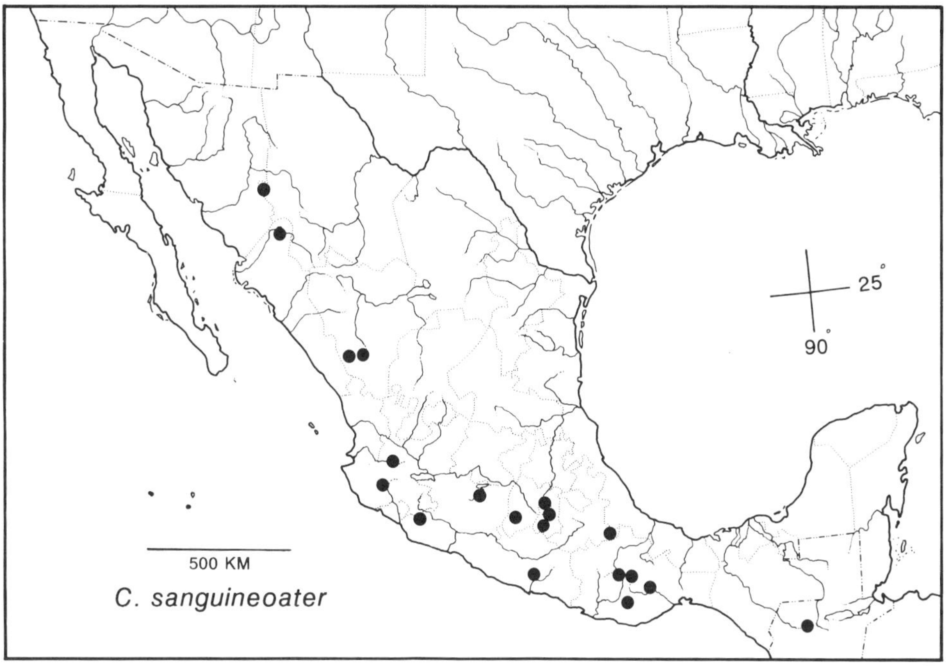

FIG. 49. Distribution of *Cyperus sanguineoater* in Mexico.

mascaltepec, Cerro Muñeca, *Hinton 1367* (F, GH, K, LCU, MO, NY, US); 2–3 km SW of Luvianos, *Roe et al. 1605* (WIS).—MICHOACÁN: vicinité de Morelia, *Arsène 5316* (MO, US); nearly W of Aquililla, "Cerritos de Agua," *McVaugh 17842* (MICH, NY).—MORELOS: near Cuernavaca, *Pringle 13231* (CU, K, MICH, US, VT).—NAYARIT: between Santa Gertrudis and Santa Teresa, *Rose 3377* (US).—OAXACA: 25 air km NE of Oaxaca on Hwy 175, *Conrad 3071* (GH, MO); 15 km by rd SE of Miahuatlán, *Marcks 1035* (DAV, WIS); Mpio. Santiago Laxopa: S. Laxopa, *Maldonado V. & Martin 328* (MO).—PUEBLA: Chinantla, *Liebmann 14378* (C, HAL).—SINALOA: Ocurahui, Sierra Surotato, *Gentry 6441* (ARIZ, LCU, MICH).

Cyperus sanguineoater is distinguished from other Mexican species of *Cyperus* by its red scales and its dense, rayless inflorescence. It might be confused with red-scaled forms of *C. manimae*, but the latter have blunt (though occasionally mucronulate or mucronate) instead of acute or subacute scale apices, and shorter anthers and styles. Kükenthal (1935–36) described *C. sanguineoater* var. *floribundus* based on a robust collection with longer than average spikelets, 7–10 mm long. In *C. sanguineoater*, the length of the spikelets shows a continuous range from 5–11 mm. Kükenthal's variety simply represents an extreme of variation in one character, and Benedict Ayers (1946) rightly did not maintain it.

Cyperus sanguineoater has not been previously reported from South America but has been collected in Peru: Depto. Pasco, Prov. Oxabampa, 5–8 km SW of Huancabamba, *Smith & Canne 5766* (MO).

The specific epithet of this species has often been spelled "sanguineo-ater," presumably because someone objected to three vowels in a row; however, Boeckeler in the protologue used no hyphen and his spelling is reactivated here.

56. Cyperus dipsaceus Liebmann, Mexic. halvgr. 31. 1850. *Mariscus dipsaceus* (Liebmann) C. B. Clarke, Bull. Misc. Inform. Kew, add. ser. 8: 16. 1908.—TYPE: MEXICO. Veracruz: Potrero de Consoquitla, Oct 1841, *Liebmann 14357* (lectotype, designated by Tucker, 1986: C!, photos: MICH! NY! F!; isolectotypes: B-fragment! C! F! GH! K!).

Cyperus wrightii Britton, Bull. Torrey Bot. Club 13: 215. 1886.—TYPE: U.S.A. New Mexico, 1852, *Wright 1947* (holotype: NY!; isotype: GH!).

Perennial, (30–) 45–80 cm tall. Rhizomes 0.5–3.0 cm long, proximally 1–2 mm in diameter, distally 2–4 mm in diameter, tuberlike, thickened, 3–6 mm in diameter, densely clothed with reddish brown scales. Culms (0.8–) 1.2–2.2 mm in diameter, trigonous. Leaves (1–) 2–3, 15–30 cm long, 0.5–3.0 mm wide, flat to v-shaped, becoming triquetrous apically, the margins and keel scabrellate throughout or smooth in the basal half. Inflorescence bracts 4–6, 3–24 cm long, 0.4–3.6 mm wide, the larger ones flat or v-shaped, the smallest capillary, triquetrous, the margins and keels scabrellate, the longest bract erect or nearly so, the others ascendent. Rays 2–4 (–5), (1.5–) 2.5–6 cm long; secondary rays absent. Spikes (9–) 12–22 mm long, (4–) 6–12 mm wide, densely oblong-cylindric, subtruncate at the base; rachis (6–) 10–16 mm long. Spikelets (10–) 20–60, 3–9 mm long, 2.0–3.4 mm wide, oblong to linear-lanceolate, compressed-quadrate and borne at right angles or slightly ascendent on the rachis, the lower spikelets never reflexed; rachilla ± deciduous, 0.6–0.8 mm wide, 0.1–0.2 mm thick, stramineous to dull white, successive scales (1.4–) 1.8–2.2 mm apart, wings 0.2–0.4 mm wide, whitish, opaque. Scales deciduous, 2–4 (–8), (2.0–) 2.2–2.8 (–3.0) mm long, (1.0–) 1.2–1.8 mm wide, ovate-lanceolate, apex acute, weakly tridentate, the mucro 0.2–0.6 mm long, with an apical tuft of crystalline prickles, medially 3-nerved, green, laterally 2–3-nerved, stramineous, dull whitish or pale reddish brown, often red-dotted. Stamens 3; filaments 1.8–3.0 mm long; anthers 0.4–0.9 mm long, ellipsoid, the connective apex up to 0.1 mm long, reddish. Style 1.0–1.4 mm long; stigmas 3, 1.0–1.6 mm long. Achenes 1.4–1.5 (–1.6) mm long, 0.8–0.9 mm wide, trigonous, broadly ellipsoid, apex subtruncate, apiculate, base substipitate, ca. 0.1 mm long, 0.1–0.2 mm wide, the faces concave to flat, the surface papillose, brown.

Phenology. Flowering and fruiting in August, September, and October.

Distribution. Mountains and plateaus of Arizona and New Mexico, U.S.A.; in Mexico (Fig. 50) in Chihuahua and northeastern Sonora, southern Baja California, northern Nayarit, and the mountains of southern Veracruz; uncommon, pine forests, seepage, moist ledges, and stream sides; (750–) 2000–2500 m.

REPRESENTATIVE SPECIMENS. **Mexico.** BAJA CALIFORNIA SUR: Cape region, El Taste, 12 Sep 1893, *Brandegee s.n.* (UC).—CHIHUAHUA: creek NW of Chuhuichupa, *LeSueur 1287* (ARIZ, TEX, UC); base of the Sierra Madre, *Pringle 1683* (B, M, NY, UC); by brooks, Sierra Madre, *Pringle 1397* (E, NY, PH).—NAYARIT: between Acaponeta and Pedro Paulo, *Rose 1922* (NY, US).—SINALOA: Sierra Surutato, Mpio. Badiraguato, Ocurahui, *Gentry 6435* (ARIZ, MICH, NY).—SONORA: Río de Bavispe region, NE of El Tigre, *White 4835* (MICH); Río Mayo Region, El Rayo, *Martin et al. s.n.* (ARIZ 302750).—VERACRUZ: Potrero de Consoquitla, *Liebmann 14357* (B, C, GH, K).

Cyperus dipsaceus is distinguished by the combination of erect to ascendent inflorescence bracts and rays, densely oblong-cylindric spikes, and acute, stramineous to light reddish brown scales. *Cyperus sphaerolepis* also has erect an

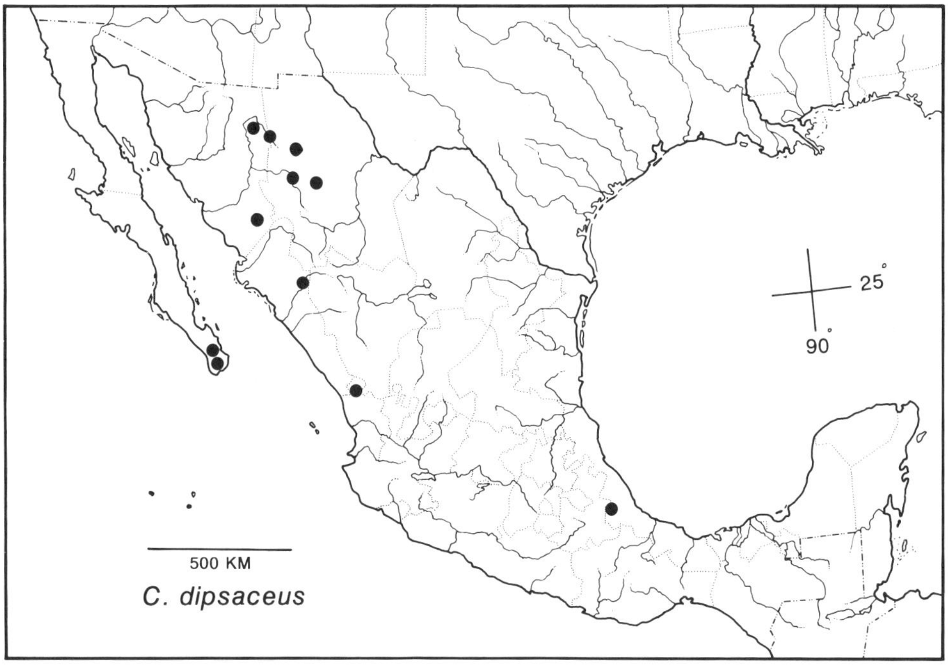

FIG. 50. Distribution of *Cyperus dipsaceus* in Mexico.

inflorescence bract, but its scales are wider than long, whereas those of *C. dipsaceus* are 1.5–2 times longer than wide.

57. Cyperus schweinitzii Torrey, Ann. Lyceum Nat. Hist. New York 3: 276. 1836. *Cyperus alterniflorus* Schweinitz in Keating, Narrat. exp. St. Peter's River, 2: 381. 1824, non *Cyperus alterniflorus* R. Brown, 1810.—TYPE: CANADA. Manitoba: Winnipeg Lake, 18–19 Aug 1823, *Say s.n.* (holotype: PH!). [See Stuckey (1970) for comments concerning the type.]

Perennial, (10–) 20–50 cm tall. Rhizomes 0.3–1.5 cm long, 3–7 mm in diameter, loosely clothed with light brown to reddish brown papery lanceolate scales. Culms (0.6–) 1.0–2.0 mm in diameter, trigonous, smooth or more often scabrellate, either in the distal half or over the entire length. Leaves 3–6, (2–) 20–35 cm long, 2–6 mm wide, flat to v-shaped, the margins and keel scabrellate; the lowest leaves bladeless, the base of the culm with 1–3 papery light brown sheaths 1–4 cm long. Inflorescence bracts 3–7, (2.5–) 8–20 (–25) cm long, (0.5–) 1.5–6 mm wide, the margins and keel scabrellate, all erect or nearly so, the longest bract usually appearing as a continuation of the culm. Rays 3–5, 1.5–8.0 (–15) cm long; secondary rays absent. Spike 1 per ray, 10–25 mm long, 7–16 mm wide, loosely ovoid; rachis 3–9 mm long. Spikelets (1–) 5–10 (–20), 7–10 (–18) mm long, (2.8–) 3.2–4.0 (–4.5) mm wide, oblong to linear-oblong, quadrangular; rachilla ± deciduous, 0.4–0.5 mm wide, 0.2–0.3 mm thick, straight, flattened, successive scales (1.9–) 2.0–2.5 mm apart, wings at most 0.1 mm wide, hyaline. Scales deciduous, (1–) 5–10 (–14), (2.3–) 2.8–3.2 mm long, (2.0–) 2.4–3.2 mm wide, broadly ovate to oblong-ovate, apex broadly rounded, often hyaline-erose, the midvein excurrent as a cusp 0.1–0.4 mm long in the proximal scales and 0.4–1.0 mm long in the distal ones, laterally 2–3-nerved, stramineous, sometimes also red-spotted, medially 3-nerved, green. Stamens 3; filaments 2.5–3.8 mm long; anthers 0.8–1.4 mm long, linear-oblong, the connective apex not prolonged. Style 0.1–0.3 mm long; stigmas 3, 1.5–3.0 mm long. Achenes 2.0–2.4 mm long, 0.9–1.4 mm wide, trigonous, broadly ellipsoid or infrequently obovoid, apex obtuse to broadly rounded, not apiculate but the remnants of the style usually persistent, base cuneate to substipitate, the faces concave to flat, the surface papillose, light to dark brown or nearly black. Fig. 47B.

Phenology. In Mexico, fruiting collections have been made in October.

Distribution. In Canada in southern Ontario and southern Manitoba; in the U.S.A., in Massachusetts (adventive), New York, Idaho, and western Washington south to northern Kentucky (adventive?), Missouri, Texas, and New Mexico. Rare in Mexico, known only from two collections from northern Chihuahua, both from sand dunes. In the U.S.A. and Canada, this species grows on river banks, sand bars, lake shores, sand dunes, and sandy openings in woods.

ADDITIONAL SPECIMENS EXAMINED. **Mexico**. CHIHUAHUA: Samalayuca Dunes area, Oct 1935, *LeSueur 1724* (BKL, GH, POM, UC), *2003* (TEX).

Cyperus schweinitzii is morphologically similar to *C. sphaerolepis*, from which it may be distinguished by its cuspidate, broadly ovate to oblong-ovate scales and its broadly ellipsoid achenes. *Cyperus sphaerolepis* has broadly ovate to suborbiculate, mucronate scales and ovoid to obovoid achenes. In addition, the anthers of *C. schweinitzii* are 0.8–1.4 mm long; those of *C. sphaerolepis* are only 0.4–0.6 mm long.

58. Cyperus sphaerolepis Boeckeler, Linnaea 35: 609. 1868.—TYPE: "North America," "*Cyp. ex Amer. boreali no. 102*, communicavit Hornemann" (holotype: B!).

Cyperus rusbyi Britton, Bull. Torrey Bot. Club 11: 29. 1884.—TYPE: U.S.A. New Mexico: near Silver City, 1880, *Rusby s.n.* (holotype: NY!; isotype: GH!).

Cyperus schweinitzii var. *debilis* Britton, Bull. Torrey Bot. Club 13: 208. 1886. *Cyperus fendlerianus* var. *debilis* (Britton) Kükenthal, Pflanzenreich 4(20): 465. 1936.—TYPE: MEXICO. Tacubaya près Mexico, *Bourgeau 529* (lectotype, here designated: NY!; isolectotypes: C! GH! P).

Cyperus leucolepis Boeckeler, Linnaea 35: 610. 1868. *Cyperus fendlerianus* var. *leucolepis* (Boeckeler) Kükenthal, Pflanzenreich 4(20): 466. 1936.—TYPE: MEXICO. *Coulter 1611 p.p.* (holotype: B, not located; isotype: GH!).

Perennial, (8–) 15–40 (–60) cm tall. Rhizomes 1–2 cm long, 2–5 mm in diameter, densely clothed with light brown to reddish brown ovate scales or their fibrous remnants, producing single culms 1–2 cm apart. Culms (0.4–) 0.8–1.5 (–2.0) mm in diameter, trigonous, smooth or sparsely to densely scabrellate to scabrous on the angles or generally on the surface, either just below the bracts or over the upper few centimeters. Leaves (1–) 3–4, (7–) 15–25 (–40) cm long, 1–3 (–5) mm wide, v-shaped, sparsely scabrellate on the margins and keel in the distal third, otherwise smooth. Inflorescence bracts 2–3 (–7), (1–) 3–20 (–26) cm long, (0.5–) 1–3 (–5) mm wide, scabrellate on the margins and keel, the smallest bracts linear, trigonous, rather densely scabrellate, the longest bract ± erect, appearing as a continuation of the culm, the remaining bracts erect to ascendent. Rays (1–) 2–6 (–11), (1–) 2–7 (–12) cm long; secondary rays absent; in some depauperate plants, the rays less than 1 cm long, the inflorescence composed of 1–3 sessile globose spikes, occasionally a single spike borne on a peduncle 5–20 cm long produced from the axil of the uppermost leaf. Spike 1 per ray, (9–) 15–30 (–45) mm long, (3–) 8–15 (–18) mm wide, loosely cylindric to oblong-ovoid; rachis (5–) 10–30 mm long, the spikelets at an oblique angle, oriented toward the summit of the spike. Spikelets (4–) 8–25 (–35), (2.0–) 7–11 (–15) mm long, (2.0–) 2.4–3.2 (–3.5) mm wide, oblong-elliptic to linear, compressed-quadrate; rachilla ± deciduous, 0.5–0.8 mm wide, (0.1–) 0.2 mm thick, geniculate, stramineous to greenish, sometimes red-dotted, successive scales (1.2–) 1.4–1.9 mm apart, wings (0.1–) 0.2 mm wide, hyaline. Scales deciduous, (1–) 4–8 (–14), (1.6–) 1.8–2.4 mm long, (1.8–) 2.2–2.8 mm wide, broadly ovate to suborbiculate, apex obtuse, with a straight mucro 0.1–0.2 (–0.3) mm long, bearing at its tip a tuft of crystalline prickles, laterally strongly 3 (–4) -nerved, dull whitish, stramineous, or light brown, or reddish brown to deep red, often red-dotted as well, the margins hyaline erose especially apically, medially weakly 1–3-nerved, greenish. Stamens 3; filaments 2–3 mm long; anthers 0.4–0.6 mm long, ellipsoid, the connective apex scarcely prolonged. Style 0.4 (–0.8) mm long; stigmas 3, 0.6–1.0 mm long. Achenes 1.4–1.6 (–1.9) mm long, 1.0–1.3 (–1.4) mm wide, trigonous, ovoid to obovoid, apex obtuse, slightly apiculate, base cuneate to substipitate, 0.1–0.2 mm long, ca. 0.2 mm wide, the faces flat to concave, the surface finely reticulate with isodiametric cells, this outer layer gradually wearing away to reveal an inner papillose brown to reddish brown layer.

Phenology. Flowering and fruiting from July to September (–October).

Distribution. Western Texas, southern New Mexico, and Arizona, U.S.A., south through the mountains of northern Mexico to the Distrito Federal and Puebla (Fig. 51); pine and oak-pine forests, meadows, swales, dry streambeds, roadsides, and disturbed soils; 1500–2600 m.

Representative Specimens. **Mexico.** Chihuahua: SE of Creel, *Bye 2436, 2437* (ASU); Sierra del Nido, *Christman 161* (UC); Culebra Mts, *LeSueur 1097* (ARIZ, F, K, TEX); Mpio. Chihuahua, 30 mi SW of Chihuahua, *Muller 3343* (GH, LL, UC).—Coahuila: Sierra del Carmen, *Hendrickson 11406* (LL); Sierra del Pino, *Johnston & Muller 540* (GH, LL, TEX); Mpio. Múzquiz, Hac. La Babía, *Wynd & Mueller 654* (ARIZ, GH, MICH, MSC, NY).—Distrito Federal: Tacubaya près Mexico, *Bourgeau 529* (C, GH, NY).—Durango: 20 km NE of El Salto, *Marcks 1262B* (MICH, TEX, WIS); Durango and vicinity, *Palmer 544* (C, F, GH, NY, UC, US).—Hidalgo: near El Salto, *Pringle 9332* (F, GH, LL, US, VT).—México: Mpio. Texcoco, 1.5 km W of of Huexotla, *Galván 685-B* (IBUG); Mpio. Ixtapaluca, Cerro del Pino, *Rzedowski 36347* (ENCB, IBUG); Río Hondo, 8000 ft, *Pringle 3415* (VT).—Nuevo León: Mpio. Galeana, Hac. Cieneguillas on Cerro Potosí, *Schneider 975* (ARIZ, F, GH, LCU, MICH, NY).—Puebla: vicinity of Puebla, *Arsène 1024* (GH).—Querétaro: near San Juan del Río, *Rose et al. 9559* (NY).—San Luis Potosí: region of San Luis Potosí, *Parry & Palmer 907* (E, F, GH, K, PH, US); circa Morales, *Schaffner 561* (GH, K).—Sonora: Fronteras, *Hartman 37* (GH); region of Río de Bavispe, 4 mi E of El Bilito, *White 4782* (MICH).—Tlaxcala: Mpio. Santa Cruz Tlaxcala, vicinity of San Bernabe Amaxac, *Hernández 362* (MICH).—Zacatecas: 29 mi W and S of Fresnillo on rd to Valparaíso, *Anderson & Laskowski 3594A* (MICH).

Cyperus sphaerolepis and *C. fendlerianus* are the only species in subg. *Cyperus* with obovoid to ovoid achenes and with scales that are wider than long. *Cyperus*

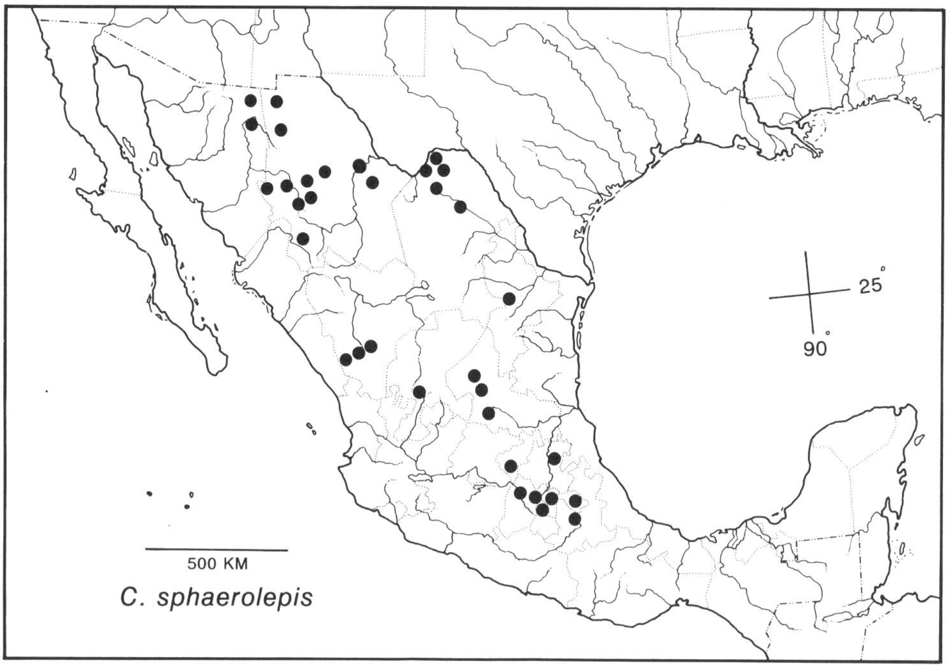

FIG. 51. Distribution of *Cyperus sphaerolepis* in Mexico.

sphaerolepis is distinguished from *C. fendlerianus* by its having at least one pedunculate or rayed spike in the inflorescence, by its shorter scales with shorter, straighter mucros, and also by the erect leading bract of the inflorescence. Some small depauperate plants of *C. sphaerolepis* have an inflorescence consisting of a single sessile spike. Such plants can be distinguished from *C. fendlerianus* by the erect inflorescence bract. A brief review of the treatments of these two taxa by Kükenthal and later authors follows.

Kükenthal (1935–36) and Benedict Ayers (1946), neither of whom saw a large number of specimens, both treated *C. sphaerolepis* as *C. fendlerianus* var. *debilis*. Benedict Ayers distinguished the two varieties as follows:

Culms scabrellate at the apex (rarely smooth); rays 0 (–2), at most 3–5 cm long; spikelets 3–5-flowered; mucronate apex of the scales 0.3–0.5 mm long, excurved. *C. fendlerianus* var. *fendlerianus*.

Culms smooth at the apex (rarely scabrellate); rays 1–8, the longest 6–13 cm long; spikelets 8–16-flowered; mucro straight, at most 0.3 mm long. *C. fendlerianus* var. *debilis*.

Correll and Johnston (1970), in their *Manual of the vascular flora of Texas*, noted that var. *debilis* has erect inflorescence bracts, whereas those of var. *fendlerianus* are borne horizontally, a distinction previous workers had failed to appreciate.

Kearney and Peebles (1951) in *Arizona Flora* recognized *C. sphaerolepis* (under the name *C. rusbyi*) at the specific level, separating it from *C. fendlerianus* thus:

1. Rays absent, the 1–4 spikes crowded into a dense, sessile, 3–4-lobed head; spikelets 3–8-flowered; mucro of the scales spreading, often curved, up to 0.5 mm long. *C. fendlerianus*.

1. Rays well developed; spikelets commonly 8–16-flowered; mucro erect or nearly so, straight, not more than 0.3 mm long. *"C. rusbyi"* [=*C. sphaerolepis*].

Martin and Hutchins (1981), in *A Flora of New Mexico*, also recognized both species but provided a simpler key:

1. Spikes sessile or if pedunculate, the rays much less than 5 mm in length; inflorescence a dense capitate cluster. *C. fendlerianus.*
1. Spikes pedunculate, the rays 5 mm or more long; inflorescence loose. *"C. rusbyi"* [=*C. sphaerolepis*].

Each difference between the two taxa noted by these authors has been critically examined. Plants of *C. fendlerianus* always have both sessile spikes and horizontal to reflexed inflorescence bracts, whereas in *C. sphaerolepis* the leading bract of the inflorescence is always erect, and there is always at least one elongate ray. Close examination of the plants yielded another difference; the scales of *C. fendlerianus* are longer than those of *C. sphaerolepis*, 2.4–2.8 (–3.4) mm vs. (1.6–) 1.8–2.4 mm. Care must be taken to compare scales subtending mature or nearly mature achenes. At or before anthesis, the scales are often ca. 0.5 mm shorter.

Kükenthal (1935–36) and Benedict Ayers (1946) noted two differences which I could not confirm: *C. sphaerolepis* supposedly has smooth culms, whereas those of *C. fendlerianus* are said to be rough; however, both rough and smooth culms occur in both species. Also, there is no abrupt distinction between the rough and smooth condition. Many plants were observed which have very few prickles on the culm and are nearly smooth. Some plants are scabrellate, whereas others are scabrous. A few collections have prickles on only one of the three angles of the culm, whereas the other angles are smooth. In all cases the prickles, if any, are confined to the upper few centimeters of the culm. Thus, in this pair of species, the roughness of the culms is not taxonomically useful.

A second difference was reported by Benedict Ayers (1946) in the number of scales per spikelet. He stated that *C. sphaerolepis* has 8–16-flowered spikelets, whereas in *C. fendlerianus* they are only 3–5-flowered. I found that *C. sphaerolepis* does have a greater maximum number of scales per spikelet (14 vs. 8), but the most frequent value for both species is the same (6). Thus, these two species cannot be distinguished on the basis of scale number.

The difference in length of the mucros of the scales has proved to be a reliable means of distinguishing the two species. Here, care must be taken to measure the mucros of the distal scales of a spikelet. The basal scales consistently have shorter mucros than the distal scales. [This has been noted in other mucronate-cuspidate species of this subgenus, e.g., *C. panamensis* (no. 85).] The scales of *C. sphaerolepis* are entire or have short straight mucros 0.1–0.2 (–0.3) mm long, whereas the mucros of *C. fendlerianus* are longer (0.3–0.6 mm long) and additionally are stouter and conspicuously excurved.

In summary, *C. sphaerolepis* and *C. fendlerianus* are distinct species, differing in at least four characters: presence of rays, orientation of the inflorescence bracts, length of the scales, and length of the mucros.

Plants of *C. sphaerolepis* often produce a single spike borne on a long peduncle originating in the axil of the uppermost basal leaf. Apparently, no other species of this genus produces such basal spikes.

59. Cyperus fendlerianus Boeckeler, Linnaea 35: 520. 1868. *Chlorocyperus fendlerianus* (Boeckeler) Palla, Österr. Bot. Zeitschr. 58: 392. 1908.—TYPE: U.S.A. New Mexico: Santa Fe, foot of mountains near irrigating ditches, 25 June–28 Jul 1847, *Fendler 865* (holotype: B!; isotypes: GH! NY! UC!).

Perennial, (7–) 20–70 (–85) cm tall. Rhizomes 1–2 cm long, 2–6 mm in diameter, indurate, densely clothed with ovate brown to reddish brown scales or their fibrous remnants (stout, curved downward from the base of the culm). Culms (0.5–) 1–1.8 (–2.6) mm in diameter, trigonous, smooth or sparsely to densely scabrellate on the surfaces or only on the angles just below the inflorescence in the upper 3–6 cm. Leaves 2–3, (12–) 20–40 (–55) cm long, (2–) 3–5 (–7) mm wide, flat to v-shaped, sparsely scabrellate on the margins and keel. Inflorescence bracts (2–) 3–6 (–10), (2–) 6–16 (–25) cm long, (1.5–) 2–4 (–7) mm wide, the margins and keel scabrellate to scabrous, the upper and lower surfaces often also scabrous, ± horizontal to reflexed downward parallel to the culm. Rays absent. Spikes (1–) 3–4 (–6), (12–) 18–30 mm long, (6–) 12–20 mm wide, loosely to densely cylindric to ovoid-cylindric; rachis (8–) 12–20 (–24) mm long. Spikelets (8–) 15–30, 5–8 (–10) mm long, (2.0–) 2.4–3.4 (–4.2) mm wide, spreading to ascendent-appressed, oblong-lanceolate, quadrate, slightly compressed; rachilla ± deciduous, 0.5–0.6 mm wide, 0.2–0.3 mm thick, geniculate, pale greenish white to stramineous, successive scales (1.3–) 1.6–1.8 mm apart, wings absent. Scales deciduous, (2–) 4–6 (–8), 2.4–2.8 (–3.4) mm long, 2.4–2.8 (–3.4) mm wide, ovate-orbiculate, apically obtuse, the mucronate to cuspidate apex stout, (0.1–) 0.3–0.6 mm long, straight to excurved, laterally (2–) 3–4 (–5) -nerved, yellowish, reddish, or greenish brown, medially (1–) 3 (–5) -nerved, greenish, smooth or rarely scabrellate distally. Stamens 3; filaments 2–3 mm long; anthers 0.4–0.8 mm long, oblong to slenderly ellipsoid, the connective apex scarcely prolonged. Style 0.7–1.3 mm long; stigmas 3, 0.7–1.2 mm long. Achenes 1.6–1.9 mm long, 1.1–1.4 (–1.6) mm wide, trigonous, obovoid, apex obtuse, apiculate, base substipitate to cuneate, 0.1–0.2 mm long, ca. 0.2 mm wide, the faces concave, the surface smooth to finely reticulate, brown to reddish brown. Fig. 47C.

Phenology. Flowering and fruiting during July, August, and September, one fruiting collection from December.

Distribution. Cordillera of the western U.S.A. and northern Mexico, from southeastern Wyoming, southern Colorado, and western Texas to southeastern Arizona, northeastern Coahuila, the Valley of Mexico, and western Durango, also in the mountains of Tamaulipas and Nuevo León, and in the Sierra San Pedro Mártir of northern Baja California (Fig. 52); pine forests, juniper woodlands, montane pastures and meadows, and borders of croplands; 1600–2800 (–3300) m.

REPRESENTATIVE SPECIMENS. **Mexico.** BAJA CALIFORNIA: Sierra San Pedro Mártir, *Moran 15452* (ARIZ, ASU, LL, MICH, RSA, UC), *Moran & Thorne 14273* (ARIZ, MICH, RSA, SD, UC).—CHIHUAHUA: Majalca, *LeSueur 228* (GH, TEX); Culebra Mts, *LeSueur 1097* (SMU, TEX, UC); Chuhuichupa, *LeSueur 1109* (CAS, NYS, PH, RM, RSA, TEX, UC); La Bufa Mtn above Cusihuiriachic, *Pringle 1391* (E, GH, LL, MICH, NY, PH, RSA, VT); near Colonia García in the Sierra Madres, *Townsend & Barber 149* (E, K, POM, UC, Z); 27.2 mi W of Vieja Casas Grandes, *J. Tucker*

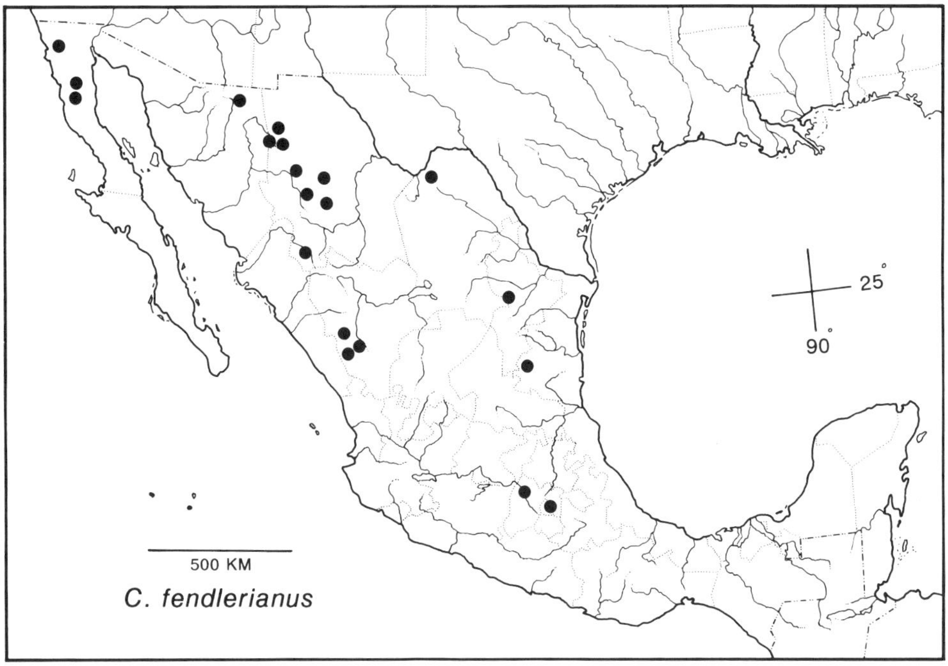

FIG. 52. Distribution of *Cyperus fendlerianus* in Mexico.

2458 (DAV, SMU, UC).—COAHUILA: Villa Acuña, Sierra del Carmen, Cañón de Sentenela, *Wynd & Mueller 512* (ARIZ, K, MICH).—DURANGO: Laguna del Progreso, *Maysilles 7921* (MICH); Otinapa, *Palmer 336* (GH, K, NY, UC).—MÉXICO: Mpio. Texcoco, 1.5 km W of Huexotla, *Galván 685-b* (ARIZ, ENCB); Flor de María, *Pringle 3420* (MICH, NY).—NUEVO LEÓN: La Jolla on Cerro Potosí, *Schneider 1003* (ARIZ, GH, LCU, MICH).—SONORA: Sierra de El Tigre, *White 3414* (ARIZ, GH, MICH).—TAMAULIPAS: en route to Peña Nevada out of Hermosa, *Stanford et al. 2526* (US).

Cyperus fendlerianus is distinguished from other New World species of *Cyperus* by its horizontal to reflexed inflorescence bracts, sessile, cylindric spikes, and ovate-orbiculate, cuspidate scales. For a comparison of *C. fendlerianus* and *C. sphaerolepis*, see the discussion of *C. sphaerolepis* (no. 58).

60. Cyperus ciliatus Junghuhn in Schlechtendal & Chamisso, Linnaea 6: 25. 1831.—TYPE: MEXICO. Veracruz: between Santa Fé and Plan del Río, Jul 1828, *Schiede 850* (holotype: HAL!; isotype: B!).

Perennial (or annual), (15–) 20–50 (–80) cm tall. Rhizomes ca. 1 cm long, 2–4 mm in diameter, indurate, oblique, clothed with fibrous scale remnants. Culms 1–12, 1.0–2.5 mm in diameter, trigonous, ciliate-scabrous on the angles, especially distally. Leaves 2–6, (10–) 20–30 (–60) cm long, 3–6 (–8) mm wide, v-shaped, the margins and keel scabrous. Inflorescence bracts 3–7, (2–) 6–20 (–27) cm long, (1–) 2–5 (–6) mm wide, v-shaped, the margins and keel scabrous, ascendent at 30–45°. Rays 3–8, (1.5–) 3–6 (–12) cm long; secondary rays absent. Spikes (12–) 15–20 (–24) mm long, loosely ovoid; rachis (3–) 5–13 mm long, ciliate-scabrellate on the angles. Spikelets (4–) 20–35 (–50), 6–14 mm long, 2.4–3.5 mm wide, oblong-

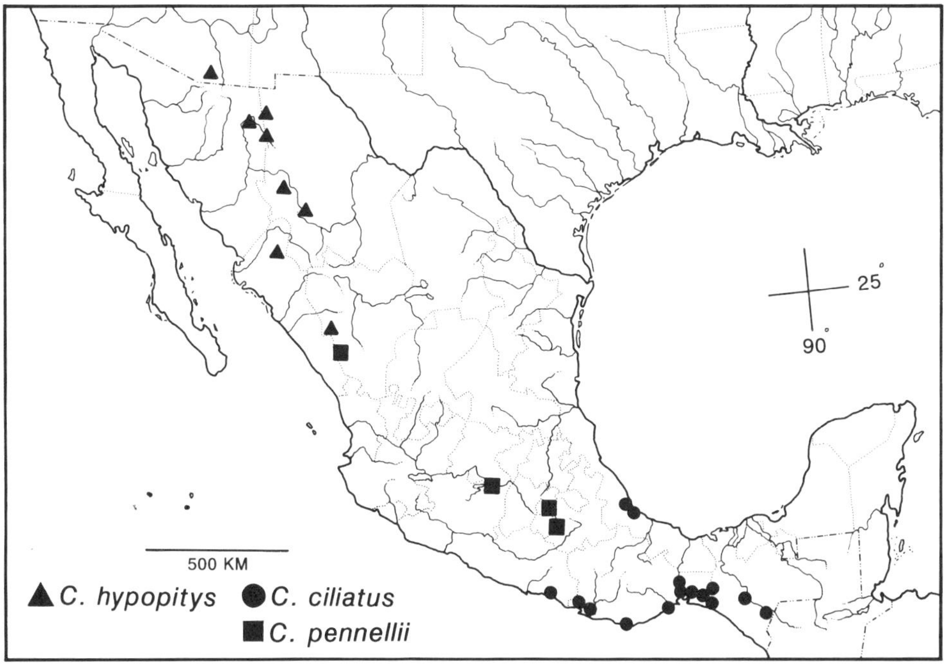

FIG. 53. Distribution of *Cyperus ciliatus*, *C. hypopitys*, and *C. pennellii*.

lanceolate, compressed-quadrate; rachilla deciduous, ca. 0.2 mm wide, 0.1 mm thick, slightly flexuous, stramineous to greenish, successive scales 2.0–2.6 mm apart, wings 0.3–0.4 mm wide, hyaline. Scales deciduous, (4–) 5–8 (–12), (3.8–) 4.0–4.5 (–4.7) mm long, 2.2–2.8 mm wide, ovate-lanceolate, apex acute, mucronulate, laterally 3–4-nerved, stramineous to bronze-colored, medially 3–5-nerved, ciliate-scabrellate along the midvein, green. Stamens 3; filaments 4.0–4.7 mm long; anthers 0.6–0.8 mm long, linear, the connective apex not prolonged. Style 0.8–1.2 (–1.6) mm long; stigmas 3, 1.6–2.6 mm long. Achenes 1.8–2.3 mm long, 1.0–1.2 mm wide, trigonous, ovoid to very broadly ellipsoid, apex broadly obtuse, apiculate, base cuneate to stipitate, 0.1–0.2 mm long, (0.2–) 0.3–0.4 mm wide, the faces flat or slightly concave, the surface finely papillose, dull brown to reddish brown.

Phenology. Flowering and fruiting from June to September.

Distribution. Mexico (Fig. 53), in the lowlands from southern Veracruz and southern Guerrero to central and southern Chiapas, and also in El Salvador; pastures, grassy ditches, sandy disturbed roadsides, and openings in thorny scrub savannas; sea level to 200 (–1000) m.

REPRESENTATIVE SPECIMENS. **Mexico.** CHIAPAS: Mpio. Acala, 27 km N of Acala, *Breedlove 36889* (CAS); 47 km N of Arriaga, *Davidse 9552* (BM, MO); ca. 5 km S of Arriaga, *Tucker 2712* (DUKE, ENCB); 7 km NW of Arriaga, *Tucker 2719* (DUKE, ENCB); 3.7 mi SE of Arriaga, *Wilbur 35953* (DUKE, ENCB); 3.5 mi NW of Arriaga, *Wilbur 35940* (DUKE, ENCB).—GUERERRO: along Hwy 200, ca. 6 km E of Puerto Copala, *Tucker 2929* (DUKE, ENCB).—OAXACA: 17 km NW of La Ventosa, *King 588* (MICH); 7 km E of Salina Cruz, *King 1286* (MICH, UC, US); 9 rd mi W of Zanatepec, *Morley 689* (UC); Trans-Ithmian hwy, ca. 35 km N of Juchitán, *Tucker 2202* (DUKE); ca. 26 km W of Arriaga, *Tucker 2724* (DUKE, ENCB, MICH); Tepanatepec, *Tucker 2735* (DUKE, ENCB); 30 km E of

Niltepec, *Tucker 2742* (DUKE, ENCB); 13 km E of Niltepec, *Tucker 2747* (DUKE, ENCB); near Santa Elena ca. 30 km E of Puerto Escondido, *Tucker 2900* (DUKE, ENCB, MICH); 2 km NW of Pinotepa Nacional, *Tucker 2913* (DUKE, ENCB); ca. 15 km W of Pinotepa Nacional, *Tucker 2917* (DUKE, ENCB); 18 mi NW of Tepanatepec, *Wilbur 35931* (DUKE, ENCB, MICH).—VERACRUZ: 10 km N of La Piedra on Hwy 180, SE of Veracruz, *Tucker 2491* (CAS, DUKE, ENCB, F, GH, MICH).

Cyperus ciliatus is distinguished by its ciliate-scabrellate leaves, culms, and rachis. It superficially resembles both *C. compressus* and *C. wilburii*, but both these species have persistent spikelets and glabrous leaves and culms. *Cyperus dentoniae* and *C. ciliatus* are the only Mexican species in which the rachis is scabrellate. The former species is easily distinguished from *C. ciliatus* by its cylindric spikes and subterete spikelets with persistent scales.

Cyperus ciliatus was one of the most poorly known of the Mexican species of *Cyperus*. Kükenthal (1935–36) saw only an isotype; Benedict Ayers (1946) saw no specimens and excluded *C. ciliatus* from his treatment. I located five modern collections of this species and made 20 collections in 1983 and 1984. Study of these leads me to agree with Kükenthal's placement of this species following *C. manimae*. The narrow, quadrangular, deciduous spikelets and closely appressed, deciduous scales indicate such a placement. *Cyperus ciliatus* resembles *C. hypopitys* in having scabrellate culms, winged rachillas, and ovate-lanceolate scales. The two species are allopatric and differ conspicuously in scale dimensions and color.

61. Cyperus hypopitys G. C. Tucker, sp. nov.—TYPE: MEXICO. Sinaloa: Ocurahui, 27–30 Aug 1941, *H. S. Gentry 6175* (holotype: MICH!; isotypes: DS! GH! LCU!).

Culmi triquetri, in angulis scabrosi, approximati, basibus incrassatis, bulbiformibus. Radii simplices; spicae laxae, globosae vel subglobosae. Spiculae lineares, compressae, 8–18 mm longae, (2.4–) 2.8–4 mm latae; rachilla conspicue alata. Squamae oblongo-lanceolatae, rufae, 2.8–3.3 mm longae, (1.4–) 1.6–1.8 (–2) mm latae, 5- vel 7-nervosae, viridi-carinatae. Achenia trigona, oblongo-ellipsoidea, 1.6–2 mm longa, 0.9–1 mm lata, papillosa, brunnea. Affinis Cypero spectabili Link.

Perennial, (20–) 30–80 (–90) cm tall, tending to form small clumps by basal offshoots. Rhizomes 1–3 cm long, 2–5 mm in diameter, horizontal, covered with fibrous scale remnants. Culms 0.7–1.2 mm in diameter, trigonous below, triquetrous above, smooth below, scabrous on the angles above, bulbous-thickened and clothed with persistent reddish fibers at the base. Leaves 3–5, the blades 20–45 cm long, 2–4 mm wide, flat to v-shaped, the margins and keel scabrellate. Inflorescence bracts (2–) 3, 1–15 (–21) cm long, 0.4–3.5 mm wide, v-shaped to inversely w-shaped, the keel and margins scabrous, borne approximately horizontally. Rays (1–) 2–4, 1–6 cm long; secondary rays rarely present, up to 2 cm long. Spikes (1.5–) 2.5–7 cm long, (1–) 1.5–3.5 cm wide, loosely ovoid to ovoid-ellipsoid; rachis (4–) 10–30 mm long. Spikelets (3–) 8–26, 8–18 mm long, (2.4–) 2.8–4.0 mm wide, linear, compressed-quadrate; rachilla deciduous, ca. 0.4 mm wide, 0.1–0.2 mm thick, quadrate, light reddish to yellowish brown, successive scales (2.4–) 2.6–3.0 mm apart, wings 0.3–0.4 mm wide, translucent whitish, often reddish tinged. Scales deciduous, 6–16, (2.6–) 2.8–3.3 mm long, (1.4–) 1.6–1.8 (–2.0) mm wide, ovate-lanceolate, apex mucronulate, tipped with a tuft of crystalline prickles, laterally

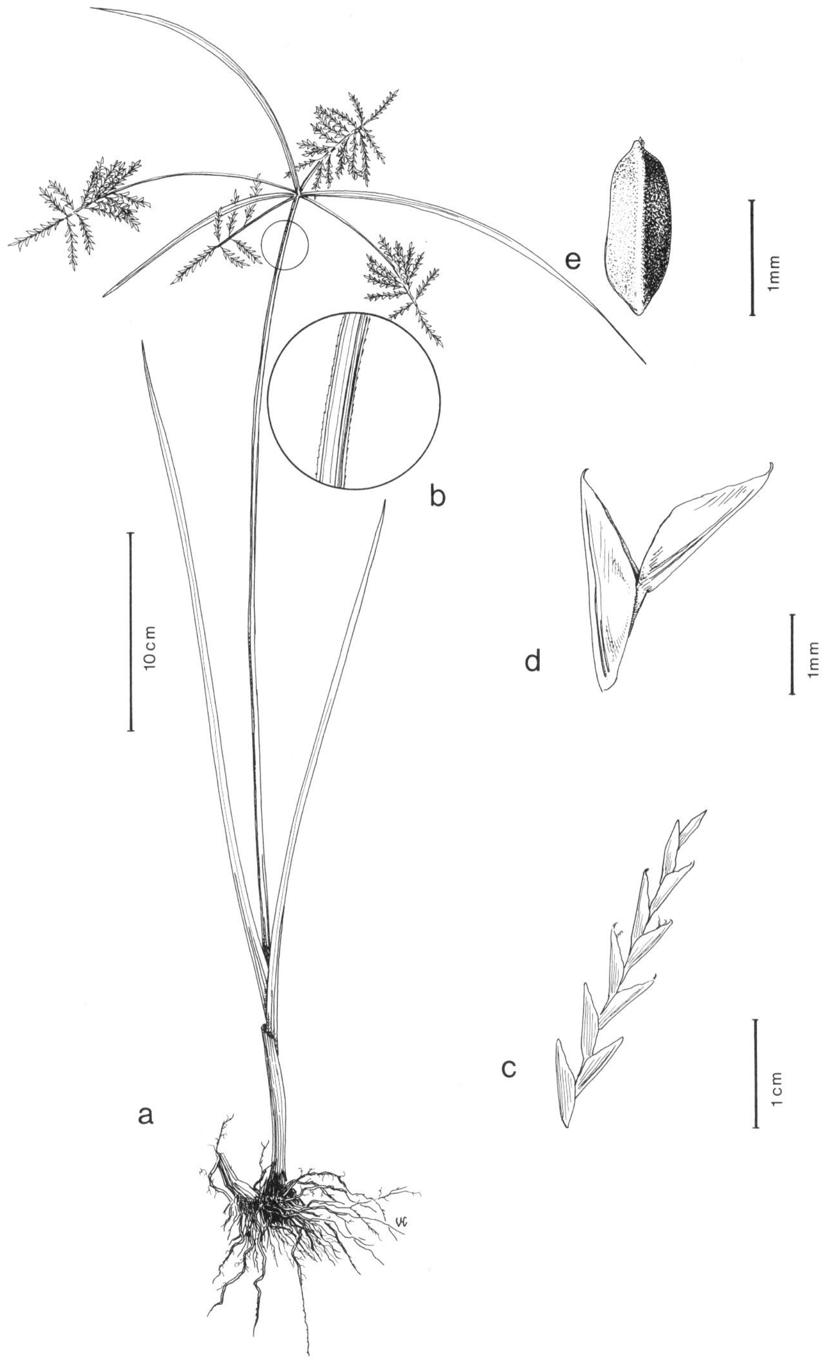

FIG. 54. *Cyperus hypopitys*. a. Habit. b. Portion of culm showing scabrous edges (×5). c. Spikelet, showing spreading, broadly lanceolate scales. d. Two scales, attached to rachilla segment. e. Achene, abaxial view. Drawn by Virginia Hoyt Cantarella.

strongly 3 (–4) -nerved, reddish to reddish brown, medially 1–3-nerved, green. Stamens 3; filaments 2.8–3.4 mm long; anthers 0.8–1.1 mm long, linear-ellipsoid, the connective apex not prolonged. Style 1.0–1.4 mm long; stigmas 3, 1.0–1.4 mm long. Achenes 1.6–2.0 mm long, (0.7–) 0.9–1.0 mm wide, trigonous but slightly dorsiventrally compressed, ellipsoid, apex obtuse, slightly apiculate, base substipitate, 0.1–0.2 mm long, whitish, the faces slightly concave, the surface papillose, brownish. Fig. 54.

Phenology. Flowering and fruiting from August through October.

Distribution. Southern Arizona, U.S.A., and northwestern Mexico (Fig. 53), along the Sierra Madre Occidental from northeastern Sonora to southwestern Durango; moist slopes, canyons, swales, and disturbed ground in pine forests; 1700–2700 m.

ADDITIONAL SPECIMENS EXAMINED. **U.S.A.** ARIZONA: Santa Cruz County, Madera Canyon, Santa Rita Mountains, 5500 ft, 7 Sep 1988, *Bertelsen s.n.* (ARIZ, NYS). **Mexico.** DURANGO: Laguna del Progresso, 10 Aug 1955, *Maysilles 8408* (MICH); Quebrada de San Juan, 10 Aug 1955, *Maysilles 8356* (MICH).—CHIHUAHUA: "Sierra Madre," 11 Oct 1888, *Pringle s.n.* (ASU, CAS, MSC, SMU, VT, YU); Chuhuichupa, Aug–Sep 1936, *LeSueur 1107* (F); Mojarachic, 5 Sep 1938, *Knobloch 5474* (F, MSC); Sierra Madre Occidental, 82 mi (by rd) W of Vieja Casas Grandes, 25 Aug 1952, *J. M. Tucker 2523* (ARIZ, DAV, RSA); Mpio. Batopilas, vic. La Bufa, Nov 1973, *Bye 5934* (UCR); Cascada Basaseachic, 26–27 Aug 1986, *Holm s.n.* (ARIZ 302749).—SINALOA: Ocurahui, Sierra Surotato, 27–30 Aug 1941, *Gentry 6175* (DS, GH, LCU, MICH, PH).—SONORA: Río de Bavispe region, between Las Tierritas and El Tigre, 22 Aug 1940, *Phillips 676* (MICH).

Cyperus hypopitys is morphologically most similar to *C. spectabilis*, as indicated by the fibrous remnants of the leaf sheaths at the bases of the culms, the compressed-quadrate spikelets with 3 (–4) nerves per side, and the size and shape of the achenes. The new species is strikingly similar in habit and scale color to *C. spectabilis*, from which *C. hypopitys* is quickly separated by its winged rachillas. The epithet "hypopitys" (Greek, under pine) denotes the occurrence of this species in pine forests.

62. Cyperus spectabilis Link, Hort. berol. 1: 318. 1827.—TYPE: based on a cultivated plant from the Berlin Botanical Garden (holotype: not located at B; probable isotype: BR!).

Cyperus scaberrimus Nees, Linnaea 19: 697. 1847. *Cyperus spectabilis* var. *scaberrimus* (Nees) Boeckeler, Linnaea 35: 606. 1868.—TYPE: MEXICO. *Aschenborn 683* (holotype: B, destroyed; isotype: B! WRSL!).

Cyperus spectabilis var. *filiformis* Boeckeler, Linnaea 35: 607. 1868.—TYPE: "C. dissitiflorus Hort. Vratislav. et Neesii herb." (not located).

Cyperus buckleyi Britton ex Coulter, Contr. U.S. Natl. Herb. 2: 461. 1894.—TYPE: U.S.A. Texas, *Buckley s.n.* (lectotype, designated by Tucker & McVaugh, 1993: NY!; isolectotype: GH!).

Cyperus parryi Britton ex C. B. Clarke, Bull. Misc. Inform. Kew, add. ser. 8: 9. 1908. *Cyperus spectabilis* var. *parryi* (C. B. Clarke) Kükenthal, Pflanzenreich 4(20): 462. 1936.—TYPE: MEXICO. San Luis Potosí, *Parry & Palmer 906* (lectotype, here designated: NY!; isolectotypes: GH! US!).

Cyperus mohrii Britton ex C. B. Clarke, Bull. Misc. Inform. Kew, add. ser. 8: 10. 1908.—TYPE: MEXICO. [Veracruz:] Orizaba, *Baker 33*, herb. Mohr (holotype: NY, not located).

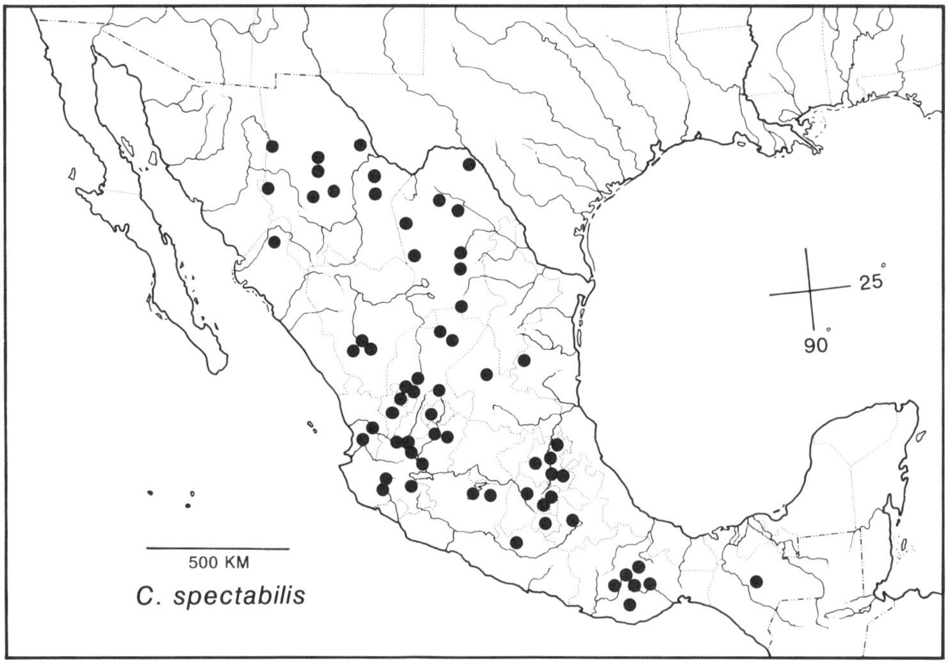

FIG. 55. Distribution of *Cyperus spectabilis* in Mexico.

Perennial, (12–) 30–80 (–90) cm tall, tending to form small clumps by basal offshoots. Rhizomes 1–5 cm long, 2–5 mm in diameter, covered with fibrous brown scale remnants. Culms 0.7–1.7 mm in diameter, trigonous, smooth or essentially so. Leaves 3–5, 20–45 cm long, 2–4 mm wide, flat to v-shaped, the margins and keel scabrellate. Inflorescence bracts 2–4, 1–15 (–24) cm long, 0.4–3.5 mm wide, v-shaped, the keel and margins scabrous, borne approximately horizontally. Rays 1–4, 1–15 cm long; secondary rays absent. Spikes (1.5–) 2.5–7.0 cm long, (1.0–) 1.5–3.5 cm wide, loosely ovoid to ovoid-ellipsoid; rachis (4–) 10–30 mm long. Spikelets (3–) 8–40, 8–18 mm long, (2.4–) 2.8–4 mm wide, linear, compressed-quadrate; rachilla ± deciduous, ca. 0.4 mm wide, 0.1–0.2 mm thick, flat, greenish, whitish or stramineous, successive scales (2.0–) 2.4–3.0 mm apart, wings absent or essentially so. Scales deciduous, (2–) 4–12, 1.6–2.6 mm long, (1.6–) 1.8–2.4 (–2.7) mm wide, oblong-ovate, apex obtuse or sometimes mucronulate, laterally weakly 2–3-nerved, light brown to reddish brown, medially 1 (–3) -nerved, green or light brown. Stamens 3; filaments 1.6–2.8 mm long; anthers 0.4–0.6 mm long, linear-ellipsoid, the connective apex not prolonged. Style 0.3–0.6 mm long; stigmas 3, 0.6–1.0 mm long. Achenes 1.7–2.2 mm long, 0.8–1.1 mm wide, trigonous, ellipsoid, apex obtuse, slightly apiculate, base substipitate, 0.1–0.2 mm long, whitish, the faces slightly concave, the surface papillose, dark brown to black. Fig. 47D.

Phenology. Flowering and fruiting from early July to early September or early November.

Distribution. Oklahoma to Arizona, U.S.A., south to Chiapas, Mexico (Fig. 55); also in the mountains of Bolivia and northwestern Argentina; pine and pine oak woods, pinyon-juniper woodlands, grasslands with cacti, chaparral, damp arroyos in semidesert scrub, and sometimes a weed of croplands; (1000–) 1500–2000 (–2400) m.

REPRESENTATIVE SPECIMENS. **Mexico.** AGUASCALIENTES: 4 km W of Asientos, *Rzedowski 25077* (ENCB, MICH).—CHIAPAS: San Cristóbal, *Breedlove 37964* (DS).—CHIHUAHUA: vic. Pirámide, *I. M. Johston 8130A* (GH); near Chihuahua, *Pringle 311* (F, GH, MICH, MO, PENN, PH, RSA); Sierra del Pino, vic. La Noria, *Johnston & Muller 455* (F, GH, LL).—COAHUILA: 16 km S of Orteaga, *Kenoyer & Crum 2738* (A); Mpio. Castaños, Cañón Bocatoche, *Müller 3134* (GH, LCU, LL, MICH, UC).—DISTRITO FEDERAL: près Santa Fe, *Bourgeau 431* (C, GH); near Mexico City, *Pringle 13233* (ARIZ, C, CU, F, GH, K, MICH).—DURANGO: Durango and vicinity, *Palmer 377* in 1896 (C, F, G, GH, K, MO, UC, US); Mpio. Santiago Papasquiaro, 3.5 km W of La Soledad, *Díaz 468* (DUKE, UTEP).—GUANAJUATO: between Guanajuato and Santa Rosa, *Moore 1356* (BH, GH).—GUERRERO: Dtto. Mina, Zihuaqueo, *Hinton 9323* (GH, K, LCU, US).—HIDALGO: near Trinidad Iron Works, *Pringle 8921* (ARIZ, ASU, CAS, DAV, DUKE, F, GH, LL, MSC, SD, TEX, WIS); Dtto. Zacualtipán, 3 mi from Zaculatipán on rd to Tianquistengo, *Moore 3321* (BH, GH, UC); Mt. Ixmiquilpan, *Purpus 5066* (F, GH, MO, UC).—JALISCO: 4.4 mi N of Guadalajara, *Denton 2090* (MICH); 10 km WSW of Cd. Guzmán, *Iltis et al. 576* (IBUG, WIS); 2 mi NW of Tequila, *McVaugh 18654* (DUKE, MICH); 54 km N of Guadalajara, Carr. Saltillo, *Puga 7599* (IBUG).—MÉXICO: Río Hondo Canyon, *Pringle 3159* (ARIZ, ASU, C, DUKE, F, GH, LL, MICH, MSC, TEX, UC, Z); hills near Lecheria, *Pringle 13234* (C, CU, F, GH, K, MICH, MSC).—MICHOACÁN: vic. Morelia, *Arsène 3169* (GH), *5122* (F, GH, K, MO), *5239* (GH), *9860* (GH, MO); 1 km E of Tuxpan, *Iltis & Doebley 37* (WIS).—MORELOS: Parque Nac. Laguna Zempoala, *Wyatt 57* (MICH).—NAYARIT: 9 mi W of Compostela, *McVaugh 16561* (MICH, TEX); 18 mi SE of Tepic, *Stevens & Fairhurst 1802* (DUKE, MICH, MO, MSC).—NUEVO LEÓN: Mpio. Galeana, Cerro Potosí, *Mex. Biol. Exped. [Schneider] 987* (ARIZ, F, GH, LCU, MICH, MO).—OAXACA: Portillo San Dionisio, *Tucker 2781* (CAS, DUKE, ENCB, MICH); 30 km S of San José del Pacífico, *Tucker 2884* (DUKE, ENCB).—PUEBLA: vic. de Puebla, *Arsène 1430* (MO).—SAN LUIS POTOSÍ: Charcas, *Lundell 5094* (GH, LCU, LL, MICH).—SINALOA: Ocurahui, Sierra Surotato, *Gentry 6205* (ARIZ, GH, LCU, MICH).—TAMAULIPAS: 3 mi N of Miquihuana, *Stanford et al. 2448* (DS).—ZACATECAS: 7 mi SE of Pinos, *I. M. Johnston 7488* (GH); 29 mi SW of Fresnillo, *Anderson & Laskowski 3594B* (MICH).

Kükenthal (1935–36) and Benedict Ayers (1946) both accepted *C. spectabilis*, which they distinguished from *C. manimae* only by its lax spikes usually borne on rays; *C. manimae* was described as having dense spikes and often lacking rays. According to Kükenthal, the geographic range of *C. spectabilis* is very nearly the same as that of *C. manimae*. In an earlier work (Tucker 1984), I reluctantly combined *C. spectabilis* with *C. manimae*, because I could not find any quantitative characters to distinguish the two species. McVaugh (pers. comm.) encouraged me to reinvestigate the matter. I have done so following his suggestions and find that the density of spikelets can be quantified as follows: in *C. spectabilis* the spikelets average 12–16 per cm of rachis [range: 3–28 per cm of rachis]; in *C. manimae* the spikelets average 50–65 per cm of rachis [range: 36–70 per cm of rachis]. In addition, McVaugh noted and I have confirmed that *C. spectabilis* has broader scales than *C. manimae*.

63. Cyperus calderoniae S. González Elizondo, Phytologia 57: 384. 1985, ('calderonii').—TYPE: MEXICO. Hidalgo: Mpio. Epazoyucan, 1 km N of Nopalillo, 2750 m, *Medina C. 1461* (holotype: ENCB, not seen but examined by Rogers McVaugh).

Perennial, (1–) 5–20 (–35) cm tall, tending to form small clumps by basal offshoots. Rhizomes 1–2 cm long, 2–5 mm in diameter, covered with fibrous brown scale remnants. Culms 0.7–1.7 mm in diameter, trigonous, smooth or essentially so. Leaves 3–5, (6–) 15–20 cm long, 1.5–3.5 mm wide, flat to v-shaped, the margins and keel scabrellate. Inflorescence bracts 2–3 (–4), 1–6 (–10) cm long, 0.4–3.0 mm wide, v-shaped, the keel and margins scabrous, borne approximately horizontally. Rays 1–3, 1–4 cm long, or absent and the inflorescence a sessile terminal spike; secondary rays absent. Spikes 1.5–3.0 cm long, 1.0–2.0 cm wide, rather loosely ovoid-ellipsoid; rachis (4–) 10–20 mm long. Spikelets (6–) 10–40, 3.5–6.0 (–9.0) mm long, 2.0–2.5 mm wide, linear, compressed-quadrate; rachilla ± deciduous, ca. 0.4 mm wide, 0.1–0.2 mm thick, flat, greenish, whitish or stramineous, successive scales 1.4–1.8 mm apart, wings absent or essentially so. Scales deciduous, 4–7, 1.7–2.0 mm long, 1.6–2.0 mm wide, ovate, apex obtuse, muticous or sometimes minutely mucronulate, laterally weakly 3–5-nerved, whitish, speckled with reddish brown, medially 3–5-nerved, green or greenish white. Stamens 3; filaments 1.6–2.0 mm long, whitish; anthers ca. 0.3 mm long, ellipsoid, the connective apex not prolonged. Style 0.3–0.6 mm long; stigmas 3, 0.5–0.8 mm long. Achenes 1.3–1.5 mm long, 0.8–1.0 mm wide, trigonous, broadly ellipsoid, apex obtuse, slightly apiculate, base substipitate, ca. 0.1 mm long, whitish, the faces slightly concave, the surface papillose, light yellowish brown.

Phenology. Flowering and fruiting from July to August.

Distribution. Endemic to Mexico, in western San Luis Potosí, Hidalgo, Querétaro, northwestern Guanajuato, and northeastern Jalisco; grassy slopes, desert flats, brushy pastures, and openings in pine forests; 1800–2900 m.

ADDITIONAL SPECIMENS EXAMINED. **Mexico**. GUANAJUATO: 16 km N of San Felipe, rd to Ocampo, *Rzedowski 43684* (IEB).—HIDALGO: Mpio. Epazoyucan, 1 km N of Nopalillo, *Medina C. 1461* (ENCB).—JALISCO: Paso de la Troje, km 36 SW of Ojuelos, *McVaugh 16767* (MICH); 3 km E of Ojuelos, *Rzedowski 16158* (ENCB).—SAN LUIS POTOSÍ: 30 mi S of San Luis Potosí, *Reeder & Reeder 2958* (NY); 28.1 km S of San Luis Potosí, 3 km S of jct. to Villa de Reyes, *Reeves 6284* (ASU).—ZACATECAS: Mpio. Concepción del Oro, Sta. Rosa, Cañada El Astillero, *Robert & Passini 63-5874* (ANSM).

Cyperus calderoniae was compared by its author to *C. spectabilis* and *C. sphaerolepis*. Although there are similarities in the dimensions of rachilla and scales, *C. sphaerolepis* is quickly distinguished by its erect inflorescence bracts and rays and mucronate scales. *Cyperus calderoniae* differs from *C. spectabilis* in the narrower spikelets, shorter rachilla internodes, scales, anthers, and achenes.

I have seen few specimens of *C. calderoniae* and think it is genuinely rare. The type at ENCB was unavailable for loan because of recent earthquake damage to the building, but it and the specimen cited from IEB were examined by Rogers McVaugh, whose generous assistance is much appreciated.

64. Cyperus pennellii O'Neill & Benedict Ayers, Leafl. W. Bot. 4: 34. 1944.—TYPE: MEXICO. Durango: Remudo between El Salto and Cueva, 2850–2900 m, 30 Aug 1934, *Pennell 18470* (holotype: PH!).

Dwarf perennial, 0.5–3.0 (–10) cm tall. Rhizomes 0.5–1.5 cm long, 2–4 mm in diameter, covered with brown scales or their fibrous remnants. Culms single, 0.8–1.4 mm in diameter, trigonous, smooth, bulbous-thickened basally. Leaves 2–5, (1.0–) 2.0–5.5 cm long, 1.0–4.0 (–5.0) mm wide, v-shaped, the margins and keel scabrellate. Inflorescence bracts (2–) 3–4, (1–) 2–15 cm long, (0.5–) 1–3.5 mm wide, v-shaped, the margins and keel scabrellate, borne approximately horizontally. Rays absent, the inflorescence a loose hemispherical cluster of 1–3 sessile spikes. Spikes 12–18 mm long, (12–) 16–24 mm wide, loosely ovoid; rachis (3–) 5–12 mm long, smooth. Spikelets (5–) 20–35, (4–) 7–10 mm long, 1.8–2.6 mm wide, narrowly ellipsoid to linear-oblong, strongly compressed, nearly flat; rachilla ± deciduous, ca. 0.4 mm wide, 0.1–0.2 mm thick, zigzag, greenish to light brown, successive scales 1.2–1.8 mm apart, wings absent. Scales deciduous, (5–) 7–9 (–12), 2.6–3.2 mm long, 1.5–2.2 mm wide, oblong-elliptic, apex obtuse, laterally 2–3-nerved, light brown, medially 3-nerved, greenish or greenish brown. Stamens 3; filaments 2.4–4.2 mm long; anthers 0.4–0.6 mm long, narrowly ellipsoid to linear, the connective apex not prolonged. Style 0.8–1.3 mm long; stigmas 3, 0.8–1.5 mm long. Achenes 1.7–2.1 mm long, 0.7–0.9 mm wide, trigonous, ellipsoid, the apex broadly rounded to acute, base sessile, the faces flat, the surface puncticulate or essentially smooth, brown or reddish brown.

Phenology. Flowering and fruiting in July, August, and September.

Distribution. Mexico and Peru. In Mexico (Fig. 53) in the region of Mexico City, in northeastern Michoacán, and in southwestern Durango; rare or rarely collected in clearings in pine and pine-fir forests; ca. 2500–3230 m.

ADDITIONAL SPECIMENS EXAMINED. **Mexico**. DURANGO: Remudo, 30 Aug 1934, *Pennell 18470* (PH).—MÉXICO: 55 km SE of Mexico City, 13 Jul 1942, *Weaver 778* (NY); Mpio. Ixtapaluca, 8 km S de Río Frio, 5 Aug 1975, *Koch 75383* (CHAPA), 11 Aug 1978, *Vega Aviña 402* (CHAPA).—MICHOACÁN: 12 km due N of Cd. Hidalgo, at SW side of Cerro San Andrés, *Beaman 4355* (MSC, UC).

Cyperus pennellii is distinguished from other species of subg. *Cyperus* by its extremely short culm. It shares with the much more common montane species *C. spectabilis* deciduous scales and rachillas, open spikes (10–15 spikelets per centimeter of rachis), elliptic to linear spikelets, oblong elliptic scales, and trigonous ellipsoid achenes. *Cyperus pennellii* differs from *C. spectabilis* by its very short culms, bulbous-thickened base, and its nearly flat spikelets. The only other montane species with such short culms is *C. seslerioides*, readily distinguished by its capitate, whitish inflorescences.

Benedict Ayers (1946) knew of this species from its type locality only. The additional collections cited here suggest that *C. pennellii* might be more widespread at higher elevations in central Mexico as well as Durango. *Cyperus pennellii*, thought to be endemic to Mexico, is now also known from Peru: Depto. Puno, Juliaca, *Stafford 471* (F).

65. Cyperus manimae H. B. K., Nov. gen. sp. 1 [quarto]: 209. 1816. *Mariscus manimae* (H. B. K.) C. B. Clarke, Contr. U.S. Natl. Herb. 10: 452. 1908. *Cyperus spectabilis* var. *coarctatus* Boeckeler, Linnaea 35: 607. 1868.—TYPE: VENEZUELA. "Orinoco River at Atabapo," *Humboldt & Bonpland s.n.* (holotype: P; isotype: B-W no. 1392, IDC microfiche!). [See note in Tucker & McVaugh, 1993.]

Perennial, (4–) 30–70 (–95) cm tall. Rhizomes 1–10 cm long, 2–5 mm in diameter, densely covered with reddish brown scales or their fibrous remnants, producing single culms 1–3 (–10) cm apart. Culms 0.5–2.3 mm in diameter, trigonous, densely scabrous or hirtellate (or sometimes smooth), either just below the apex or over the upper half of the culm. Leaves 3–5 (–8), 7–30 (–45) cm long, 1.5–3.5 (–6) mm wide, flat to v-shaped, often becoming involute in drying, smooth except for scattered prickles on the keel and margins in the apical third, or sometimes both the upper and lower surfaces scabrous throughout. Inflorescence bracts (1–) 3–5 (–10), 3–12 (–36) cm long, 0.4–3.0 (–5.0) mm wide, v-shaped, the margins and keel and often both surfaces scabrous, borne approximately horizontally to reflexed downward parallel to the culm. Rays often absent, the inflorescence of 1–3 (–4) sessile spikes, occasionally so densely congested that the individual spikes are not discernible; or rays 2–5 (–10), 1–6 (–16) cm long, smooth or scabrous like the culm; secondary rays absent but rarely spikes produced from the leaf sheaths on peduncles up to 15 cm long. Spikes (15–) 20–35 (–50) mm long, 10–30 (–40) mm wide, densely ovoid to oblong-ovoid; rachis 8–20 (–25) mm long. Spikelets 25–50 (–80), 4–12 (–22) mm long, 2–3 (–4) mm wide, elliptic to linear, compressed-quadrate; rachilla deciduous, 0.4–0.5 (–0.6) mm wide, 0.1–0.2 mm thick, flat, greenish to stramineous, sometimes reddish speckled, successive scales 1.2–1.8 (–2.0) mm apart, essentially wings absent or with inconspicuous hyaline wings 0.1 (–0.2) mm wide. Scales deciduous, (2–) 3–10 (–14), (1.5–) 1.8–2.4 (–2.7) mm long, (1.2–) 1.4–2.0 (–2.5) mm wide, oblong-elliptic to broadly lanceolate, apex blunt to mucronulate, the mucro 0.1 (–0.3) mm long, laterally 3 (–4) -nerved, brownish to hyaline distally, often erose apically, or light brown, reddish brown or deep red throughout, marginally hyaline, medially 1–3-nerved, smooth or rarely scabrellate, green and often also reddish striate. Stamens 3; filaments 1.8–2.5 mm long; anthers 0.4–0.6 mm long, oblong-ellipsoid, the connective apex not prolonged. Style (0.2–) 0.4–0.8 mm long; stigmas 3, 0.7–1.1 mm long. Achenes 1.3–1.8 (–2.2) mm long, (0.6–) 0.7–0.9 (–1.1) mm wide, trigonous, oblong-ellipsoid, apex obtuse, scarcely to distinctly apiculate, base substipitate to stipitate, 0.1–0.2 mm long, the faces concave, the surface finely to distinctly papillose, brown to reddish brown.

Phenology. Flowering and fruiting from July to October.

Distribution. Western Texas (Big Bend region), southern Arizona, U.S.A.; in Mexico in northeastern Sonora and southern Baja California Sur, south through the Sierra Madre and the High Plateau to Guatemala; mountains of Honduras and Nicaragua, Meseta Central of Costa Rica, Volcán Chiriquí in western Panama, and northwestern South America; pine, oak-pine, and fir forests, grasslands, pastures, semidesert scrub, canyons, roadsides, and disturbed areas; 1500–2700 (–2900) m (a few collections from 700–1000 m in Nayarit).

Infraspecific variation in *C. manimae* has confounded many taxonomists (Kükenthal 1935–36; Benedict Ayers 1946; Tucker, 1984). Kükenthal recognized four varieties in *C. manimae*, which were also accepted by Benedict Ayers, who thought there was "bewildering" intergradation among the varieties, particularly between the red and brown-scaled taxa. Representatives of var. *manimae* have smooth culms and blunt brown scales. According to Benedict Ayers, the other varieties are distinguished as follows: var. *asperrimus*, culms rough; var. *divergens*, scales mucronate; and var. *phaeocephalus*, scales red. None of the last three varieties, possessing supposedly distinguishing features, has a distinct range—all occur more or less throughout the range of the species.

Benedict Ayers did not state the mucro length of the scales of var. *divergens*. In specimens cited by him the mucros exceed 0.3 mm. The lengths of mucros in collections studied show a continuous distribution from less than 0.1 mm to ca. 0.7 mm, if one measures all the scales of the spikelet together as I did in an earlier study (Tucker 1984). I erred by comparing apples with oranges. In *C. manimae,* as in other species of this subgenus, the distal scales of a spikelet have longer mucros or cusps than the basal scales. Thus, I redid the measurements used for the plot of mucro length (Tucker 1984, fig. 1) and found that if one compares the two distal scales of the spikelet, two distinct classes are evident: 0–0.2 (–0.3) mm corresponding to var. *manimae* and var. *asperrimus,* and (0.5–) 0.7–1.1 mm corresponding to var. *divergens,* as accepted by Kükenthal and Benedict Ayers.

Brown- and red-scaled plants can be found within any of the varieties of *C. manimae*; there is no evidence to support recognition of a red-scaled variety, var. *phaeocephalus*.

A very few collections of *C. manimae* are intermediate among these three varieties here recognized: Culms rough and scales cuspidate (intermediate between var. *asperrimus* and var. *divergens*); Jalisco: 2 km S of Ixtlahuacán, *Vázquez G. 317* (IBUG). Culms roughened on angles only or just below inflorescence bracts (intermediate between var. *asperrimus* and var. *manimae*); Distrito Federal: San Gregorio, *Matuda 21179* (LL), Pedregal de Tlalpan, *Matuda 21378* (LL), La Cima, *Martínez s.n.*, 19 Aug 1964 (DS, MICH, MSC), Ajusco, *Fisher s.n.*, 18 Aug 1926 (DS, MICH, WVA); Guatemala: *Standley 61276*, *61277* (F), *Steyermark 51644* (F). Culms with few scattered prickles all over; mucronate apex of scales 0.2–0.4 mm long (combining features of all three varieties): Honduras, *Molina R. 12832* (F).

Both Kükenthal (1935–36) and Benedict Ayers (1946) recognized the Mexican species *C. apiculatus*. Benedict Ayers distinguished *C. apiculatus* from *C. manimae* by the former species having fewer inflorescence bracts, wider scales, and longer, iridescent achenes. He cited ten collections of *C. apiculatus*, eight of which I studied: Distrito Federal: *Bourgeau 432* (C, E, GH); Guerrero: Manchón, *Hinton 9244* (GH, K, LL, MICH); Jalisco: Guadalajara, *Pringle 3844* (K, LL, MSC), Huejotitán, *Diguet s.n.*, Oct 1912 (K); Michoacán: vic. Morelia, *Arsène 2692a* (GH), *5579* (NY), *5751* (MO), *5844* (GH, MO). Examination of these collections did not confirm that *C. apiculatus* differed from *C. manimae* in any of the characters listed by Benedict Ayers. For the these specimens, the average number of inflorescence bracts per plant was 3; the average for *C. manimae* (based on examination of some 550 collections from throughout its range) was 4. For scales, the average measurements were 2.2 mm long and 1.6 mm wide for "*C. apiculatus*" vs. 2.1 mm long and 1.8 mm wide for *C. manimae*, and for achenes 1.5 mm long and 0.8 mm wide for "*C. apiculatus*" vs. 1.6 mm long and 0.8 mm wide for *C. manimae*.

Several collections cited by Benedict Ayers (for example, *Pringle 3844*) have glossy achenes. Benedict Ayers used the inappropriate word "iridescent." Kükenthal (1935–36) used the term "nitida" to describe the achenes of both *C. manimae* and *C. apiculatus*. This is accurate, although both dull and glossy achenes were observed in *C. manimae*.

To summarize, I found no evidence to support the recognition of *C. apiculatus*. None of the differences used by Benedict Ayers could be confirmed by the large sample of specimens of *C. manimae* examined nor could any other consistent differences be detected. Thus, the type of *C. apiculatus* is included in *C. manimae* var. *divergens*.

Key to the Varieties of Cyperus manimae

1. Culms densely scabrous or hirtellate, especially just below the inflorescence bracts. *Cyperus manimae* var. *asperrimus*.
1. Culms smooth.
   2. Distal scales of spikelets with cusps (0.4–) 0.7–1.1 mm long. *Cyperus manimae* var. *divergens*.
   2. Distal scales of spikelets blunt to mucronulate (mucros 0–0.2 (–0.3) mm long). *Cyperus manimae* var. *manimae*.

65a. Cyperus manimae var. **manimae.**

Cyperus inaequalis Link, Jahrb. Gewächsk. 3: 88. 1820. *Cyperus anisos* Roemer & Schultes, Syst. Veg. Mantissa 2: 131. 1824, nom. superfl.—Type: "America meridionalis." *Humboldt & Bonpland s.n.* (holotype: B-W no. 1392, IDC microfiche!).

Cyperus spectabilis var. *filiformis* Boeckeler, Linnaea 35: 607. 1868.—Type: Mexico. *Schmitz 808* (not located).

Cyperus phaeocephalus Grisebach, Abh. Königl. Ges. Wiss. Göttingen 19: 216. 1874. *Mariscus phaeocephalus* (Grisebach) C. B. Clarke, Bull. Misc. Inform. Kew, add. ser. 8: 16. 1908. *Cyperus manimae* var. *phaeocephalus* (Grisebach) O'Neill & Benedict, Leafl. W. Bot. 4: 36. 1944.—Type: Ecuador. "in Andinibus Ecuadoriensibus," 1857–1859, *Spruce 5904* (holotype: S!, photo: NY!).

Cyperus manimae var. *pseudospectabilis* Kükenthal, Pflanzenreich 4(20): 463. 1936.—Type: Mexico. Puebla: Chinantla, *Liebmann 14460* (lectotype, here designated: C!).

Culms 0.5–2.3 mm in diameter, trigonous, smooth or essentially so. Scales deciduous, (3–) 6–10 (–20), 1.7–2.2 mm long, 1.4–2.0 mm wide, oblong-elliptic, apex blunt to mucronulate (the mucro up to 0.3 mm long), laterally 3 (–4)-nerved, brownish to hyaline distally, often erose apically, or light brown, reddish brown or deep red throughout, marginally hyaline, medially 1–3-nerved, green and often also reddish striate.

Phenology. Flowering and fruiting from July to October.

Distribution. Mountains of Mexico from Durango to Chiapas, Guatemala, and northwestern South America; grassy or rocky slopes in pine or pine-oak woods; (1000–) 1800–2700 m.

Representative Specimens. **Mexico.** Chiapas: Amatenango del Valle, *Ton 1091* (DS, MICH); 50 km E of San Cristóbal on Hwy 190, *Tucker 2675* (DUKE, ENCB).—Distrito Federal: Ajusco, *Fisher s.n.*, 18 Aug 1926 (CAS).—Durango: Mpio. Pueblo Nuevo, 0.5 km E. of Hac. Coyotes, *Clausen 76-26* (CU).—Jalisco: ca. 15 km S of Aguascalientes, E slopes of Cerro de los Gallos, *McVaugh 17091* (NY, TEX).—México: Amecameca, *Kuntze 23662* (NY); 18.2 km W of Toluca, *Lane 2661* (TEX); Toluca Rd near Mexico City, *MacDaniels 507* (BH, F); Río Hondo Canyon, *Pringle 3160* (LL, M, MSC); Mpio. Huehuetoca, Cerro Sincoque, *Zarate F. 98* (IBUG).—Tamaulipas: vic. San José above La Vegonia, *Bartlett 10042* (MICH).—Veracruz: near Orizaba, *Balls B5384* (E, UC).

65b. Cyperus manimae var. **asperrimus** (Liebmann) Kükenthal, Pflanzenreich 4(20): 463. 1936. *Cyperus asperrimus* Liebmann, Mexic. halvgr. 30. 1850.—TYPE: MEXICO. Veracruz, Consoquitla, Aug 1841, *Liebmann 14416* (holotype: C!; isotype: HAL!).

Cyperus triceps Nees, Linnaea 19: 697. 1847.—TYPE: MEXICO. *Aschenborn 123* (holotype: B!; isotype: B, Herb. Bothe!).

Cyperus asperrimus var. *multiflorus* Liebmann, Mexic. halvgr. 31. 1850.—TYPE: MEXICO. Veracruz: Mirador, Oct 1841, *Liebmann 14415* (holotype: C!; isotype: K!).

Culms 0.5–2.3 mm in diameter, trigonous, densely scabrous or hirtellate, either just below the apex or over the upper half of the culm. Scales deciduous, (2–) 3–10 (–14), (1.5–) 1.8–2.4 (–2.7) mm long, (1.2–) 1.4–2.0 (–2.5) mm wide, oblong-elliptic to broadly lanceolate, apex blunt to mucronulate [the mucro 0.1 (–0.3) mm long], laterally 3 (–4) -nerved, brownish to hyaline distally, often erose apically, or light brown, reddish brown or deep red throughout, marginally hyaline, medially 1–3-nerved, smooth or rarely scabrellate, green and often also reddish striate.

Phenology. Flowers and fruits from July to October.

Distribution. Southern Arizona, U.S.A., south through the mountains of Mexico (Fig. 56) and Central America to Volcán Chiriquí, Panama, from Colombia to northern Argentina; openings in pine or pine-oak forests, pastures, seepage slopes, and in croplands as a weed; 1000–2000 (–2500) m.

REPRESENTATIVE SPECIMENS. **Mexico.** BAJA CALIFORNIA SUR: Cape Region, Sierra de San Francisquito, *Brandegee s.n.*, 1 Oct 1899 (UC, with *C. dipsaceus*).—CHIAPAS: NW edge of Teopisca, *Breedlove 10515* (DS, F, MICH); S side of San Cristóbal, *Wilbur 35721* (DUKE, ENCB).—CHIHUAHUA: Mpio. Cuauhtémoc, 12 mi E of San Antonio, *Muller 3366* (GH, LL); La Bufa Mtn above Cusihuiriachic, *Pringle 1392* (F, GH, K, MICH, MSC, NY); "foothills of Sierra Madre," *Pringle 1394* (F, GH, MICH, NY).—COAHUILA: Cañón de Centinela just S and W of Pico de Sentinela, *M. C. Johnston et al. 11978A* (LL).—DISTRITO FEDERAL: Villa de Obregón, *Fisher 35220* (CAS, F, NY, SMU, TEX); El Rosario, *MacDaniels 693* (BH, CU, F).—DURANGO: 5 1/2 mi E of El Salto, *Waterfall & Wallis 13646* (F, GH, TEX).—HIDALGO: Dtto. Jacala, Puerto de la Zorra, *Moore 1699* (GH).—JALISCO: 21 mi SW of Guadalajara, *Denton 2053* (MICH); 7 mi WNW of Tototlán, *McVaugh 17284* (MICH); NW of Poncitlán, *Puga 7689* (IBUG).—MÉXICO: Río Hondo Canyon, *Pringle 3163* (ARIZ, CAS, E, GH, K, LL, M, MSC, NY, PH, UC, Z); Mpio. Juchitepec, Pedregal el Púlpito, *Ventura A. 1450* (WIS).—MICHOACÁN: vic. Morelia, *Arsène 2692a* (GH), *5579* (NY); Dtto. Zitacuaro, Loma Larga, *Hinton 13112* (K, LL, MICH, UC, US).—MORELOS: near Cuernavaca, *Pringle 13237* (C, CU, F, GH, K, MICH, MO).—NAYARIT: 12 mi SE of Tepic, *Feddema 603* (DS, MICH, TEX), *604* (MICH, TEX).—OAXACA: 7 mi N of Oaxaca, *Kenoyer 1664* (GH); 8 km N of San José del Pacífico, *Tucker 2871* (DUKE, ENCB).—PUEBLA: Puente del Emperador, *Sharp 44541* (GH); vic. Puebla, *Arsène 171* (GH).—SAN LUIS POTOSÍ: San Luis Potosí, *Virlet 1178* (F).—SONORA: Río de Bavispe, Cañón de Huépari, N of Aribabi, *White 2640* (ARIZ, GH, MICH).—TAMAULIPAS: ca. 40 km NNW of Aldama, *Dressler 2298* (GH, MICH).—VERACRUZ: Mt. Orizaba, Loma Grande, *Balls 5384* (GH, K); 5.4 mi N of Jalapa on rd to Misantla, *Tucker 2079* (DUKE, ENCB).

65c. Cyperus manimae var. **divergens** (H. B. K.) Kükenthal, Pflanzenreich 4(20): 463. 1936. *Cyperus divergens* H. B. K., Nov. gen. sp. 1 [quarto]: 208. 1816.—TYPE: MEXICO. [Michoacán:] Jorullo, *Humboldt & Bonpland s.n.* (holotype: P!; isotypes: B-W no. 1316, IDC microfiche!).

Cyperus apiculatus Liebmann, Mexic. halvgr. 32. 1850. *Cyperus manimae* var. *apiculatus* (Liebmann) C. D. Adams, Ann. Missouri Bot. Gard. 78: 254.

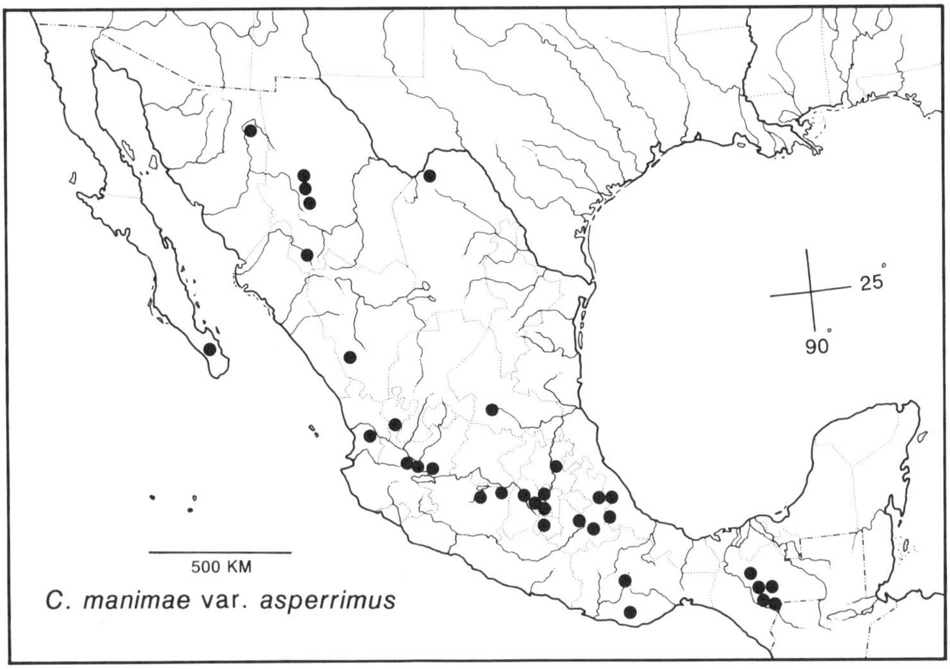

FIG. 56. Distribution of *Cyperus manimae* var. *asperrimus* in Mexico.

1991.—TYPE: MEXICO. Veracruz: San Antonio Huatusco, Aug 1841, *Liebmann 14352* (lectotype, designated by Tucker, 1986: C!; isolectotype: C!).

Cyperus apiculatus var. *latifolius* Kükenthal, Pflanzenreich 4(20): 463. 1936.—SYNTYPES: MEXICO. Monterey, *Abbon s.n.* (not located); Michoacán, Morelia, *Arsène 5640* (not located). [Neither syntype at Berlin; no duplicates found; assignment of name based on Kükenthal's protologue.]

Culms 0.5–2.3 mm in diameter, trigonous, smooth or essentially so. Scales deciduous, (3–) 6–10 (–20), (1.5–) 1.8–2.4 (–2.7) mm long, (1.2–) 1.4–2.2 (–2.5) mm wide, oblong-elliptic, distinctly cuspidate, the ± excurved cusp of the distal scales (0.5–) 0.6–0.8 (–1.1) mm long and bearing a tuft of crystalline prickles at the apex, laterally 3 (–4) -nerved, brownish to hyaline distally, often erose apically, or light brown, reddish brown or deep red throughout, marginally hyaline, medially 1–3-nerved, smooth, green and often also reddish striate.

Phenology. Flowering and fruiting from mid-July to early October.

Distribution. Northwestern Mexico (Fig. 57) south to the mountains of Costa Rica; openings in pine or pine-oak forests, pastures, seepage slopes, sand dunes, and in croplands as a weed; 1000–2000 (–2500) m.

REPRESENTATIVE SPECIMENS. **Mexico.** CHIAPAS: on Hwy 195 at the turnout to Waltic, *Tucker 2657* (DUKE, ENCB).—CHIHUAHUA: Mpio. Temosachic, 10 mi S of Madera, *Muller 3462* (GH, LL, MICH); 25–30 km SW of Minaca, *Pennell 18869* (GH, PH); base of Sierra Madre, *Pringle 1682* (M, MO, NY, UC).—COAHUILA: ca. 54 air mi SE of Big Bend Park, *Henrickson 15030* (LL).—DISTRITO FEDERAL: *Bourgeau 432* (C, E, GH).—GUANAJUATO: near San Miguel Allende, *Kenoyer 2164*

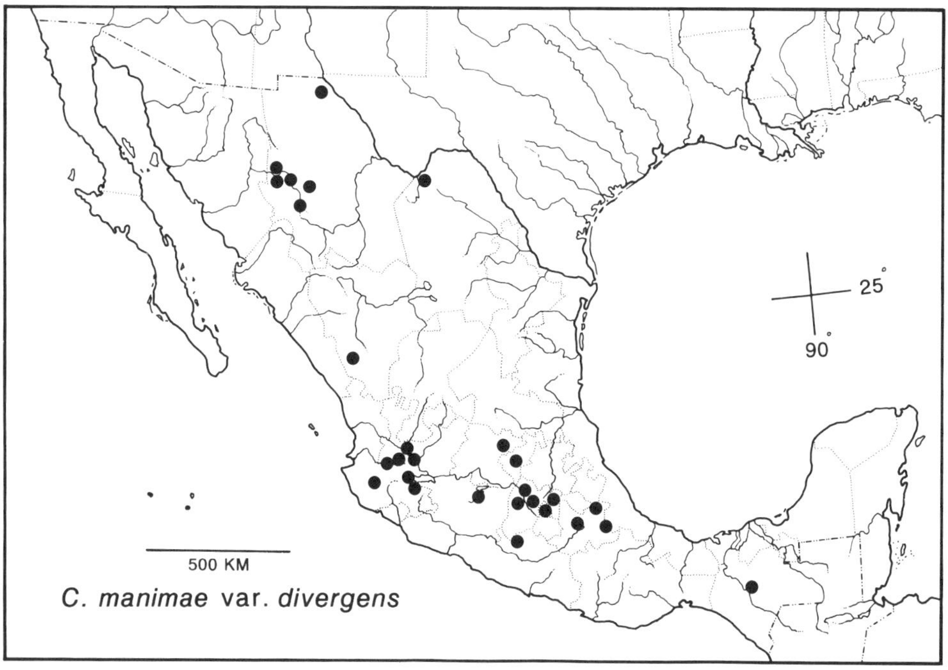

FIG. 57. Distribution of *Cyperus manimae* var. *divergens* in Mexico.

(GH).—DURANGO: Llano Grande, rd from Durango to Mazatlán, *Beetle M-7731* (MICH).—GUANAJUATO: near San Miguel Allende, *Kenoyer 2164* (GH).—GUERRERO: Dtto. Mina, Manchón, *Hinton 9244* (GH, K, LL, MICH, UC).—HIDALGO: Mpio. Tepeapulco, Cerro Santa Ana, *Ventura A. 172* (ARIZ, ENCB, NY, SD).—JALISCO: near Zapopan, *Iltis & Lasseigne 628*, *631* (WIS); Sierra del Tigre, 2 mi NE of Mazamitla, *McVaugh 13022* (MICH, SMU); 14 km N of Ixtlahuacán, *Puga 8647* (IBUG).—MÉXICO: Río Hondo Canyon, *Pringle 3161* (E, GH, MICH); 2 km N of Tepotzotlán, *Cruz Cisneros 269* (MICH, MO).—MICHOACÁN: vic. Morelia, *Arsène 2692* (GH), *5317* (BH, GH), *5751* (GH, MO), *5848* (GH).—PUEBLA: vic. Puebla, *Arsène & Nicolas 5861* (GH).—QUERÉTARO: San Juan del Río, *Rose 9558* (GH, NY).—VERACRUZ: Orizaba, *Botteri 761* (GH); near Jalapa, *MacDaniels 355* (BH, F).

Cyperus manimae var. *divergens* is distinguished from other varieties of *C. manimae* by its cuspidate scales. It is evidently less common in Mexico than var. *asperrimus*, but in Central America the two varieties are about equally common.

66. Cyperus oxylepis Nees ex Steudel, Syn. pl. glum. 2: 25. 1854.—TYPE: GUAYANA. Collector unknown (holotype: not located).

Coarse, caespitose annual or short-lived perennial, 10–50 cm tall; culms, leaves, bracts, and rays viscid. Rhizomes absent. Culms 0.9–2.4 mm in diameter, roundly trigonous, smooth. Leaves 1–5, 10–46 cm long, 1.5–4 mm wide, involute below, becoming flat to triquetrous apically, the margins smooth or sparsely scabrous. Inflorescence bracts 3–5, 2.5–25 cm long, 1.2–4.0 mm wide, involute, margins scabrous, vertical to ascendent at 45°; secondary bracts 0–2, 5–20 mm, linear, margins and keel scabrous. Rays 3–6, smooth, 0.5–5.0 cm long; or, in small plants, rays sometimes absent and the inflorescence a congested head of spikelets, 1.0–3.5 cm in diameter; secondary rays 0–3, 1–3 cm long. Spikes 1 (–3) per ray, 1.0–5.5 cm in diameter, ovoid, loose; rachis 8–20 mm long. Spikelets 5–24, 7–20 (–30) mm long, 2.5–4.0 (–6.0) mm wide, oblong to linear-lanceolate, strongly compressed; rachilla ± deciduous, ca. 0.2 mm wide, ca. 0.1 mm thick, flat, slightly flexuous, pale green to stramineous to golden-brown, successive scales 0.6–0.8 mm apart, wings absent. Scales deciduous, 10–20 (–40), 3.1–4.0 mm long, 1.5–2.4 mm wide, ovate-lanceolate, apex with a cusp (0.2–) 0.4–0.8 mm long, laterally 2–3-nerved, pale green to stramineous, medially 1–3-nerved, green to brown, smooth. Stamens 3; filaments 2.8–4.0 mm long; anthers 0.6–0.8 mm long, ellipsoid, the connective apex not prolonged. Style 0.8–1.6 mm long; stigmas 3, 1.0–1.4 mm long. Achenes 2.0–2.4 mm long, 0.5–0.8 mm wide, trigonous, ellipsoid, apex acute, the persistent style forming a beak 0.5–1.2 mm long, the base cuneate, the stipitate portion 0.2–0.3 mm long, 0.2–0.4 mm wide, spongy, whitish, the faces flat, the surface smooth or finely papillose, light to dark brown, rarely somewhat reddish. Fig. 58D–F.

Phenology. Flowering and fruiting from August to February.

Distribution. Primarily in South America, apparently a recent introduction in North America; in the U.S.A., in South Carolina, Louisiana, and Texas (Tucker 1987); in Mexico, in the Pacific lowlands of Sonora, Sinaloa, Oaxaca, and Chiapas; margins of pools, swales between dunes, and moist disturbed sites, generally in saline or alkaline soils; sea level to 200 m.

Representative Specimens. **Mexico.** Chiapas: Mpio. Tonalá, east shore of Mar Muerto, N of Paredón, *Breedlove & Thorne 20803* (DS, RSA), *20828* (DS, DUKE, MICH, RSA), *20832* (DS, MICH, RSA).—Colima: Laguna de Cuyutlán, *Díaz Luna 20111* (MEXU).—Oaxaca: Salina Cruz, *Kral 25313* (MO, US).—Sinaloa: Mazatlán, *Eyerdam & Beetle 8693* (G, UC); Altata, *Gentry 5446* (ARIZ, F, LCU).—Sonora: 1.25 km N of Huatabampito, *Sanders 9219* (NYS, UCR).

Cyperus oxylepis is easily recognized by its involute leaves, its sticky leaves, culms, and bracts, and its golden-brown spikelets. The ovate-lanceolate scales and the ellipsoid, brownish achene with a persistent beak distinguish this species from others with deciduous scales. Both Kükenthal (1935–36) and Benedict Ayers (1946) placed this species and *C. elegans* in sect. *Viscosi* C. B. Clarke [sect. *Glutinosi* (Boeck.) Kük.]. They believed this section was most closely related to sect. *Luzeoloidei* of subg. *Pycnostachys*. However, both *C. oxylepis* and *C. elegans* have the spikelets borne on a rachis, not in the capitate or digitate clusters that characterize subg. *Pycnostachys*. In addition, anatomical examination of fresh material of *C. oxylepis* and other species of this section (*C. elegans* and *C. gardneri* Nees) shows the presence of kranz anatomy, a characteristic of subgenus *Cyperus*. Subgenus *Pycnostachys*, contrarily, has non-kranz anatomy and C_3 photosynthesis. Thus, I have placed *C. elegans* and *C. oxylepis* in subg. *Cyperus*.

67. Cyperus elegans L., Sp. pl. 1: 45. 1753. *Papyrus elegans* (L.) G. Don, Loudon's Hort. Brit. 22. 1830, non *Papyrus elegans* Schrader ex Nees, 1842.—Type: Sloane, Voy. Jam. Nat. Hist. 1: 117, t.75, fig. 1, 1707 (lectotype, designated by Tucker & McVaugh, 1993).

Cyperus trachynotus Torrey, Rep. U.S. Mex. bound. 2: 227. 1859.—Type: U.S.A. Texas: Painted Caves, 1850, *Bigelow s.n.* (lectotype, here designated: NY!; isolectotypes: GH! YU!).

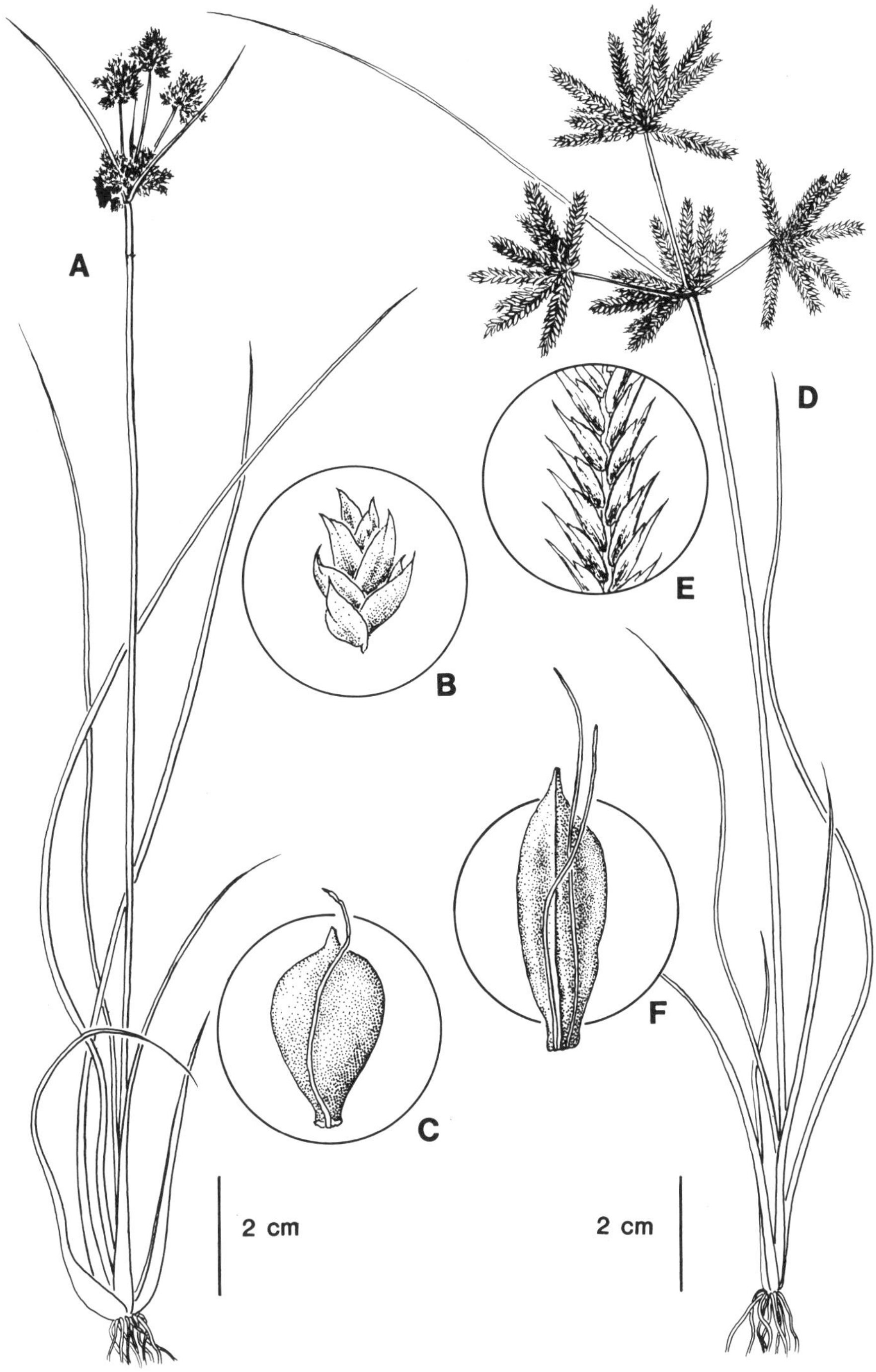

FIG. 58. *Cyperus elegans* and *C. oxylepis*. A–C, *C. elegans*. A. Habit. B. Spikelet. C. Achene with persistent filament. D–F, *C. oxylepis*. D. Habit. E. Portion of spikelet. F. Achene with two persistent filament. Drawn by Patricia M. Kernan.

Coarse, caespitose perennial, 15–80 cm tall; culms, leaves, bracts, and rays viscid. Rhizomes 3–6 mm in diameter, erect to oblique, or absent. Culms 2–4 mm in diameter, round to roundly trigonous, smooth or occasionally distantly scabrous with sparse crystalline prickles. Leaves 3–6, 15–75 cm long, 1–4 mm wide, involute, the adaxial face concave, becoming triquetrous apically, the margins distantly but prominently scabrous. Inflorescence bracts 3–7, 2.5–5.5 cm long, 1–4 mm wide, involute, margins scabrous, more so than the leaf blades, horizontal to vertical; secondary bracts 0–3, linear, triquetrous, scabrous, 1–3 cm long. Rays 3–8 (–10), smooth or sparsely scabrous, 0.5–12 cm long; secondary rays 0–3, 1–3 cm long. Spike 1 per ray, 10–20 mm in diameter, ovoid to capitate, ± dense; rachis 3–10 mm long, often concealed by spikelets. Spikelets 3–20 (–30), 3–15 mm long, 2.5–3.5 mm wide, oblong-elliptic to oblong-lanceolate, strongly compressed; rachilla persistent, ca. 0.2 mm wide, ca. 0.1 mm thick, flat, slightly flexuous, green to stramineous or brown, successive scales 0.6–0.8 mm apart, wings absent. Scales deciduous, 6–22, (2.0–) 2.8–3.6 mm long (the slightly excurved mucro 0.2–0.3 mm long), (2.0–) 2.4–2.6 mm wide, ovate-lanceolate, laterally 2-nerved, pale greenish white, stramineous, or reddish brown, medially 3–5-nerved, pale green to yellowish green, apically scabrous. Stamens 3; filaments 2.7–3.2 mm long; anthers 0.6–1.4 mm long, oblong, the connective apex less than 0.1 mm long, red. Style 1.2–1.5 mm long; stigmas 3, 1.0–1.5 (–3.0) mm long. Achenes (1.4–) 1.8–2.0 mm long, 0.9–1.0 mm wide, trigonous, obovoid, apex truncate, apiculate, base cuneate, the stipitate portion 0.1–0.2 mm long or sometimes absent, 0.2–0.4 mm wide, spongy, the faces concave, flat, or subtly convex, the surface coarsely papillose, dark reddish brown to black. Fig. 58A–C.

Phenology. Flowering and fruiting from July to October, sometimes in the remaining months.

Distribution. Florida and New Mexico, U.S.A., south to Nicaragua, in Trinidad, Peru, and Suriname. In Mexico (Fig. 59) frequent along the Pacific and Gulf Coasts; borders of mangrove swamps, ditches, water holes (tinajas), lagoons, shores, and river flats; near sea level to ca. 150 (–1500) m.

Representative Specimens. **Mexico.** Baja California: 2 km NE of San Isidro, *Moran 7466* (ARIZ, NY, SD); 1.5 mi W of Rancho La Presa, *Wiggins 15492* (DS, NY, SD).—Colima: Laguna de Cuyutlán, Colomos, Manzanillo, *Díaz Luna 4180* (IBUG, MICH).—Guerrero: near Acapulco, *Jones 22475* (WIS).—Jalisco: San Patricio Melaque, *Rzedowski 24955* (ENCB, LL, NY); Mpio. La Huerta, La Manzanilla, *Puga 12669* (IBUG).—Michoacán: El Tícuiz, on Hwy 200 ca. 6 km E of the Colima state line, *Tucker 2956* (DUKE, ENCB).—Nayarit: Mpio. Tepic, Ingenio de Puga, 10 Aug 1976, *Castaneda s.n.* (IBUG).—Nuevo León: 12 mi S of China, *Johnston 6067* (NY, TEX).—Oaxaca: Laguna Colorada, *Liebmann 14462* (C); 12 km SW of Juchitán, *Tucker 2764* (DUKE, ENCB, MICH); on Hwy 175, 55 km N of Pochutla, *Tucker 2889* (DUKE, ENCB); San Antonio, *Pringle 5946* (VT).—Puebla: near Coxcatlan, Tehuacán area, *Smith et al. 4131* (F, TEX).—Quintana Roo: San Miguel, Cozumel Island, *Swallen 2882* (LCU).—San Luis Potosí: 21.5 mi E of Tamuín, *Tucker 2004* (DUKE, ENCB).—Sinaloa: Imala, *Gentry 5094* (ARIZ, LCU, NY).—Sonora: Isla Tiburón, *Felger & García 2750* (ARIZ); Km 90 on Hwy 15, SE of Guaymas, *Gibson 2045* (BM, CAS, SD).—Tabasco: 15 km S of Villahermosa, *Tucker 2568* (DUKE, ENCB).—Tamaulipas: Soto la Marina, *Tucker 1974* (DUKE, ENCB); 20 mi N of Mante at Cinco de Mayo, *Tucker 1991* (DUKE, ENCB).—Veracruz: between Pánuco and Tampico, *Tucker 2412* (DUKE, ENCB); on Hwy 180 ca. 1 km from Río Tonalá, *Tucker 2545* (DUKE); 5 mi E of Coatzacoalcos, *Wilbur 35529* (DUKE, ENCB).—Yucatán: Izamal, *Gaumer 2485* (C, G, LCU, NY).

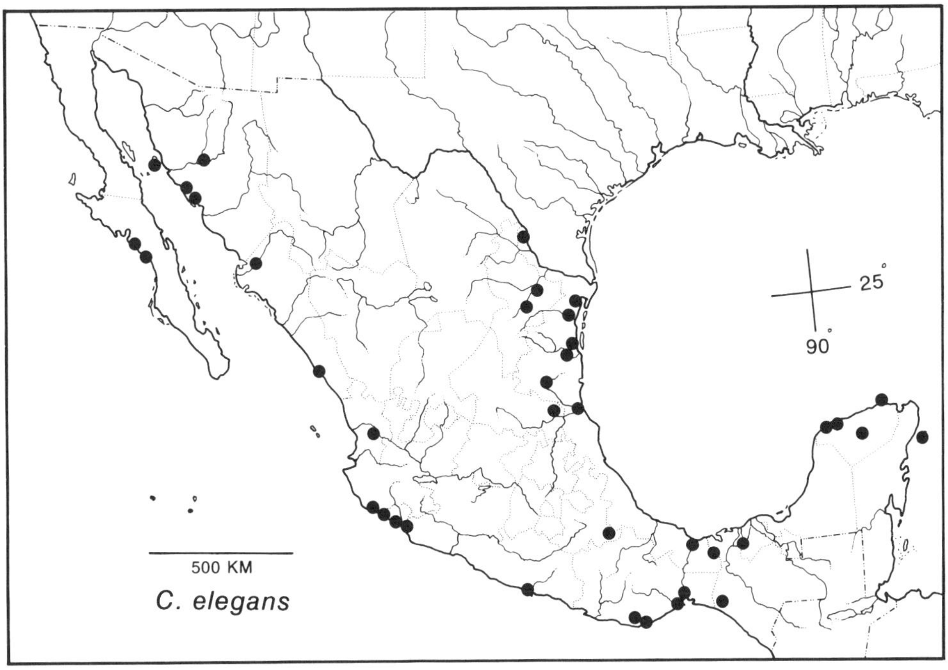

FIG. 59. Distribution of *Cyperus elegans* in Mexico.

The combination of involute leaves and black, truncate achenes distinguishes *C. elegans* from other Mexican species of the genus. Living plants are sticky, a useful character for recognizing this species in the field.

68. Cyperus planifolius L. C. Richard, Actes Soc. Hist. Nat. Paris 1: 106. 1792.—TYPE: VIRGIN ISLANDS. St. Croix, *Richard s.n.* (holotype: P).

Cyperus brunneus Swartz, Fl. Ind. occid. 1: 116. 1797. *Cyperus planifolius* var. *brunneus* (Swartz) Kükenthal, Repert. Spec. Nov. Regni Veg. 23: 189. 1926.—TYPE: West Indies. *Swartz s.n.* (holotype: S!; isotype: M!).

Cyperus brizaeus Vahl, Enum. pl. 2: 316. 1805.—TYPE: FRENCH GUIANA. Cayenne, *Richard s.n.* (lectotype, here designated: C!).

Cyperus purpurascens Vahl, Enum. pl. 2: 359. 1805.—TYPE: VIRGIN ISLANDS. St. Croix, *West s.n.* (holotype: C!).

Caespitose perennial, 25–50 (–80) cm tall. Rhizomes 5–10 cm long, 3–7 mm in diameter, erect. Culms 1–5 (–20), 0.8–2.6 mm in diameter, trigonous, smooth. Leaves 5–10 (–15), (20–) 30–80 cm long, 3–10 mm wide, flat to v-shaped, margins extrorsely scabrellate to smooth. Inflorescence bracts 4–8, the longest (10–) 15–35 cm long, 4–8 mm wide, the shortest 2–5 cm long, 1.5–3 mm wide, flat to v-shaped, margins smooth or scabrellate, horizontal to ascendent at 30°. Rays 4–7, 1–10 cm long; secondary rays 0.5–2 cm long, occasionally present; rays often not elongate, the inflorescence then capitate, 2–5 cm in diameter. Spikes 1–3, 1.5–3.0 cm (–4.5) cm long, 1.5–3.0 cm wide, loosely to densely

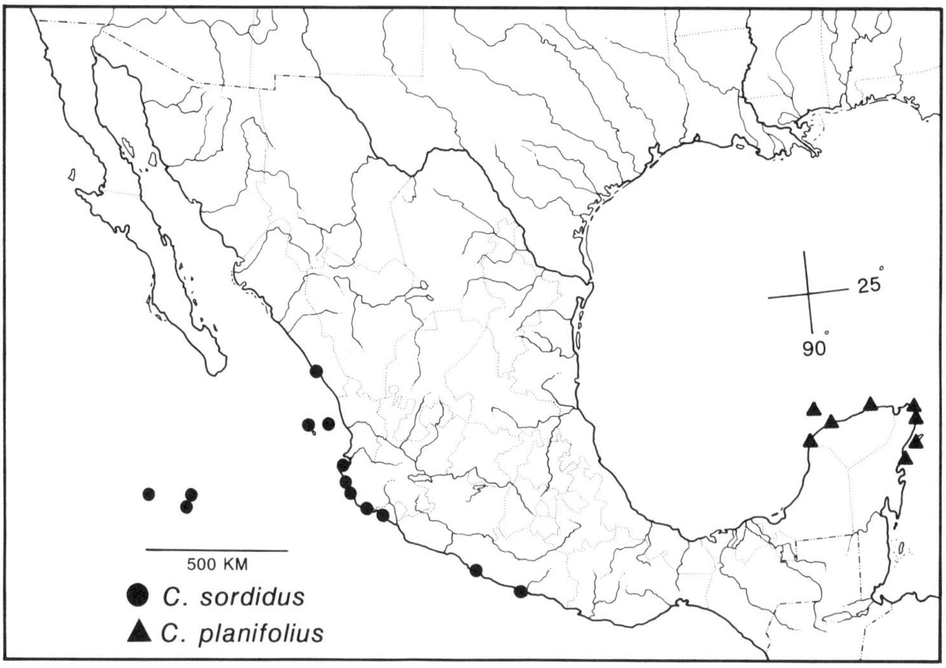

FIG. 60. Distribution of *Cyperus planifolius* and *C. sordidus* in Mexico.

ovoid; rachis 1–4 cm long. Spikelets (10–) 20–60, 6–22 cm long, (1.0–) 2.0–3.0 (–4.0) mm wide, linear-lanceolate, flattened-elliptic in cross section; rachilla persistent, 0.7–1.2 mm wide, 0.4–0.6 mm thick, straight to geniculate, stramineous, spongy-thickened, successive scales 1.0–1.6 mm apart, wings 0.4–0.6 mm wide. Scales deciduous, 6–16, 2.5–3.0 mm long, 1.4–1.8 mm wide, loosely imbricate, ovate-lanceolate to suborbiculate, apex acute to obtuse and mucronulate, medially greenish, 3–5-nerved, laterally reddish brown to brown, 3–5-nerved. Stamens 3; filaments (1.2–) 2.2–3.9 mm long; anthers 0.7–1.8 mm long, narrowly ellipsoid, the connective apex ca. 0.1 mm long, blunt, reddish. Style (0.3–) 1.2–2.8 mm long; stigmas 3, (1.0–) 2.0–3.0 mm long. Achenes 1.3–2.0 mm long, 0.75–1.0 (–1.1) mm wide, trigonous, dorsiventrally compressed, ellipsoid to obovoid or oblong, obtuse, apex apiculate or not, base stipitate or not, the adaxial face concave or flat, the abaxial faces concave to strongly convex, angles rounded, the surface coarsely punctate, brownish to black.

Phenology. Flowering and fruiting from March to November.

Distribution. Primarily in the Caribbean region; southern Florida, U.S.A., the Greater and Lesser Antilles, from the Yucatan Peninsula (Fig. 60) to Honduras, also in northeastern Brazil; coral islands (cays), sandy beaches, and grassy areas near salt water; at or near sea level.

REPRESENTATIVE SPECIMENS. **Mexico.** CAMPECHE: Mpio. Champotón, ca. 38 km S of Campeche on Hwy 180, *Sanders 9574* (NYS, UCR).—QUINTANA ROO: camino a Nizuc (entronca a la carr. Cancún–Tulum), *Moreno 106* (MEXU); Tancah, *Swallen 2798* (F, US); Cozumel Island, San Miguel, *Swallen 2890* (F, LL, MICH, US).—YUCATÁN: Alacrán Atoll, *Fosberg 41863* (DUKE); Progreso, *Millspaugh 1697* (F); Celestún, *Schott 412* (BM, F).

Kükenthal (1935–36) recognized several varieties of *C. planifolius*, but these seem to be trivial variants. Koyama (1978) separated *C. planifolius* and *C. brunneus* based on the shape of scales, the width of spikelets, and the shape of achenes. I have studied specimens at DUKE (25 sheets) and GH (60 sheets) and found nearly continuous variation in the characters used by Koyama. Therefore, it seems prudent to consider *C. brunneus* a synonym of *C. planifolius* until a comprehensive examination of *C. planifolius* and its relatives can be made.

69. Cyperus sordidus J. Presl & K. Presl in K. Presl, Reliq. Haenk. 1: 171. 1828.—TYPE: "California ad Monte-Rey," 1791, *Haenke s.n.* (holotype: PR!).

Cyperus planifolius f. *decolorans* Kükenthal, Pflanzenreich 4(20): 448. 1936.—TYPE: MEXICO. Colima: Manzanillo, *Palmer 1095* in 1890 (lectotype, designated by Tucker & McVaugh, 1993: US; isolectotypes: B! BM! K! UC!).

Cyperus howellii O'Neill & Benedict Ayers, Leafl. W. Bot. 4: 3 5. 1944.—TYPE: MEXICO. Sinaloa: Mazatlán, 2 Aug 1932, *Howell 10560* (holotype: CAS!; isotype: LCU!).

Caespitose perennial, 25–45 (–60) cm tall; leaves and stems papillate throughout. Rhizomes very short, erect. Culms 1.2–2.5 (–5.0) mm in diameter, trigonous to roundly trigonous, somewhat bulbous-based, papillose but not at all scabrous. Leaves 5–10, the blades (10–) 20–50 cm long, 3–5 mm wide, flat to broadly v-shaped, arctuate, coriaceous, the margins coarsely scabrous. Inflorescence bracts 3–5 (–7), (1.5–) 8–25 (–45) cm long, 1.0–4.0 (–6.5) mm wide, the margins and keel scabrellate, horizontal to ascendent to 45° or rarely the longest bract nearly vertical. Inflorescence of (1–) 3–6 spikes aggregated into a loose or dense head (10–) 20–40 (–60) mm wide, or of 1–4 (–6) spikes borne singly or 2–3 together on rays 5–25 (–45) mm long. Spikes ovoid to broadly cylindric, (7–) 12–20 mm long, 6–12 mm wide; rachis 3–14 mm long. Spikelets (10–) 25–70 (–100), 3–7 mm long, 2.0–3.2 mm wide, rather closely spaced and concealing the rachis, oblong-ellipsoid to oblong-lanceolate, compressed, obscurely quadrate to elliptic in cross section; rachilla ± deciduous, 0.4–0.6 mm wide, (0.1–) 0.2 (–0.3) mm thick, quadrate, geniculate, reddish brown, successive scale scars 1.0–1.4 mm apart, wings absent. Scales deciduous, (8–) 10–16 (–18), 2.2–3.0 mm long, (1.6–) 2.0–2.2 mm wide, ovate, apex mucronulate, laterally 5–7-nerved, reddish to grayish brown, the nerveless margins often conspicuously lighter, red-glandular-dotted, medially 1–3-nerved, greenish. Stamens 3; filaments 2.0–3.0 mm long; anthers 0.8–1.1 mm long, linear-oblong, the connective apex ca. 0.1 mm long, subulate, red. Style 0.8–1.6 mm long. persistent on the achene; stigmas 3, 1.3–1.7 (–2.2) mm long. Achenes 1.2–1.4 (–1.6) mm long, 0.7–0.8 (–0.9) mm wide, trigonous, ellipsoid to obovoid, apex acute to subacute, base cuneate to substipitate, 0.1–0.2 mm long, ca. 0.2 mm wide, the angles rounded, the faces flat, equal, the surface papillose, brown to reddish brown.

Phenology. Flowering and fruiting from June to November on the mainland and during March and April in the Islas Revillagigedo.

Distribution. Endemic to Mexico (Fig. 60), Pacific Coast from Mazatlán to Acapulco, Islas Tres Marías and Islas Revillagigedo; rocky cliffs, coastal bluffs, and arroyos; sea level to ca. 100 m.

REPRESENTATIVE SPECIMENS. **Mexico.** COLIMA: Manzanillo, *Palmer 1095* in 1890 (BM, K, UC, US).—GUERRERO: Hwy 200, ca. 200 km NW of Acapulco and ca. 2.5 km S of Puente Cayacal, *Tucker 2935* (CAS, DUKE, ENCB, F, GH, MICH, NY, US).—JALISCO: Mpio. La Huerta, Est. Biol. Chamela, *Lott 653* (DUKE, MEXU, MICH); Puerto Vallarta, *Rzedowski 17736* (ENCB, MT).—NAYARIT: Islas Tres Marías, San Juanito, *Howell 10477* (BKL, CAS, K).—SINALOA: Mazatlán, *Howell 10560* (CAS, LCU). Islas Revillagigedo: Isla Socorro, Braithewaite Bay, *Howell 8415*, (CAS); N of Benner's Cove, *Mason 14594* (CAS, UC, US); Isla San Benedicto, Volcán Barcena, *Cruz Cisneros 1621* (IBUG).

Cyperus sordidus is distinguished from all other species of the genus in Mexico by the combination of papillose culms and deciduous scales. It may be separated from both *C. ligularis* and *C. planifolius* by its wingless rachilla. O'Neill and Benedict Ayers (1944) believed "*C. howellii*" [=*C. sordidus*] to be closely related to *C. ligularis* because of its papillose culms; however, the geniculate wingless rachilla and deciduous, firmly conduplicate scales are unlike those of *C. ligularis*. These characters suggest rather that *C. sordidus* is related to *C. planifolius*.

Kükenthal (1935–36) did not see the type of *C. sordidus* and listed it among "*Species a me non visae*." He noted that Index Kewensis placed *C. sordidus* near *C. virens* (subg. *Pycnostachys*), a placement with which he disagreed. Kükenthal cited only the type collection, *Palmer 1095* in 1890, from Manzanillo, when he published *C. planifolius* f. *decolorans*. O'Neill and Benedict Ayers (1944), also unaware of the identity of *C. sordidus,* believed that this species of western Mexico was undescribed and published for it the name "*C. howellii*." Examination of the type of *C. sordidus* from PR shows it to be the same species as *C. howellii*. Presl and Presl (1828) stated that the type of *C. sordidus* came from "California ad Monte-Rey," in agreement with the label on the type, but this species does not occur in California. Probably the type was collected at Acapulco or San Blas, coastal localities visited by the Malaspina Expedition in 1791.

70. Cyperus pycnostachyus (H. B. K.) Kunth, Enum. pl. 2: 77. 1837. *Mariscus pycnostachyus* H. B. K., Nov. gen. sp. 1 [quarto]: 215. 1816.—TYPE: MEXICO. Michoacán: between Puerto de Andaracuas and Jurirapandaro, Sep 1803, *Humboldt & Bonpland s.n.* (holotype: P-HBK, IDC microfiche!).

Cyperus schmitzianus Boeckeler, Allg. Bot. Zeitschr. Syst. 2: 1. 1896.—TYPE: MEXICO. Valle de Mexico, 1855, *Schmitz s.n.* (holotype: B!; isotype: BM!).

Perennial, 30–60 cm tall. Rhizomes 2–3 cm long, ca. 1.5 cm in diameter, erect; horizontal rhizomes also produced, up to 10 cm long, 3–5 mm in diameter, indurate, clothed with scattered ovate-lanceolate brownish scales 1–1.5 cm long. Culms 2.0–4.5 mm in diameter, trigonous, smooth. Leaves (3–) 5–7, (12–) 20–30 cm long, 5–11 mm wide, lanceolate, acuminate, the margins distally scabrellate, the keel smooth to sparsely scabrellate. Inflorescence bracts 4–7, 3–15 (–22) cm long, 2–11 mm wide, the margins and keel scabrellate to scabrous, ascendent at 45–75° to nearly vertical. Rays 6–10, 0.5–5.5 cm long; secondary rays absent. Spikes usually 1 per ray, (7–) 12–25 (–30) mm long, (4–) 7–11 mm wide, densely cylindric or less often ovoid-cylindric to ovoid, the base subtruncate, the apex rounded; rachis (5–) 15–42 mm long, not visible when pressed. Spikelets (20–) 50–100, 2.8–4.2 mm long, 1.2–1.4 (–1.8) mm wide, ellipsoid to oblong-ellipsoid, subterete; rachilla deciduous, ca. 0.4 mm wide, 0.1–0.2 mm thick, stramineous to light green, successive scales 1.6–2.8 mm apart, wings 0.3–0.4 mm wide, hyaline. Scales persistent,

(2–) 3, 2.4–3.2 mm long, 1.5–2.0 mm wide, broadly elliptic, obtuse, apex mucronulate owing to the excurrent tip, laterally 2–3-nerved, red or reddish brown, medially 3-nerved, greenish. Stamens 3; filaments light brown with reddish speckles, 2.8–4.4 mm long; anthers (0.8–) 1.2–1.4 (–1.6) mm long, ellipsoid, the connective apex 0.1–0.15 mm long, linear-subulate, reddish. Style 0.6–1.5 mm long; stigmas 3, 1.2–2.2 mm long. Achenes 1.2–1.5 mm long, 0.6–0.7 mm wide, trigonous, oblong-ellipsoid to ellipsoid-obovoid, apex obtuse, apiculate, base cuneate to stipitate, up to 0.2 mm long, 0.1–0.2 mm wide, the angles rounded, the faces flat to slightly concave, the surface finely papillose to nearly smooth, reddish brown to brown.

Phenology. Flowering and fruiting from June to September.

Distribution. Endemic to Mexico; rare and local, marshes, canal banks, and wet places; 1800–2300 m.

Representative Specimens. **Mexico.** Distrito Federal: Mexico City, *Aschenborn 118* (B); Canal de México, *Bourgeau 209* (BR, GH, K); Valley of México, *Pringle 6313* (CAS, E, GH, MICH, MO, MSC, MTMG, UC, US, VT, Z); swamps of Tlalpam, *Pringle 9500* (CU, GH, K, NY, VT, Z); Valle de México, *Schaffner 2* (NY).—México: Mpio. Texcoco, Chapingo, 6 Jun 1967, *Palacios s.n.* (DS, MICH, MSC).—Michoacán: between Puerto de Andaracuas and Jurirapandaro, Sep 1803, *Humboldt & Bonpland s.n.* (P).—Puebla: vicinité de Puebla, Rancho San Miguel, *Arsène 272* (MO).—Tlaxcala: 8 km SW of Tlaxcala, *Weaver 854* (NY).

Cyperus pycnostachyus is distinguished from other Mexican species of subg. *Cyperus* by its narrowly cylindric, dense reddish spikes borne singly at the ends of the rays. This species is morphologically similar to *C. duripes* (no. 71). Both have subulate anther connective apices, an uncommon condition in this subgenus. Also, in both species, the spikes are nearly always borne singly at the ends of the rays. They are easily distinguished by the colors of the scales. Additionally, the two species are separated geographically and ecologically. *Cyperus duripes* is endemic to the Revillagigedo Islands, where it grows on arid rocky slopes and in brackish swales; *C. pycnostachyus* is hydrophytic and occurs in the valleys of central Mexico. Vegetative differences of these two species seem to be correlated with their habitats: *C. pycnostachyus* has wide leaves of a soft thin texture, but *C. duripes* has narrow, coriaceous leaves with involute tips.

71. Cyperus duripes I. M. Johnston, Proc. Calif. Acad. Sci. 20: 53. 1931.—Type: Mexico. Revillagigedo Islands, Clarion Island, 25 May 1897, *Anthony s.n.* (holotype: UC-200907!; isotypes: CAS! K!).

Caespitose perennial, 25–70 cm tall. Rhizomes 1–2 cm long, oblique, indurate. Culms 1–10 grouped, 5–15 mm in diameter, indurate basally, clothed with leaf bases and their remnants to a height of as much as 25 cm, trigonous above the leaf bases, smooth or sparsely scabrellate on the angles apically. Leaves 10–25 (–60), densely covering the lower culm in older plants, (15–) 25–70 cm long (about as long as the culms), 6–15 mm wide, coriaceous, the margins recurved apically and forming long involute tips, the margins and keels scabrous. Inflorescence bracts 6–12, 4–20 (–35) cm long, 5–8 mm wide, the tips involute, the margins and keels scabrous, borne ascendent at 45–60°; secondary bracts absent. Rays 7–14, 2–6.5 cm long, smooth or very sparsely scabrellate distally; secondary rays absent. Spikes 1 (–3) per ray, (11–) 15–25 mm long, 6–10 mm wide, densely cylindric; rachis (7–) 10–20 mm long, not visible when pressed. Spikelets (50–) 100–150, 2.6–3.8 mm long, 0.7–1.1 mm wide, oblong-ellipsoid, compressed-quadrate to subterete; rachilla deciduous, ca. 0.4 mm wide, 0.1–0.2 mm thick, flexuous, pale to stramineous, often red-dotted, successive scales 1.6–2.8 mm apart, wings 0.4–0.5 mm wide, hyaline, clasping the angles of the achene. Scales persistent, (1–) 2, 2.3–3.2 mm long, 1.1–1.6 mm wide, oblong-elliptic, apex obtuse, mucronulate, laterally 3 (–4) -nerved, stramineous to reddish brown, medially 3-nerved, greenish brown. Stamens 3; filaments 2.2–3.6 mm long; anthers 1.3–1.6 mm long, narrowly ellipsoid, the connective apex 0.3–0.5 mm long, linear-subulate, smooth, red. Style 0.8–1.6 mm long; stigmas 3, 1.4–3.8 mm long. Achenes 1.5–1.8 (–1.9) mm long, 0.6–0.7 mm wide, trigonous, oblong-ellipsoid, often slightly wider toward the base, apex obtuse, apiculate or not, base sessile to cuneate, the angles broadly rounded, the adaxial faces flat or concave, the surface coarsely and distinctly papillose, brown.

Phenology. Flowering and fruiting from March to June (one collection from November).

Distribution. Endemic to the Revillagigedo Islands, Mexico; known from at least one collection from each of the three islands: Clarión, Socorro, and San Benedicto; swales at sea level, presumably in saline brackish conditions [Johnston (1931) recorded no fresh water on the islands], hillsides, gullies, and ocean bluffs; nearly at the summit of Cerro Evermann, 980 m, the main peak of Isla Socorro and the highest elevation in the island group.

Additional Specimens Examined. **Mexico.** Islas Revillagigedo. Isla Clarión: above Sulphur Bay, *Howell 8369* (CAS); north slopes, *Mason 1590* (CAS, F). Isla Socorro: *Cruz C. 1614* (IBUG); Ocean bluffs, North Anchorage, *Howell 8433* (CAS, DS); Cerro de los Dientes, *Moran 5789* (DS); Mt. Evermann, *Moran 5836* (DS, UC); vicinity of Academy Bay, *Moran 5894* (DS). Isla San Benedicto: Volcán Barcena, *Cruz 993* (IBUG); N end of island, *Mason & Hanna 14558* (K, UC), *Barkelew 173* (UC).

For discussion, see comments under *C. pycnostachyus* (no. 70).

72. Cyperus dentoniae G. C. Tucker, Syst. Bot. Monogr. 2: 56. 1983. *Mariscus pubescens* J. Presl & K. Presl in K. Presl, Reliq. Haenk. 1: 181. 1828, non *Cyperus pubescens* Steudel, 1855.—Type: "California," 1791, *Haenke s.n.* (holotype: PRC!).

Cyperus flavomariscus var. *peduncularis* Britton, Bull. Torrey Bot. Club 13: 215. 1886. *Cyperus flavus* var. *peduncularis* (Britton) Kükenthal, Pflanzenreich 4(20): 532. 1936.—Type: Mexico. Chihuahua, Aug 1885, *Palmer 49* in 1885 (lectotype, here designated: NY!; isolectotypes: BM! GH! US!).

Caespitose perennial, (2–) 20–50 (–80) cm tall. Rhizomes 1–2.5 cm long, 3–6 mm thick, indurate, horizontal. Culms 1–4, (0.8–) 1.5–2.0 (–2.8) mm in diameter, trigonous, smooth or nearly so basally, hispidulous distally and especially so just below the bracts. Leaves 2–5 (–7), (5–) 15–30 (–45) cm long, (1.5–) 2.5–5.0 (–8.0) mm wide, v-shaped, both surfaces scabrous to hispidulous, the margins and keels scabrellate. Inflorescence bracts 4–8 (–10), (2–) 10–25 (–45) cm long, (1–) 3–6

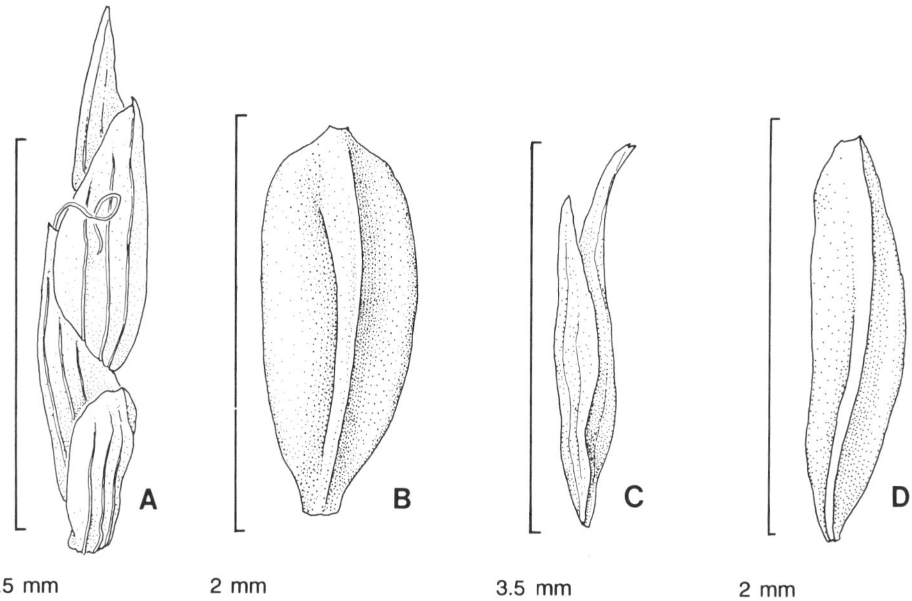

FIG. 61. *Cyperus dentoniae* and *C. tempeae*. A, B, *C. dentoniae*: A, spikelet; B, achene. C, D, *C. tempeae*. C, spikelet; D, achene. Drawn by Patricia M. Kernan.

(–8) mm wide, v-shaped, both surfaces hispidulous, the margins and keel scabrellate, borne ascendent at 30–60°. Rays 3–6 (–10), (1–) 3–6 (–18) cm long, hispidulous; secondary rays absent. Spike 1 per ray or occasionally with 1–2 smaller sessile spikes at the base, (7–) 11–20 (–25) mm long, (5–) 7–10 (–12) mm wide, densely cylindric; rachis 5–18 mm long, hispidulous, not visible when pressed. Spikelets (20–) 70–120 (–140), (2.2–) 3.4–4.5 (–5.6) mm long, (0.8–) 0.9–1.2 (–1.3) mm wide, oblong-ellipsoid to oblong-lanceolate, subterete to quadrate; rachilla deciduous, ca. 0.4 mm wide, 0.1–0.2 mm thick, greenish to pale stramineous, nearly straight to slightly flexuous, successive scales (1.6–) 2.0–2.5 (–3.0) mm apart, wings 0.3–0.4 mm wide, hyaline and sometimes also red-glandular-dotted. Scales persistent, 1 (–3), (2.4–) 2.8–3.3 (–3.6) mm long, 1.6–2.0 mm wide, ovate to elliptic-ovate, apex acute, mucronulate, laterally strongly 3 (–4) -nerved, stramineous, red-spotted, medially weakly 3-nerved, green. Stamens 3; filaments (2.4–) 2.8–3.2 mm long; anthers (0.3–) 0.4–0.6 mm long, ellipsoid, the connective apex less than 0.1 mm long, reddish. Style 0.4–0.6 (–0.8) mm long; stigmas 3, 1.0–1.4 mm long. Achenes (1.6–) 1.8–2.0 mm long, (0.7–) 0.8–0.9 mm wide, trigonous, ellipsoid, apex subtruncate, apiculate, base stipitate, 0.1–0.2 mm long, ca. 0.2 mm wide, the adaxial face flat, the abaxial faces concave, the surface puncticulate to essentially smooth, brown. Fig. 61A, B.

Phenology. Flowering and fruiting from July to September and sporadically as late as March.

Distribution. Southern Arizona, U.S.A., southward along the Pacific Coast of Mexico (Fig. 62) and Central America to western Panama; occasional in various dry open situations, most often among grasses in sandy or gravelly soils, also at borders of cultivated fields, wet roadsides, and shaded river banks; sea level to ca. 1000 m.

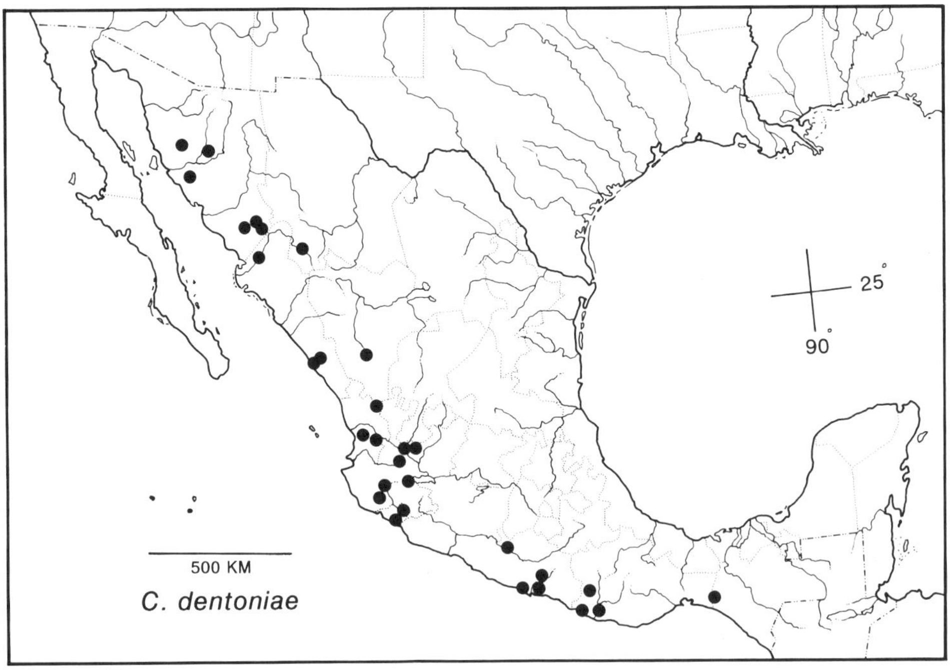

FIG. 62. Distribution of *Cyperus dentoniae* in Mexico.

REPRESENTATIVE SPECIMENS. **Mexico.** CHIAPAS: 3.5 mi NW of Arriaga, *Wilbur 35941* (DUKE).—CHIHUAHUA: "Southwestern Chihuahua," *Palmer 49* in 1885 (BM, LCU, NY, US).—COLIMA: Colima, *Orcutt 4544* (BM, DS); Colima, *Palmer 2* in 1897 (ARIZ, BM, C, MICH, UC).—DURANGO: "State of Durango," *Ibaña García 594* (US).—GUERRERO: 1 mi E of Acapulco, *Barkley 14092* (F, TEX, US, WIS); between Acapulco and Tierra Colorada, *Tucker 2260* (DUKE).—JALISCO: Mpio. Amatitán, Barranca Sta. Rosa, *Puga 6678* (IBUG); ca. 12 km N of La Huerta, *Tucker 2973* (DUKE).—NAYARIT: 15 km W of Ahuacatlán, *Feddema 479* (DS, MICH, TEX).—OAXACA: 33 km S of Putla, *Roe et al. 474* (WIS); ca. 2 km NW of Pinotepa Nacional, *Tucker 2912* (DUKE).—SINALOA: La Noria, *Mexia 128* (C, CAS, UC, US).—SONORA: Quiricoba, Dtto. Alamos, *Gentry 793M* (ARIZ, DS, MICH); 1.2 mi N of El Oasis, *Reichenbacher 1008, 1489* (ARIZ); 19 mi N of Colorado on rd to Mazatán, *Wiggins & Rollins 358* (ARIZ, DS, NY, US).

The combination of hispidulous culms, leaves, and bracts, densely cylindric spikes, and caespitose habit distinguish *C. dentoniae* from all other species of *Cyperus* occurring in Mexico. Among the New World species of *Cyperus*, a hispidulous rachis like that of *C. dentoniae* is known only in *C. ciliatus,* which can be distinguished by its flattened spikelets and deciduous scales.

Benedict Ayers (1946) treated this species as "*Cyperus asper* (Liebm.) O'Neill," a name based on *Mariscus asper*. Examination of Liebmann's type (Tucker 1983) showed it to be a specimen of *C. mutisii*, as treated by Kükenthal (1935–36). Thus, reports of *C. asper* in various floras of Mexico, Central America, and the southwestern U.S.A. should probably be referred to *C. dentoniae*.

Horvat (1941) and Benedict Ayers (1946) placed *C. dentoniae* (as *Cyperus asper*) near *C. manimae*, but presented no evidence for such placement. *Cyperus dentoniae* is decidedly out of place in such a position. *Cyperus manimae* is a rhizomatous species with compressed spikelets and loosely imbricate scales, whereas *C.*

dentoniae is caespitose and has subterete spikelets with closely appressed scales. *Cyperus dentoniae* appears to be possibly related to *C. aggregatus*. These two species are similar in having densely cylindric spikes, oblong-ellipsoid, subterete, 1–2-flowered spikelets, and golden-brown, closely appressed, persistent scales.

73. Cyperus tempeae G. C. Tucker, sp. nov.—TYPE: MEXICO. Tamaulipas: Mpio. Aldama, 40 km NNW of Aldama, Las Yucas, 23°14′N, 98°10′W, 9 Oct 1957, *Dressler 2314* (holotype: MICH!; isotypes, GH! MO!).

Culmi trigoni, apice hispiduli. Radii simplices, breves; spicae cylindricae. Spiculae oblanceolatae vel subulatae, 2.8–4.2 mm longae, 0.4–0.5 mm latae; rachilla alata. Squamae ovato-lanceolatae vel anguste ellipticae, stramineae, 2.4–2.9 mm longae, 0.8–1.1 mm latae, valde nervosae. Achenia trigona, anguste ellipsoidea, 1.7–2.1 mm longa, 0.4 mm lata, puncticulata, brunnea. Affinis Cypero dentoniae.

Caespitose perennial, 20–40 cm tall. Rhizomes not seen. Culms 1–4 (–10), 1.5–2.0 mm in diameter, trigonous, smooth basally, hispidulous distally and especially so just below the bracts. Leaves 4–7, 10–20 (–28) cm long, (1.5–) 3–6 mm wide, inversely w-shaped or v-shaped, upper surfaces scabrous to hispidulous, the margins and keels scabrellate. Inflorescence bracts 6–10, 3–10 (–20) cm long, (2–) 3–6 (–8) mm wide, inversely w-shaped, upper surfaces hispidulous especially distally, the margins and keel scabrellate, borne ascendent at ca. 45°. Rays 6–11, 1–4 (–8) cm long, hispidulous or smooth; secondary rays absent. Spikes 1–4, 15–25 (–32) mm long, 6–8 mm wide, cylindrical; rachis 18–25 mm long, ± hispidulous, visible when pressed. Spikelets 50–100, 2.8–4.2 mm long, 0.4–0.5 mm wide, oblong-lanceolate to subulate, subterete to quadrate; rachilla deciduous, ca. 0.3 mm wide, 0.1–0.2 mm thick, greenish to pale stramineous, nearly straight to slightly flexuous, successive scales 1.6–2.0 mm apart, wings 0.3 (–0.4) mm wide, hyaline. Scales persistent, 1 (–2), 2.4–2.9 mm long, 0.8–1.1 mm wide, ovate to elliptic-ovate, acute, mucronulate, laterally strongly 3 (–4)-nerved, stramineous, red-spotted, medially strongly 1 (–3)-nerved, pale green. Stamens 3; filaments 1.8–2.4 mm long; anthers ca. 0.6 mm long, narrowly ellipsoid, the connective apex less than 0.1 mm long, reddish. Style 0.5–0.6 mm long; stigmas 3, 1.0–1.3 mm long. Achenes 1.7–2.1 mm long, about 0.4 mm wide, trigonous, narrowly ellipsoid, apex acute, apiculate, base stipitate, 0.1–0.2 mm long, 0.1–0.2 mm wide, the adaxial face flat, the abaxial faces concave, the surface puncticulate, brown. Fig. 61C, D.

Phenology. Collected in early anthesis in July, with immature to mature achenes in September and October.

Distribution. Lowlands of northeastern Mexico; remnant grasslands (potreros), grassy disturbed roadsides, and lightly grazed oak woodlands; ca. 100–300 m.

ADDITIONAL SPECIMENS EXAMINED. **Mexico.** SAN LUIS POTOSÍ: carr. Río Verde–Cárdenas, Km 39, 10 Sep 1961, *Gómez G. 258* (ANSM).—TAMAULIPAS: on rd to Lauro Aguirre, 1.5 km E of Hwy 180 and ca. 10 km N of Aldama, 16 Jul 1984, *Tucker 2401* (DUKE).—VERACRUZ: Hwy 127, ca. 30 km S of Pánuco, 17 Jul 1984, *Tucker 2431* (DUKE).

This new species is morphologically similar to *C. dentoniae*, most obviously differing in its narrower achenes and narrower scales. The two species are allopatric; *C. dentoniae* occurs along the Pacific Coast and *C. tempeae* along the Atlantic Coast. *Cyperus mutisii* occasionally has scabridulous upper culms like *C. tempeae*, but it differs in its broad scales and broad achenes.

Cyperus tempeae appears to be genuinely rare. At both sites at which I found it in 1984, I saw only a solitary, several-stemmed plant.

This species is named for Christine L. Tempe, botany collection manager at the New York State Museum, with appreciation for her assiduity in mounting and filing sedges and other (less tedious) plants.

74. Cyperus svensonii G. C. Tucker, Rhodora 88: 510. 1986.—TYPE: MEXICO. Chiapas: just E of Colonia Aztlán on Hwy 195 ca. 1.7 km N of the Pan American Hwy, ca. 1000 m, 24 Jul 1984, *Tucker 2661* (holotype: DUKE!; isotypes: CAS! ENCB! F! GH! K! MEXU! MICH! MO! NY! US! WIS!).

Caespitose perennial, (15–) 25–40 (–55) cm tall. Rhizomes 1–5 cm long, 3–6 mm in diameter, covered with dark brown scales, indurate, oblique. Culms (0.8–) 1.0–1.3 (–2.0) mm in diameter, trigonous, smooth. Leaves 3–9, (10–) 15–25 (–35) cm long, 2.5–4.0 mm wide, inversely w-shaped, scabrous on the margins and veins below and on the distal adaxial surface. Inflorescence bracts (3–) 4–5 (–6), (3–) 10–14 (–24) cm long, (1.0–) 2.0–4.0 (–4.5) mm wide, v-shaped to inversely w-shaped, scabrous on the margins and veins, horizontal to reflexed downward parallel to the culm. Rays absent. Spikes (2–) 4–5 (–8), (3–) 10–14 (–24) mm long, (5–) 7–9 (–12) mm wide, densely oblong-ellipsoid to oblong-ovoid; rachis 3–12 (–20) mm long, not visible when pressed. Spikelets (30–) 60–120, (2.2–) 3.6–4.2 (–4.5) mm long, (0.7–) 0.8–1.1 (–1.2) mm wide, ellipsoid to oblong-ellipsoid, subterete; rachilla deciduous, ca. 0.1 mm thick, 0.2–0.4 mm wide, flexuous, light green, successive scales (1.2–) 1.3–1.6 (–1.9) mm apart, wings 0.3–0.4 mm wide, hyaline. Scales persistent, 2 (–4), (2.2–) 2.6–3.0 (–3.2) mm long, (1.2–) 1.4–1.8 (–2.0) mm wide, obovate, apex mucronulate, laterally 3–4-nerved, whitish to whitish green, usually with reddish streaks between the veins, medially 3-nerved, greenish or rarely off-white. Stamens 3; filaments (2.4–) 3.2–4.0 mm long; anthers 0.4–0.5 (–0.7) mm long, oblong to ellipsoid-oblong, the connective apex up to 0.1 mm long, whitish or pale brown. Style (0.4–) 0.6–0.8 (–1.2) mm long; stigmas 3, 0.8–1.4 mm long. Achenes 1.5–1.7 (–1.8) mm long, 0.6–0.7 mm wide, trigonous, oblong-ellipsoid to ellipsoid, apex acuminate, strongly apiculate owing to the persistent style base ca. 0.2 mm long, base stipitate to substipitate, 0.2 mm long, 0.2 mm wide, brownish or whitish, the faces concave, the surface puncticulate, brown to reddish brown. Fig. 2D–F.

Phenology. Flowering and fruiting from late June until November.

Distribution. Uplands of Chiapas, Mexico, and Guatemala, Honduras, and Nicaragua, also in the mountains of Ecuador; frequent throughout the range, especially in the upland valleys of Chiapas, in moist open situations: meadows, pastures, brushy slopes, irrigation ditches, and open slopes in oak or pine forests; (800–) 1200–2300 m.

REPRESENTATIVE SPECIMENS. **Mexico.** CHIAPAS: Mpio. Tenejapa, paraje of Sibanil Ha', *Breedlove 6073D* (DS, NY); just W of San Cristóbal, *King 3001* (MICH, TEX, UC); S. Benito, *Miranda 1889*

(MEXU); on rd to Ocosingo, NE of San Cristóbal, near Huixtan, *Tucker 2208* (DUKE, ENCB); ca. 20 km S of Tapilula on Hwy 195, *Tucker 2630* (DUKE).

Cyperus svensonii is the only Mesoamerican species of *Cyperus* that has dense oblong-ellipsoid to oblong-ovoid *whitish* spikes. *Cyperus svensonii* is morphologically most similar to *C. regiomontanus* from which it is distinguished by several characters. *Cyperus svensonii* has (3–) 4–5 (–6) inflorescence bracts, ellipsoid to oblong-ellipsoid spikelets, obovate scales that are laterally whitish to whitish green, whereas *C. regiomontanus* has 5–8 bracts, subulate-lanceolate spikelets, and ovate scales that are greenish to dingy greenish brown. The two species are allopatric; *C. svensonii* occurs in moist upland sites from (800–) 1200–2300 m from Chiapas southward, and *C. regiomontanus* occurs along the Pacific Coast at or near sea level from Sonora southward.

75. Cyperus regiomontanus Britton, Contr. U.S. Natl. Herb. 1: 362. 1895. *Mariscus haenkei* J. Presl & K. Presl in K. Presl, Reliq. Haenk. 1: 181. 1828, non *Cyperus haenkei* J. Presl & K. Presl, 1828.—TYPE: "California ad Monte-Rey," *Haenke s.n.* in 1791 (holotype: PR!).

Cyperus pittieri Boeckeler, Allg. Bot. Zeitschr. Syst. 2: 19. 1896. *Cyperus regiomontanus* var. *pittieri* (Boeckeler) Kükenthal, Pflanzenreich 4(20): 527. 1936.—TYPE: COSTA RICA. Guanacaste: zone littorale de la baie Salinas (forêts), Jul 1890, *Pittier 2711* (holotype: B, destroyed; isotypes: BR! Z!).

Caespitose perennial, 30–70 cm tall. Rhizomes 1–5 cm long, 2–5 mm in diameter, clothed with brownish scales, indurate, horizontal. Culms (0.5–) 1.0–2.0 (–2.6) mm in diameter, trigonous, smooth, basally indurate. Leaves 5–10 (–13), 8–40 cm long, 3–7 mm wide, inversely w-shaped, the margins and keel scabrellate. Inflorescence bracts 5–8, 2–30 cm long, 1–4 mm wide, v-shaped to inversely w-shaped, the margins and keel scabrellate, horizontal. Rays absent, or rarely 1 or 2 rays, 0.5–2.0 cm long. Spikes 5–10, 10–25 mm long, 5–8 (–12) mm wide, densely and narrowly cylindric; rachis 10–23 mm long. Spikelets (30–) 70–120, 3.0–4.0 (–5.5) mm long, 0.6–0.8 (–1.0) mm wide, subulate-lanceolate, subterete; rachilla deciduous, ca. 0.2–0.3 mm wide, ca. 0.1 mm thick, flexuous, greenish, successive scales 0.8–1.1 mm apart, wings ca. 0.3 mm wide, hyaline. Scales persistent, 1–3, (1.5–) 2.5–3.0 mm long, (0.8–) 1.4–1.8 mm wide, ovate, apex acute, laterally 3–4-nerved, greenish to dingy greenish brown, often red-speckled, medially 3-nerved, greenish. Stamens 3; filaments 1.8–2.4 mm long; anthers ca. 0.6 mm long, narrowly ellipsoid, the connective apex not prolonged. Style 0.4–0.6 mm long; stigmas 3, 1.0–1.6 mm long. Achenes ca. 2 mm long, 0.4–0.6 mm wide, trigonous, narrowly ellipsoid, apex acute to apiculate, base stipitate, 0.1–0.2 mm long, the faces concave to flat, the surface smooth or nearly so, reddish brown to light brown.

Phenology. Flowering and fruiting from May to December.

Distribution. Along the Pacific Coast (Fig. 63) from Sinaloa, Mexico, to northwestern Costa Rica; collected once inland in Sonora, Mexico; occasional, sandy thickets near beaches, brushy lots, and dry arroyos and canyons; sea level to 100 (–700) m.

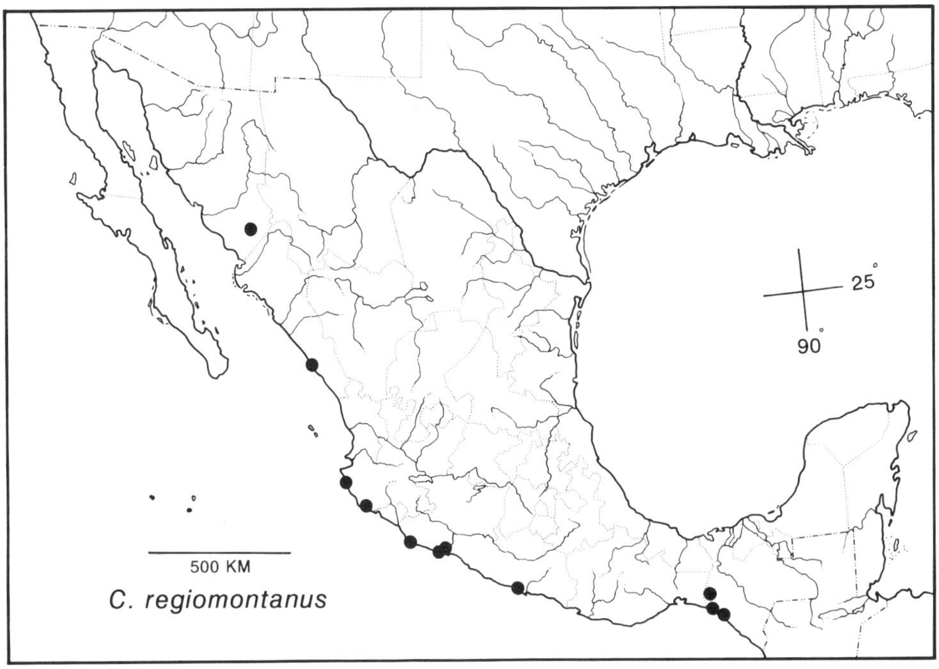

FIG. 63. Distribution of *Cyperus regiomontanus* in Mexico.

REPRESENTATIVE SPECIMENS. **Mexico.** CHIAPAS: Mpio. de Cintalapa, 23 km W of Las Cruces, *Breedlove 52811* (CAS); Puerto Arista, *Lathrop 5910* (US); 40 mi SE of Puerto Arista, *Morley 706* (UC).—COLIMA: Colima, *Palmer 1080* in 1890 (NY, US).—GUERRERO: El Tibor, *Langlassé 298* (G, US); between Acapulco and Pie de Cuesta, *MacDaniels 224* (BH).—JALISCO: Mpio. La Huerta: orilla S de la Manzanilla, *Puga & Carvajal 9771* (IBUG); Mpio. Cihuatlán, San Patricio Melaque, *Rzedowski 24935* (IBUG); Mpio. Tomatlán, Playa La Fortuna, 11 Oct 1982, *Vázquez s.n.* (IBUG); Mpio. La Huerta, Playa El Tamarindo, *Puga & Carvajal 9887* (IBUG).—MICHOACÁN: Playa Azul, *Tucker 2951* (CAS, DUKE, ENCB, MICH); Playa de Cocula, on Hwy 200, 170 km WNW of Playa Azul, *Tucker 2954* (DUKE, ENCB), *Wilbur 36632* (DUKE, ENCB).—SINALOA: 5 mi N of Mazatlán, *Gould 12245* (MICH, UC, WIS).—SONORA: 7 km SW of Alamos, *Van Devender et al. 92–768* (ARIZ, NYS).

Cyperus regiomontanus is distinguished from other species of subg. *Cyperus* in Mexico and Central America by its rayless inflorescence of densely cylindric spikes composed of 1–2-flowered greenish to brownish spikelets. It is most similar morphologically to *C. svensonii*, from which it is immediately distinguished by the color of the spikes and by its distribution.

76. Cyperus aggregatus (Willdenow) Endlicher, Cat. horti Vindob. 1: 93. 1842. *Mariscus aggregatus* Willdenow, Enum. hort. berol. 70. 1809. *Cyperus flavus* var. *aggregatus* (Willdenow) Kükenthal, Pflanzenreich 4(20): 532. 1936.—TYPE: based on a cultivated plant (holotype: B-W no. 1426, IDC microfiche!).

Kyllinga cayennensis Lamarck, Tabl. encycl. 1: 149. 1791. *Cyperus cayennensis* (Lamarck) Britton, Bull. Dept. Agric. Jamaica 5: 8. 1907, non *Cyperus cayennensis* Willdenow ex Link, 1820.—TYPE: FRENCH GUIANA. *Von Rohr s.n.* (holotype: P-LA!).

Mariscus flavus Vahl, Enum. pl. 2: 374. 1805. *Cyperus flavus* (Vahl) Nees, Linnaea 19: 698. 1847, non *Cyperus flavus* J. Presl & K. Presl, 1828.—TYPE: VIRGIN ISLANDS. St. Croix, 1757–1791, *Von Rohr 70* (lectotype, designated by Tucker & McVaugh, 1993: C!).

Mariscus laevis H. B. K., Nov. gen. sp. 1 [quarto]: 214. 1816, non *Mariscus laevis* R. Brown, 1810. *Cyperus flavus* var. *laevis* (H. B. K.) Kükenthal, Pflanzenreich 4(20): 531. 1926.—TYPE: VENEZUELA. Near Curichana and Esmeralda, *Humboldt & Bonpland s.n.* (holotype: P-HBK, IDC microfiche!).

Kyllinga squarrosa Baldwin, Trans. Amer. Philos. Soc. Philadelphia, ser. 2: 170. 1825.—URUGUAY. Banda Oriental on the Río de la Plata, Jan–Apr 1818, *Baldwin s.n.* (lectotype, here designated: PH!).

Cyperus obesus Liebmann, Mexic. halvgr. 23: 1850. *Cyperus flavus* var. *obesus* (Liebmann) Kükenthal, Pflanzenreich 4(20): 532. 1936.—TYPE: MEXICO. Veracruz: Consoquitla, Oct 1841, *Liebmann 14370* (lectotype, designated by Tucker, 1986: C!).

Mariscus latebracteatus Palla, Österr. Bot. Zeitschr. 63: 404. 1913.—TYPE: MEXICO. Michoacán: Morelia, *Arsène 5916* (lectotype, here designated: MO!). [Kükenthal (1935–36) cited *Arsène 5916*, but no duplicate could be located at B.]

Caespitose perennial, 20–100 cm tall. Rhizomes 0.5–3.5 cm long, ca. 1 cm in diameter, indurate, ± horizontal. Culms 0.8–2.5 mm in diameter, trigonous, smooth, basally indurate. Leaves 5–10, 10–70 (–90) cm long, 2–7 mm wide, v-shaped to flanged v-shaped, the margins and keel scabrellate. Inflorescence bracts 4–7, 1–16 cm long, 0.5–4.0 mm wide, v-shaped, the margins and keel scabrellate, horizontal to slightly ascendent. Rays 3–12, 0.4–5.0 (–7.0) cm long; usually only 1–2 spikes of an inflorescence borne on elongate rays, the other spikes sessile or nearly so. Spikes 1–6, 6–30 mm long, 5–11 mm wide, densely cylindric; rachis 6–25 mm long. Spikelets 20–80, 3.0–5.0 mm long, 1.0–1.4 mm wide, ellipsoid, roughly quadrate; rachilla deciduous, 0.3–0.4 mm wide, ca. 0.1 mm thick, flexuous, light green to stramineous, often red-spotted, successive scales 1.0–2.0 mm apart, wings 0.3–0.5 mm wide, hyaline. Scales persistent, 1–2 (–4), 2.4–3.4 mm long, 1.8–2.6 mm wide, elliptic to ovate, apex obtuse, laterally 3-nerved, stramineous to golden-brown, often red-speckled, medially 3-nerved, greenish. Stamens 3; filaments 2.5–3.3 mm long; anthers 0.4–0.6 mm, ellipsoid, the connective apex not prolonged. Style 0.7–1.1 mm long; stigmas 3, 1.4–2.1 mm long. Achenes 1.8–2.1 mm long, 0.8–1.0 mm wide, trigonous, broadly ellipsoid, apex apiculate, base sessile, the faces concave, the surface smooth to finely puncticulate, dark brown to reddish brown.

Phenology. Flowering and fruiting from May to February.

Distribution. Mostly in neotropical regions; Louisiana, Texas, and New Mexico, U.S.A., south through Central America to northern Argentina, also in the West Indies, adventive in Australia, collected on ballast in New Jersey, U.S.A. Frequent in the lowlands of Mexico (Fig. 64); sand dunes, gravelly roadsides, clearings, pastures, and dry open disturbed sites; sea level to 500 (–1200) m.

REPRESENTATIVE SPECIMENS. **Mexico.** CHIAPAS: Escuintla, 3 km toward El Triunfo, *Tucker 2243* (DUKE, ENCB).—GUANAJUATO: near Salamanca, *Kenoyer 1838* (GH).—JALISCO: La Barranca de

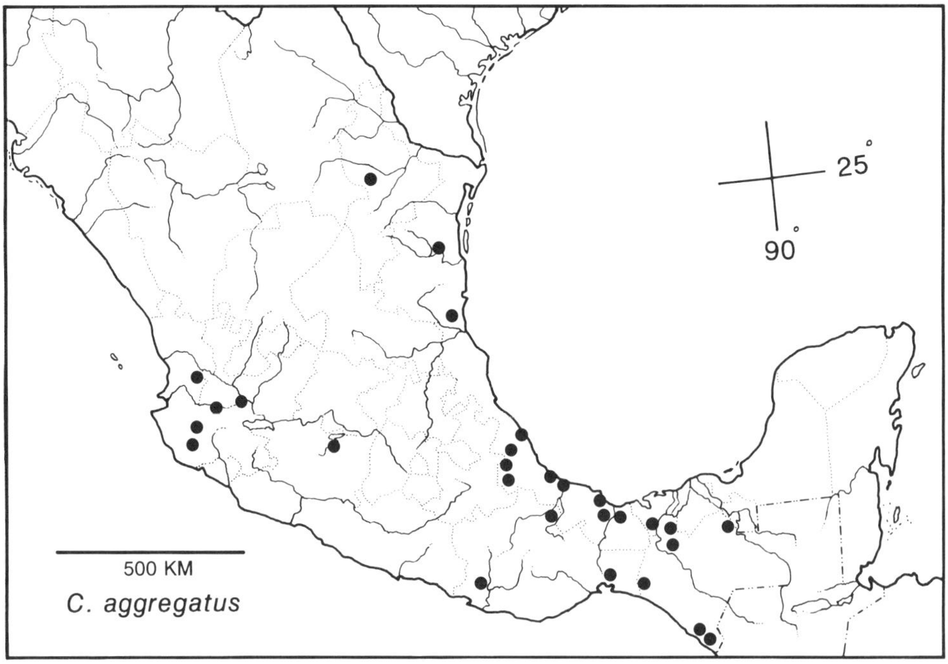

FIG. 64. Distribution of *Cyperus aggregatus* in Mexico.

Oblatos, 20 Feb 1980, *Briseño Avina s.n.* (IBUG); Unión de Tula, *Wilbur 36764* (DUKE, ENCB).—MICHOACÁN: vicinité de Morelia, *Arsène 2810* (BH, MO).—NAYARIT: 12 mi SE of Tepic, *Feddema 550* (MICH, TEX, WIS).—NUEVO LEÓN: near Monterey, *Pringle 2055* (NY, Z).—OAXACA: 33 km S of Putla, *Roe et al. 474* (WIS); 30 km E of Niltepec, *Tucker 2739* (DUKE).—SAN LUIS POTOSÍ: San Luis Potosí, *Schaffner 194* (NY).—SONORA: Santa Magdalena, *Schott III-7* (F).—TABASCO: Hwy 180, ca. 30 km E of Río Tonalá, *Tucker 2553* (DUKE).—TAMAULIPAS: 18 km S of Est. Manuel, on Hwy 180, *Tucker 2387* (DUKE, ENCB).—VERACRUZ: near Los Mangos, SE of Catemaco, *Dressler & Jones 119* (MICH, MO, NY, TENN, UC); Potrero de Consoquitla, *Liebmann 14466, 14467* (C); 10 km N of La Piedra on Hwy 180, SE of Veracruz, *Tucker 2490* (DUKE, ENCB).—YUCATÁN: Uxmal, *Swallen 2599, 2601* (MICH).

Cyperus aggregatus is recognized by its dense, cylindric, stramineous to golden-brown spikes, sessile or borne on short rays up to 7 cm long. It has denser spikes and grows at lower elevations than *C. pallidicolor*, a similar species with which it has been confused.

Three other names have been applied to *C. aggregatus*. Kükenthal (1935–36) used the name *C. flavus*, and Fernald (1950) and Tucker (1983) used *C. cayennensis*; both names are illegitimate. Correll and Johnston (1970) used the name *C. huarmensis*, which I consider a synonym of *C. odoratus*. The type of *C. huarmensis* is a plant with loosely ovoid, ragged spikes and is certainly not referable to *C. aggregatus*. The oldest available name for this species in *Cyperus* is *C. aggregatus*, based on *Mariscus aggregatus* Willd., which Kükenthal listed as a synonym of *C. flavus*. The type of *Mariscus aggregatus* exhibits the cylindric spikes characteristic of the species. Boeckeler, a monographer of the Cyperaceae, annotated this sheet in his distinctive hand as "*Mariscus flavus*."

77. Cyperus pallidicolor (Kükenthal) G. C. Tucker in Tucker & McVaugh in McVaugh, Fl. Novo-galiciana 13: 324. 1993. *Cyperus subambiguus* var. *pallidicolor* Kükenthal, Pflanzenreich 4(20): 528. 1936.—TYPE: MEXICO. Durango: Durango and vicinity, Apr–Nov 1896, *Palmer 722* (lectotype, designated by Tucker in Tucker & McVaugh, 1993: US!; isolectotypes: GH! MO! NY! UC!).

Caespitose perennial, (10–) 30–50 (–80) cm tall. Rhizomes 1 (–3) cm long, ca. 0.5–1.0 cm in diameter, clothed with fibrous remnants of leaf sheaths, indurate, ± horizontal. Culms (0.8–) 1.0–2.0 mm in diameter, trigonous, smooth or sometimes scabrellate on the angles distally. Leaves 3–7, (10–) 25–40 (–55) cm long, 2–5 mm wide inversely w-shaped or v-shaped, the margins and keel scabrellate. Inflorescence bracts (3–) 4–6 (–8), (3–) 10–20 (–30) cm long, 1–5 mm wide, inversely w-shaped, the margins and keel scabrellate, horizontal to reflexed downward nearly parallel to the culm. Rays usually absent, or occasionally 1–3, up to 4 cm long. Spikes 3–6 (–12), (6–) 10–22 (–34) mm long, (5–) 8–11 mm wide, loosely cylindric, ca. 2–3 times longer than wide, ± flattened when pressed; rachis 6–20 (–28) mm long, visible when pressed. Spikelets (10–) 25–60 (–100), 2.4–3.6 mm long, 0.8–1.2 mm wide, broadly ellipsoid to oblong, subterete; rachilla deciduous, 0.3–0.4 mm wide, ca. 0.1 mm thick, flexuous, light green to stramineous, often red-spotted, successive scales 1–2 mm apart, wings 0.3–0.5 mm wide, hyaline. Scales persistent, 1–2 (–4), 2.6–3.3 mm long, 1.8–2.4 mm wide, oblong, apex broadly rounded, laterally 3–4-nerved, golden-brown to dull grayish white, frequently streaked with red, medially 3-nerved, greenish. Stamens 3; filaments 2.2–3.2 mm long; anthers 0.6–0.8 mm long, ellipsoid, the connective apex less than 0.1 mm long. Style 0.8–1.2 (–1.4) mm long; stigmas 3, 1.4–2.0 mm long. Achenes (1.6–) 1.8–2.0 (–2.2) mm long, (0.6–) 0.7–0.9 mm wide, trigonous, ellipsoid, apex subacute to broadly rounded, base cuneate to substipitate, the faces flat or slightly concave, the surface finely papillose to essentially smooth, reddish brown to dark brown.

Phenology. Flowering and fruiting from June to October, a few collections from as late as February.

Distribution. Southwestern U.S.A., Mexico (Fig. 65), and Central America: from Arizona, northern Chihuahua, and Nuevo León south to western Nicaragua; also in Colombia, Ecuador, and Peru; frequent, clearings, trails, roadsides, disturbed slopes, and lava flows; 1500–2900 m.

REPRESENTATIVE SPECIMENS. **Mexico.** AGUASCALIENTES: ca. 10 mi SE of Calvillo, *McVaugh 18375* (MICH, US).—BAJA CALIFORNIA SUR: Sierra de la Laguna, 19 Oct 1893, *Brandegee s.n.* (UC).—CHIAPAS: Mpio. San Cristóbal, on rd to San Lucas Zapotal, *Breedlove 37303* (DS).—CHIHUAHUA: Mojarachic, *Knobloch 1250* (MICH).—COAHUILA: Sierra de Zapaliname, 3 Oct 1974, *Velasco Nuño s.n.* (IBUG).—DISTRITO FEDERAL: 4 km E of Ajusco, *López 98* (ARIZ, ENCB, LL).—DURANGO: Mpio. Santiago Papasquiaro, 3.5 km W of La Soledad, *Díaz 436* (UTEP).—HIDALGO: 18 km W of Ixmiquilpan, *González Quintero 3157* (MICH).—JALISCO: Sierra de Bolaños, *Puga 1992* (IBUG); Sierra de Mamantlán, Est. Biol. "Las Joyas," *Judziewicz & Cochrane 4864* (IBUG, WIS); Mpio. Jocotepec, Trojes Sierra del Tecuane, *Puga 695* (IBUG).—MÉXICO: Mpio. Chalco, 1 km SE of Sta. Ana, *Pineda R. 780* (ENCB, MICH); Río Hondo Canyon, *Pringle 3162* (ARIZ, BKL, E, F, M, MICH, MO, MTMG, NY, UC, VT, Z).—MICHOACÁN: 8–10 km SW of Jiquilpan, *Feddema 236* (DUKE, MICH, TEX); 3.5 mi N of Volcán Parícutin, *Turner 1893* (MICH, SMU).—MORELOS: Mpio. Huitzilac, Coajomulco, *Díaz M. 117* (ENCB, MICH, TEX).—NAYARIT: 7 mi W of Compostela, *Stevens & Fairhurst 1803* (MICH).—

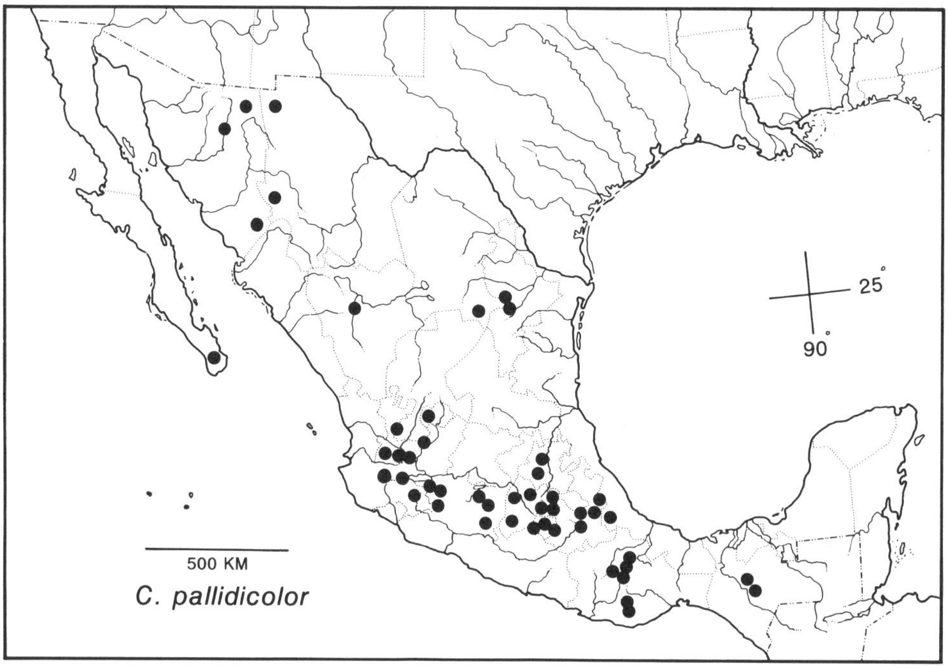

FIG. 65. Distribution of *Cyperus pallidicolor* in Mexico.

NUEVO LEÓN: Mpio. Montemorelos, La Trinidad, *Muller 2836* (F, LL, MICH, MO, NY, TENN, UC).—OAXACA: Ixtlán de Juárez, *McVaugh 22334* (MICH); 3 km S of San José del Pacífico on Hwy 175, *Tucker 2880* (CAS, DUKE, ENCB, MICH).—PUEBLA: Puente del Emperador, *Sharp 44423* (TENN).—SONORA: Río de Bavispe region, Cañón de la Mescalera, *White 4728* (MICH); Río Mayo region, 3 km W of Las Chinacas, *Martin et al. s.n.* (ARIZ 302755).—TLAXCALA: between Volcán de Malinche and Apizaco, *Hernández X. 400* (MICH).—VERACRUZ: below El Puerto, above Acultzingo, *Sharp 44792* (MO, TENN).—ZACATECAS: 18 mi S of Valparaíso, *McVaugh 17737* (MICH); Mpio. San Pedro Piedra Gorda, Cerro de los Coyotes, *del Castillo & Trujillo ZA-34* (CHAPA).

Cyperus pallidicolor is distinguished from other species with sessile spikes by its lax spikelets and horizontal to reflexed inflorescence bracts. Only two other species with sessile spikes occur at the same elevations with *C. pallidicolor*; *C. svensonii* can be immediately distinguished by its dense whitish spikes, and rayless forms of *C. mutisii* by their ascendent inflorescence bracts and broadly ellipsoid achenes.

Most collections of *C. pallidicolor* examined had been determined as "*C. flavus*" or "*C. cayennensis*" [=*C. aggregatus*]. In the field, *C. pallidicolor*, an upland species, is strikingly different from *C. aggregatus*, an inhabitant of lowlands. Particularly apparent are the lax, rayless spikes and horizontal to reflexed inflorescence bracts of *C. pallidicolor*; typical representatives of *C. aggregatus* have dense, pedunculate spikes and horizontal to slightly ascendent bracts. I also noted differences in the achenes; those of *C. pallidicolor* are ellipsoid with a subacute to broadly rounded apex and cuneate to stipitate base, whereas the achenes of *C. aggregatus* are broadly ellipsoid with an apiculate apex and sessile base.

78. Cyperus swartzii (Dietrich) Boeckeler ex Kükenthal, Repert. Spec. Nov. Regni Veg. 23: 186. 1926. *Mariscus swartzii* Dietrich, Sp. pl. 2: 343. 1833. *Kyllinga filiformis* Swartz, Prodr. 20. 1788, non *Cyperus filiformis* Swartz, 1788.—TYPE: JAMAICA. *Swartz s.n.* (holotype: S!).

Caespitose perennial, (10–) 15–50 (–65) cm tall. Rhizomes 0.5–1.5 cm long, 2 (–3) mm in diameter, soft, ± horizontal. Culms (0.5–) 1.0–1.5 mm in diameter, trigonous, smooth. Leaves 1–3, (1–) 8–20 cm long, 1–4 mm wide, flat, the margins and keel scabrellate distally. Inflorescence bracts 3–5, 1–6 (–10) cm long, 1–2 mm wide, flat, the margins and keel scabrellate, the longest erect, the others ascendent. Rays absent. Spikes 1–3, 3–7 mm long, 3–4 mm wide, densely ovoid to oblong-ovoid; rachis 1–4 mm long. Spikelets (20–) 30–45, 1.0–2.3 mm long, 0.6–1.0 mm wide, oblong, quadrate; rachilla deciduous, ca. 0.3 mm wide, ca. 0.1 mm thick, flexuous, greenish brown, successive scales 0.2–0.4 mm apart, wings ca. 0.5 mm wide, hyaline. Scales ± persistent, 1–2, 1.5–2.0 mm long, 1.4–1.7 mm wide, ovate, apex obtuse, laterally 4 (–6) -nerved nearly to the margins, light brown to tan, the margins clasping the achene, medially 3-nerved, greenish to brownish. Stamens 3; filaments 1–2 mm long; anthers 0.3–0.6 mm long, oblong to ellipsoid, the connective apex not prolonged. Style 0.2–0.5 mm long; stigmas 3, 0.4–1.0 mm long. Achenes 1.0–1.2 mm long, 0.6–0.7 (–0.8) mm wide, trigonous, ovoid, apex apiculate, base sessile or nearly so, the faces slightly concave, the surface puncticulate, dark brown.

Phenology. The only Mexican collection was made in July; in the West Indies, the species flowers and fruits throughout the year.

Distribution. The West Indies, from Cuba and Jamaica to Martinique. In Mexico, known from a single collection from eastern San Luis Potosí.

ADDITIONAL SPECIMENS EXAMINED. **Mexico**. SAN LUIS POTOSÍ: Ciudad Valles, wet area along roadsides, Jul 1937, *Lundell 7277* (LCU, NY, US, WVA).

Cyperus swartzii is distinguished from other Mexican species of *Cyperus* with dense cylindric spikes by its tiny spikes (only 3–7 mm) long and by bearing the longest inflorescence bract erect.

79. Cyperus mutisii (H. B. K.) Grisebach, Fl. Brit. W. I. 567. 1864. *Mariscus mutisii* H. B. K., Nov. gen. sp. 1 [quarto]: 216, pl. 66. 1816.—TYPE: COLOMBIA. Near Bogotá, *Mutis s.n.* (holotype: P?; isotype?: *Mutis 338*–US!). [Application of name follows Kükenthal (1935–36) and interpretation of Kunth's plate.]

Mariscus asper Liebmann, Mexic. halvgr. 41. 1850. *Cyperus mutisii* var. *asper* (Liebmann) Kükenthal, Pflanzenreich 4(20): 486. 1936. *Cyperus asper* (Liebmann) O'Neill in Tidestrom & Kittell, Fl. Ariz. New Mex. 577. 1941.—TYPE: MEXICO. Puebla: Chinautla, Aug 1842, *Liebmann 14357* (holotype: C!).

Cyperus ochreatus Boeckeler, Linnaea 36: 386. 1870.—TYPE: DOMINICAN REPUBLIC. Santo Domingo, 1856, *Meyerhoff 218* (holotype: B!).

Cyperus ehrenbergianus Boeckeler, Linnaea 36: 391. 1870. *Cyperus mutisii* var. *ehrenbergianus* (Boeckeler) Kükenthal, Pflanzenreich 4(20): 485. 1936.—TYPE: MEXICO. *Ehrenberg s.n.* (not located).

Cyperus semitribrachiatus Boeckeler, Linnaea 36: 386. 1870. *Cyperus mutisii* var. *semitribrachiatus* (Boeckeler) Kükenthal, Pflanzenreich 4(20): 485. 1936.—TYPE: DOMINICAN REPUBLIC. Santo Domingo, 1856, *Meyerhoff 217* (holotype: B!).

Cyperus pringlei Britton, Bull. Torrey Bot. Club 12: 7. 1885. *Cyperus tetragonus* var. *pringlei* (Britton) Kükenthal, Pflanzenreich 4(20): 493. 1936.—TYPE: U.S.A. Arizona: Santa Catalina Mtns, 3 Aug 1884, *Pringle s.n.* (holotype: NY!; isotypes: AC! BR! GH! VT! Z!).

Cyperus compresso-triqueter Boeckeler, Beitr. Cyper. 1: 10. 1888.—TYPE: PUERTO RICO. Farajones near Coamo, 13 Dec 1885, *Sintensis s.n.* (holotype: B!; isotype: B!).

Cyperus mutisii var. *contractus* Kükenthal, Pflanzenreich 4(20): 486. 1936.—TYPE: MEXICO. Distrito Federal: San Angel, 18 Jun 1865, *Bourgeau 428 p.p.* (lectotype, here designated: US!). [No duplicate was found at B. The duplicate at GH is *C. pallidicolor*.]

Caespitose perennial (20–) 30–120 cm tall. Rhizomes 1–3 cm long, ca. 1 cm in diameter, clothed with fibrous brownish scales, indurate, horizontal. Culms 1.4–3.5 mm in diameter, trigonous, smooth or occasionally with scattered prickles on the angles distally. Leaves 2–4 (–6), 10–60 (–70) cm long, (2–) 4–13 mm wide, inversely w-shaped, the margins, keel and adaxial nerves scabrous. Inflorescence bracts 5–8, (4–) 14–40 (–46) cm long, 1–12 mm wide, inversely w-shaped, the margins and nerves scabrellate, ascendent at ca. 45°. Rays absent or 3–6, 2–23 cm long. Spikes (1–) 3–6, borne digitately at the ends of the rays, 1–4 (–6) cm long (the terminal spike usually longer that the others), 5–11 mm wide, narrowly cylindric, 2–10 times longer than wide, not much flattened when pressed; rachis 8–40 (–50) mm long, visible when pressed. Spikelets 35–100, 3–7 mm long, (0.9–) 1.3–2.0 mm wide, oblong-ellipsoid, subterete (elliptic in cross section); rachilla deciduous, 0.2–0.3 mm wide, 0.1–0.2 mm thick, flexuous, green, successive scales (1.0–) 1.4–2.0 mm apart, wings 0.3–0.5 mm wide, hyaline, often red-dotted. Scales persistent, 1–2 (–5), 2.2–3.2 mm long, 1.5–2.1 mm wide, ovate to elliptic, apex obtuse, becoming hyaline and often erose at maturity, laterally 3–5-nerved, brownish to reddish brown, medially 3-nerved, greenish. Stamens 3; filaments 1.6–2.5 mm long; anthers 0.6–0.8 mm long, oblong-ellipsoid, the connective apex not prolonged. Style 0.4–0.8 mm long; stigmas 3, 1.2–2.2 mm long. Achenes 1.5–1.8 mm long, 0.7–0.9 (–1.0) mm wide, trigonous, slightly dorsiventrally compressed, broadly ellipsoid, apex obtuse and apiculate, base sessile, the angles rounded, the faces flat to slightly concave, the surface papillose to puncticulate, brown to reddish brown.

Phenology. Flowering and fruiting from June to November.

Distribution. Neotropical regions; Arizona, U.S.A., south to Bolivia, also in Puerto Rico. Frequent in Mexico (Fig. 66); disturbed soils of paths, roadsides, clearings, thickets, and edges of pastures; (500–) 1000–2800 (–3100) m.

REPRESENTATIVE SPECIMENS. **Mexico.** BAJA CALIFORNIA: Sierra de San Francisquito, 18 Oct 1890, *Brandegee s.n.* (UC); El Taste, 14 Sep 1893, *Brandegee s.n.* (UC).—CHIAPAS: San Cristóbal, *Breedlove*

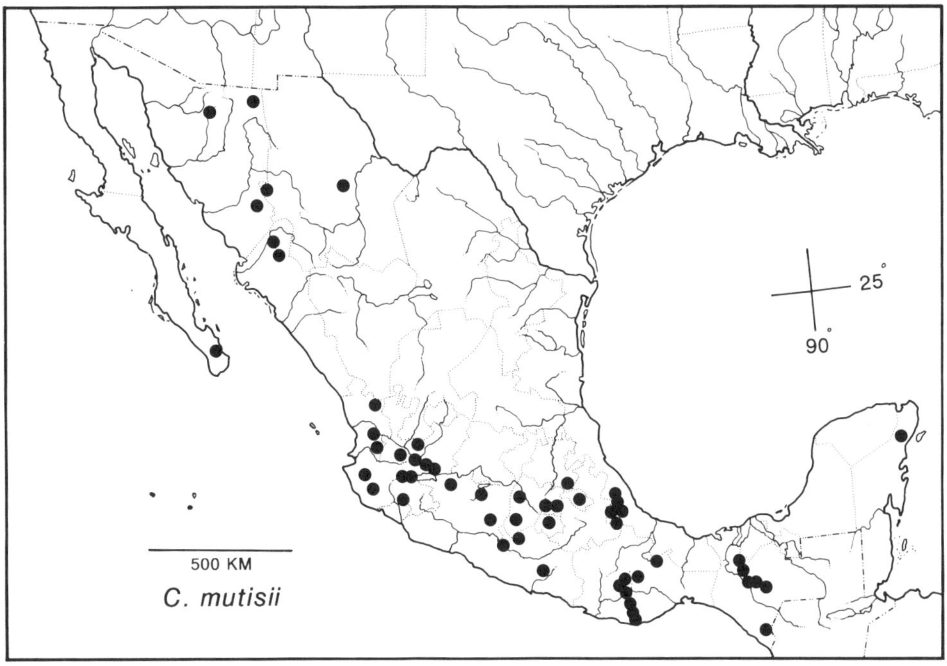

FIG. 66. Distribution of *Cyperus mutisii* in Mexico.

51443 (CAS); Pueblo Nuevo Solistahuacán, *Ton 2852* (DS, WIS); 20 km S of Tapilula, *Tucker 2632* (DUKE, ENCB).—CHIHUAHUA: Río Mayo at Guasaremos, *Gentry 2397* (LCU); near Chihuahua, *Pringle 512* (BM, GH, US), *1008* (MO, NY, UC).—DISTRITO FEDERAL: près México, *Bourgeau 428* (US); México, *Seaton 6* (GH, NY).—GUERRERO: Dtto. Adama, Temisco, *Mexia 8832* (ARIZ, BH, F, MO, NY, UC); west of Chilpacingo, *Sharp 441415* (TENN).—HIDALGO: 10 km ENE of Apam, *West G-2* (WIS); Cerro de Santa Mónica, *West P-21* (WIS).—JALISCO: 4.4 mi N of Guadalajara on Hwy 41, *Denton 2091* (MICH); near Guadalajara, *Pringle 4476* (BM, CAS, E, GH, K, M, MO, MSC, MTMG, NY, UC, VT, Z); 10 mi S of Autlán, *Wilbur & Wilbur 2187* (GH, MICH, TEX, US).—MÉXICO: Dtto. Temascaltepec, Puerto Salitre, *Hinton 1782* (K, LCU, MO, US); Mpio. Juchitepec, Pedregal el Púlpito, *Ventura 1450* (ARIZ, ENCB).—MICHOACÁN: along rd from Tsitzio to Altamirano, *Sohns 884* (US).—MORELOS: near Cuernavaca, *Pringle 13235* (CU, MICH, VT).—NAYARIT: 7 mi W of Compostela, *Norris & Taranto 13823* (MICH); between La Labor and Santa María del Oro, *Pennell 19890* (LCU, PH).—OAXACA: 35 km from Mitla on the rd to Ayutla, *Tucker 2199* (DUKE); 11 mi NE of Oaxaca on Rte 175, *Wilbur 36295* (DUKE, ENCB).—PUEBLA: vic. Puebla, *Arsène 1009* (US).—QUINTANA ROO: Coba, *Lundell 7837* (LCU, MICH).—SINALOA: Sierra Surutato, 7 mi NE of El Quintero, *Breedlove & Thorne 18123*, *18124* (CAS).—SONORA: Puerto de los Aserraderos, *White 3223* (ARIZ, MICH).—TLAXCALA: Santa Ana Chiantenopam, *Arsène 10145* (US).—VERACRUZ: Orizaba, *Bourgeau 2979* (C, GH, K, NY); Fortín, *Fisher 35344* (ARIZ, LCU, MO, NY, US); Mpio. Dos Ríos, Cerro Falso, *Ventura 4225* (ARIZ, ASU, ENCB, IBUG).

Cyperus mutisii is recognizable by its long, cylindric spikes borne digitately at the ends of the rays and by its short "plump" spikelets.

80. Cyperus ligularis L., Pl. Jamaic. Pugill. 3. 1759.—TYPE: JAMAICA. *P. Browne s.n.* (holotype: LINN 70.37, IDC microfiche!).

Mariscus rufus H. B. K., Nov. gen. sp. 1 [quarto]: 216. 1816.—TYPE: MEXICO. Michoacán: Ario et Playa de Jorullo, Apr 1804, *Humboldt & Bonpland s.n.* (holotype: P-HBK, IDC microfiche!).

Caespitose perennial, (30–) 60–130 cm tall; leaves, stems, bracts, and rays densely papillose, glaucous when fresh. Rhizomes 1–12 cm long, 1–4 cm in diameter, clothed with the fibrous brown remnants of scales, indurate, oblique. Culms 1–4 (–9) mm in diameter, trigonous, smooth, basally somewhat indurate. Leaves 6–13, 30–100 cm long, 5–13 (–20) mm wide, v-shaped, the cross-veins prominent, the margins and midvein harshly scabrous with brownish prickles. Inflorescence bracts 5–12, 4–50 (–90) cm long, 1–15 mm wide, flat to v-shaped, the margins and keels scabrous, ascendent at 30–75°. Rays 5–12, 1–16 cm long; secondary rays 5–33 mm long. Spikes 3–7, borne digitately at the ends of the rays, the terminal spike sessile, the others borne on secondary rays, 10–28 (–35) mm long, 8–12 (–15) mm wide, densely oblong-cylindric to subglobose, not flattened when pressed; rachis 8–25 (–30) mm long, not visible when pressed. Spikelets 20–80, 3–7 mm long, 1.0–2.0 (–2.7) mm wide, oblong-elliptic, subterete (slightly compressed); rachilla deciduous, 0.2–0.3 mm wide, ca. 0.1 mm thick, flexuous, greenish, successive scales 1.8–2.8 mm apart, wings 0.3–0.5 (–0.6) mm wide, hyaline. Scales persistent, (2–) 4–7, (2.0–) 2.5–3.3 mm long, 1.2–2.3 mm wide, ovate, apex acute, mucronate, laterally 3–4-nerved, hyaline to brownish, reddish streaked, medially 3-nerved, green. Stamens 3; filaments (2.4–) 3.0–3.8 (–4.1) mm long; anthers 0.6–0.8 mm long, narrowly ellipsoid, the connective apex less than 0.1 mm long, reddish. Style 0.5–1.0 mm long; stigmas 3, 1.0–2.3 mm long. Achenes (1.2–) 1.5–1.6 (–1.7) mm long, 0.6–0.8 mm wide, trigonous, obovoid to broadly ellipsoid, apex apiculate, base slightly stipitate, the angles rounded, the faces concave, the surface puncticulate, brown.

Phenology. Flowering and fruiting throughout the year.

Distribution. Southern Florida and southwestern California, U.S.A., south to Brazil, also in the Bahama Islands and tropical west Africa. Frequent in Mexico (Fig. 67); beaches, brackish marshes, mangrove swamps, disturbed soils, ditches, riverbanks, and croplands near the coasts; sea level to 100 (–700) m.

REPRESENTATIVE SPECIMENS. **Mexico.** BAJA CALIFORNIA: San José del Cabo, 24 Apr 1897, *Brandegee s.n.* (UC); San José del Cabo, *Wiggins 5687B* (DS).—CAMPECHE: Champotón, *Steere 1972* (MICH).—CHIAPAS: Mpio. Tonalá, NW of Puerto Arista, *Breedlove & Davidse 54180* (CAS); on Hwy 211, ca. 1 km N of Belisario Domínguez, *Tucker 2690* (DUKE).—COLIMA: Mpio. Tecomán, El Real, *Brizuela 4* (MSC); Socorro Island, NE Anchorage, Old-Man-of-the-Rocks, *Mason & Hanna 14608* (BH, CAS, MO, UC).—GUERRERO: Acapulco and vicinity, *Palmer 295* in 1895 (E, MICH, MO, NY, UC, US).—JALISCO: Mpio. La Huerta, Est. Biológica Chamela, *Lott 511* (DUKE, MEXU); 2 km N of Puerto Vallarta, *Rzedowski 17790* (DS, ENCB, MSC).—NAYARIT: along Mantanchen Bay 3 km SE of San Blas, *Marcks 1175* (LL, WIS).—OAXACA: Playa Abierta, just E of Salina Cruz on Pacific Coast, *Tucker 2769* (DUKE, ENCB).—QUINTANA ROO: Laguna de Coba, *Barrera et al. 279* (MO).—SINALOA: N of Mazatlán, *Mexia 82* (C, CAS, UC).—TABASCO: along coast 3 km E of Paraíso, *Conrad & González 2924* (MO, SD).—TAMAULIPAS: ca. 1 mi N of San Rafael on Hwy 180, *Tucker 1978* (DUKE, ENCB).—VERACRUZ: Morro de la Mancha, *González 1559* (IBUG); Estero de la Silvena, cerca de Tecolutla, *Rzedowski 19990* (CAS, ENCB, LL, MSC); ca. 1 km from Río Tonalá on Hwy 180, *Tucker 2541* (DUKE).—YUCATÁN: San Felipe, *Gaumer 1401* (NY, UC).

The papillose culms, rays, and leaves of *C. ligularis* distinguish it from all other New World species of *Cyperus,* except *C. sordidus*. *Cyperus sordidus* has conduplicate deciduous scales, flattened spikelets, and loosely ovoid spikes, whereas *C. ligularis* has rounded, persistent scales, subcylindric spikelets, and dense cylindric to subglobose spikes.

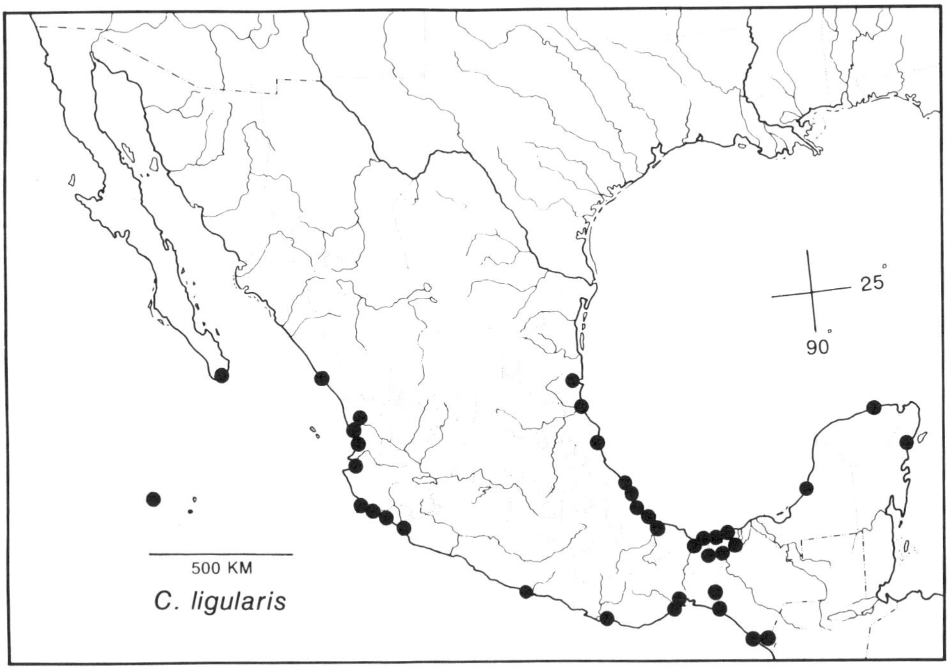

FIG. 67. Distribution of *Cyperus ligularis* in Mexico.

81. Cyperus camphoratus Liebmann, Mexic. halvgr. 28. 1850.—TYPE: MEXICO. Veracruz: Potrero de Consoquitla near Mirador, Oct 1841, *Liebmann 14354* (holotype: C!; isotypes: GH! HAL! MICH!).

Cyperus camphoratus var. *boliviensis* Kükenthal, Pflanzenreich 4(20): 422. 1936.—TYPE: BOLIVIA. Santa Cruz: Buenavista, *Steinbach 6878* (holotype: not located).

Cyperus herndoniae G. C. Tucker, Syst. Bot. 7: 345. 1982.—TYPE: COSTA RICA. Guanacaste: entrance of the Santa Rosa National Park, 21 Jul 1981, *Wilbur 31825* (holotype: DUKE!; isotypes: F! GH! MO!).

Caespitose perennial, 30–75 cm tall. Rhizomes 5–25 mm long, 1.5–2.5 mm in diameter, indurate, clothed with fibrous remains of scales, horizontal. Culms 0.5–3 mm in diameter, trigonous, smooth. Leaves 1–6, 20–40 cm long, 2–3 mm wide, v-shaped, the margins scabrellate. Inflorescence bracts 3–6, 2–25 cm long, 1–6 mm wide, v-shaped, the margins scabrellate, ascendent at 15–30°. Rays 4–8, 1–10 cm long; secondary rays absent. Spikes (5–) 15–35 mm long, 15–45 mm wide, loosely ovoid to hemispherical; rachis 3–15 mm long. Spikelets (10–) 30–60 (–100), 7–18 mm long, 1.5–2.3 mm wide, linear-lanceolate, slightly flattened; rachilla ± deciduous, 0.2–0.3 mm wide, ca. 0.1 mm thick, slightly flexuous, light brown, successive scales 1.7–2.4 mm apart, wings 0.3–0.4 mm wide, hyaline. Scales deciduous, 4–12, 2.7–3.4 mm long, 1.3–2.0 mm wide, elliptic, obtuse, apex slightly mucronulate, laterally 2–3-nerved, brown to stramineous, the margins hyaline, medially 3-nerved, greenish. Stamens 3; filaments 3.4–4.0 mm long; anthers 0.9–1.1 mm long, oblong, the connective apex not prolonged. Style 0.4–0.7 mm long; stigmas 3, 2.5–

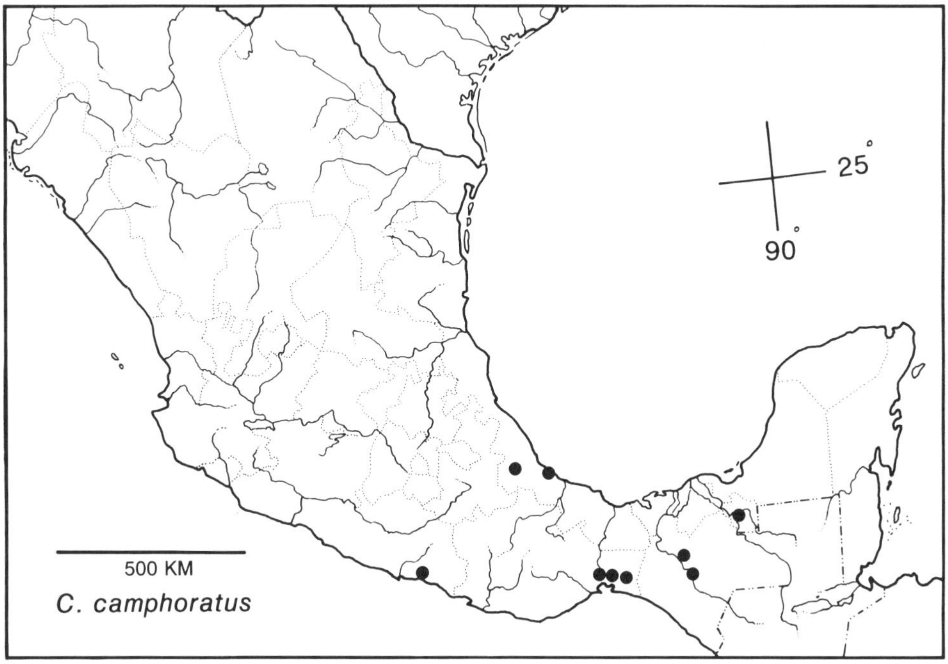

FIG. 68. Distribution of *Cyperus camphoratus* in Mexico.

3.2 mm long. Achenes 1.9–2.3 mm long, 0.6–0.7 mm wide, trigonous, oblong, apex apiculate, base substipitate, the faces flat to slightly convex, the surface finely papillose, brown. Fig. 2G–J.

Phenology. Flowering and fruiting from July to October.

Distribution. Southern Guerrero and southern Veracruz, Mexico, south along the Pacific Coast to Guanacaste, Costa Rica, also in Ecuador. Uncommon in Mexico (Fig. 68); damp roadside ditches, wet spots in pastures, and openings in thorn scrub and deciduous woodlands; sea level to 500 (–1100) m.

ADDITIONAL SPECIMENS EXAMINED. **Mexico**. CHIAPAS: Mpio. Villa Corzo, 56 km S of Hwy 190 near junction of Jerico on rd to Nueva Concordia, *Breedlove 37532* (CAS); on Hwy 195 ca. 7 km N of Hwy 190, near turnout to Waltic, *Tucker 2653* (DUKE, ENCB); Colonia Aztlán, *Tucker 2660* (CAS, DUKE, ENCB, F, GH, MEXU, MICH, MO, US).—GUERRERO: along Hwy 200 ca. 6 km E of Puerto Copala, *Tucker 2930* (CAS, DUKE, ENCB, F, GH, MICH, MO).—OAXACA: 2 km E of Zanatepec, *King 1875* (MICH, UC); ca. 27 km E of Niltepec, *Tucker 2746* (CAS, DUKE, ENCB, GH).—TABASCO: Tenosique, *Beristain 7977* (F, US).—VERACRUZ: Potrero de Consoquitla near Mirador, *Liebmann 14354* (C, GH, HAL, MICH); La Piedra 50 km SE of Veracruz, *Tucker 2495* (DUKE, ENCB).

Cyperus camphoratus (Fig. 2G–J) is difficult to place within *Cyperus*. In having both deciduous scales and rachillas, it is similar to species of sect. *Laxiglumi* (C. B. Clarke) Kük. (Tucker 1987); in its linear to lanceolate more or less deciduous spikelets, it is similar to *C. hermaphroditus* and *C. tenuis* and their North American relatives *C. strigosus* L. and *C. stenolepis* Torrey. It is not related to although similar to *C. distans,* as I once thought (Tucker 1982). Kükenthal (1935–36) placed *C. camphoratus* in sect. *Strigosi* Kük. and thought its nearest relatives among

Mesoamerican species to be *C. tenuis* and *C. hermaphroditus;* I now concur with this placement. When I published *C. herndoniae*, specimens then available to me had deciduous scales but not obviously deciduous rachillas, and thus I thought the species similar to *C. distans*. More collections and field experience convinced me that *C. herndoniae* was conspecific with *C. camphoratus*. *Cyperus camphoratus* comprises plants with deciduous and/or persistent rachillas and also exhibits some variation in deciduousness of scales. *Cyperus strigosus* and *C. stenolepis* (also placed in sect. *Strigosi* by Kükenthal, 1935–36) similarly are variable in the deciduous nature of the scales and rachillas (O'Neill 1940; Tucker 1987).

Cyperus camphoratus was recorded from Bolivia by Kükenthal (1935–36), but I have not located the specimen he cited. I did, however, find an unidentified collection from Ecuador: ad Guayaquil, Apr 1930, *Mille 1031* (F, with *C. oxylepis*).

82. Cyperus tenuis Swartz, Prodr. 20. 1788.—Type: Jamaica. *Swartz s.n.* (lectotype, designated by Tucker, 1986: S!; isolectotypes: B-W no. 1390, IDC microfiche! M! S!).

Mariscus flabelliformis H. B. K., Nov. gen. sp. 1 [quarto]: 215. 1816. *Cyperus caracasanus* Kunth, Enum. Pl. 2: 86. 1837, non *Cyperus flabelliformis* Rottbøll, 1773.—Type: Venezuela. Caracas, *Humboldt & Bonpland s.n.* (holotype: B-W no. 1433, IDC microfiche!).

Cyperus discigerus Liebmann, Mexic. halvgr. 24. 1850. *Cyperus sertularinus* var. *discigerus* (Liebmann) Kükenthal, Pflanzenreich 4(20): 411. 1936.—Type: Mexico. Oaxaca: Trapiche de la Concepción, 3000 ft, Aug 1842, *Liebmann 14356* (holotype: C!).

Cyperus breviradiatus Liebmann, Mexic. halvgr. 26. 1850. *Cyperus camphoratus* var. *breviradiatus* (Liebmann) Kükenthal, Pflanzenreich 4(20): 422. 1936.—Type: Mexico. Veracruz: Potrero de Consoquitla, 2500 ft, Oct 1841, *Liebmann 14353* (lectotype, designated by Tucker, 1986: C!; isolectotypes: C!).

Cyperus tenuis var. *brevior* Kükenthal, Pflanzenreich 4(20): 417. 1936. —Type: Costa Rica. Santo Domingo de Golfo Dulce, *Tonduz 14624* (lectotype, here designated: B!).

Cyperus tenuis var. *grandiceps* Kükenthal, Pflanzenreich 4(20): 417. 1936.—Type: Costa Rica. Guanacaste: Baie de Salinas, *Pittier & Durand 2717* (lectotype, here designated: B!; isolectotypes: BR! US! Z!).

Caespitose perennial, (4–) 10–50 cm tall. Rhizomes 3–15 mm long, 1–2.5 mm in diameter, indurate, horizontal. Culms (0.7–) 1.2–1.6 (–2.0) mm in diameter, trigonous, smooth. Leaves 3–6, 5–45 cm long, (1–) 2–3 mm wide, v-shaped, the margins and keel scabrellate. Inflorescence bracts (4–) 5–9, 3–25 cm long, (1–) 2–3 mm wide, v-shaped, the margins and keel scabrellate, horizontal to ascendent at 45°. Rays (2–) 5–12 (–14), 1.5–9.0 (–12.0) cm long, smooth; secondary rays absent. Spikes (10–) 15–25 (–35) mm long, (8–) 10–25 mm wide, loosely ovoid to ovoid-cylindric; rachis (4–) 8–12 (–23) mm long. Spikelets (10–) 30–70, 5–22 mm long, 0.7–1.1 mm wide, linear, compressed-quadrate; rachilla deciduous, ca. 0.4 mm wide, 0.1–0.2 mm thick, flexuous, greenish, successive scales 1.8–2.6 (–3.3) mm apart, wings 0.3–0.5 mm wide, hyaline. Scales persistent, (2–) 4–9 (–20), 2.4–3.4 mm long, (1.0–) 1.3–1.8 mm wide, oblong-elliptic, the broadly rounded

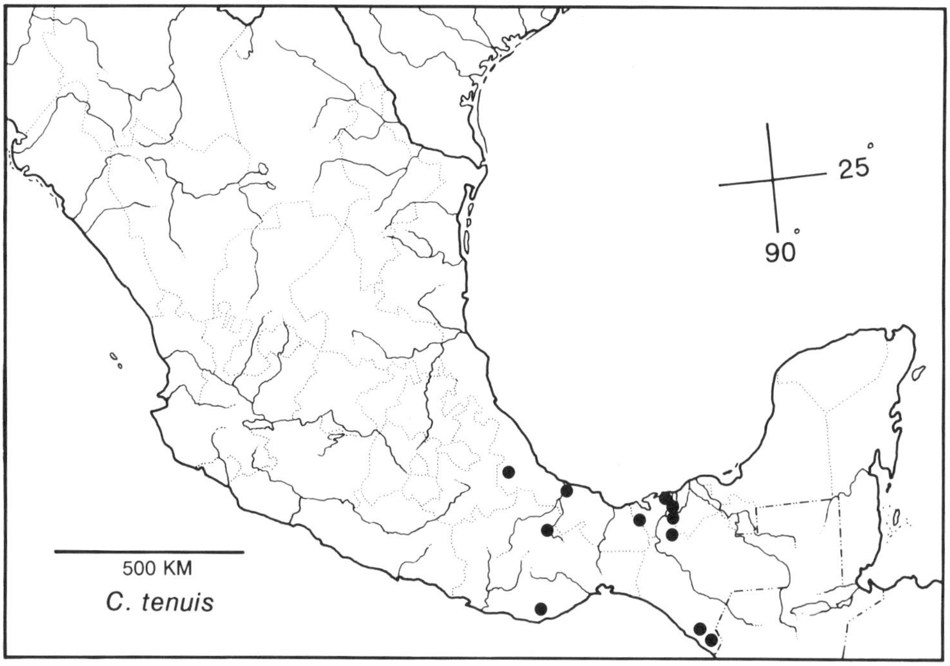

FIG. 69. Distribution of *Cyperus tenuis* in Mexico.

mucronulate apex overlapping lower 1/4–1/3 of the next scale, laterally 3–5-nerved, greenish white, often streaked with red, medially 3-nerved, green. Stamens 3; filaments 2.0–3.4 mm long; anthers 0.3–0.5 mm long, ellipsoid to oblong-ellipsoid, the connective apex not prolonged. Style 0.4–1.0 mm long; stigmas 3, 1.0–3.0 mm long. Achenes (1.4–) 1.6–1.8 (–2.0) mm long, 0.4–0.5 mm wide, trigonous, linear to narrowly oblong, apex subacute, apiculate, base stipitate, 0.1–0.2 mm long, the faces concave to barely convex, the surface punctate, brown to reddish brown.

Phenology. Flowering and fruiting throughout the year.

Distribution. Pantropical regions; in the New World from southern Mexico and Jamaica to southern Brazil, also in tropical Africa. Occasional in southern Mexico from the region of the Isthmus of Tehuantepec eastward (Fig. 69), frequent to common in Central America; edges of roadsides, coffee and banana groves, along trails and streams; sea level to 1200 m.

REPRESENTATIVE SPECIMENS. **Mexico.** CHIAPAS: Tapachula, *Ross 6–57* (US); 1 km E of Ixtacomitán, *Tucker 2610* (DUKE, ENCB).—OAXACA: along Route 175, 28 km S of Valle Nacional, *King 2132* (MICH, TEX).—TABASCO: S. J. Bautista, *Rovirosa 9* (US); 40 km S of Villahermosa, *Tucker 2575* (CAS, DUKE, ENCB).—VERACRUZ: Potrero de Consoquitla, *Liebmann 14353* (C); 1 km from Río Tonalá on Hwy 180, *Tucker 2539* (DUKE).

Cyperus tenuis is separated from others of the genus in Mexico by the combination of ovoid to ovoid-cylindric spikes, greenish, slender spikelets (at most 1.1 mm wide), and lincar to narrowly oblong achenes. Differences between this species and *C. thyrsiflorus*, with which it has been frequently confused, are summarized below:

1. Spikes densely ovoid; spikelets slightly flattened, ca. 1.5 times wider than thick; scales laterally greenish white, often red-streaked, the apex overlapping the lower 1/4–1/3 of the next scale above. *C. tenuis*.
1. Spikes loosely oblong-cylindric; spikelets subcylindric, about as wide as thick; scales laterally reddish brown or tawny, the apex barely reaching the base of the next scale above. *C. thyrsiflorus*.

The report of *C. tenuis* from Baja California by Wiggins (1980) was apparently based on misidentifications of *C. pallidicolor* and *C. hermaphroditus*, e.g., *Chambers 901* (UC).

83. Cyperus thyrsiflorus Junghuhn, Linnaea 6: 24. 1831.—TYPE: MEXICO. Veracruz: Jalapa, Aug 1828, *Schiede 849* (lectotype, here designated: HAL!).

Cyperus dissitiflorus Nees ex Torrey, Ann. Lyceum Nat. Hist. New York 3: 266. 1836. *Mariscus dissitiflorus* (Torrey) C. B. Clarke in Urban, Symb. Antill. 2(1): 45. 1900.—TYPE: U.S.A. Louisiana: New Orleans, *Ingalls s.n.* (holotype: NY!).

Cyperus anceps Liebmann, Mexic. halvgr. 25. 1850.—TYPE: CUBA. Bejucal, Apr 1843, *Liebmann 14351* (holotype: C!).

Mariscus pallens Liebmann, Mexic. halvgr. 42. 1850. *Cyperus regiomontanus* var. *pallens* (Liebmann) Kükenthal, Pflanzenreich 4(20): 527. 1936. *Cyperus pallens* (Liebmann) Standley & Steyermark, Fieldiana, Bot. 24(1): 133. 1958.—TYPE: MEXICO. Oaxaca: Trapiche de la Concepción, Aug 1842, *Liebmann 14380* (holotype: C!).

Mariscus tribrachiatus Liebmann, Mex. halvgr. 42. 1850. *Cyperus tribrachiatus* (Liebmann) Kükenthal, Pflanzenreich 4(20): 489. 1936.—TYPE: COSTA RICA. Aguacate, *Oersted 14405* (holotype: C!).

Perennial, (15–) 20–40 (–65) cm tall. Rhizomes (rarely collected) 1–2 (–3) cm long, 2–6 mm in diameter, clothed with dark purple scales, indurate, horizontal. Culms 1–2 mm in diameter, trigonous, smooth. Leaves 3–10, (10–) 20–40 cm long, 2.0–3.5 (–7) mm wide, v-shaped, the margins, keels, and veins adaxially ciliate-scabrellate. Inflorescence bracts (4–) 5–7 (–8), (4–) 10–30 (–35) cm long, (1–) 2–4 (–7) mm wide, v-shaped, the margins, keels and adaxial surface ciliate-scabrellate, ascendent at ca. 30°. Rays (5–) 6–8 (–11), 0.5–4.0 (–7.5) cm long; secondary rays absent. Spikes (15–) 25–35 (–40) mm long, 12–18 (–22) mm wide, loosely oblong-cylindric, the largest rarely with 1–3 small spikes borne at the base; rachis 13–25 (–30) mm long. Spikelets (20–) 40–80 (–100), (4.5–) 6.0–9.0 (–11.0), (0.5–) 0.6–0.7 (–1.0) mm wide, linear to narrowly ellipsoid, subcylindric, scarcely compressed; rachilla deciduous, 0.2–0.25 mm wide, ca. 0.1 mm thick, flexuous, greenish to stramineous, successive scales (1.8–) 3.0–4.0 mm apart, wings 0.4–0.5 mm wide, hyaline. Scales persistent, (1–) 2–3 (–4), (2.6–) 2.8–3.2 (–3.4) mm long, (1.0–) 1.2 (–1.4) mm wide, oblong, apex acute, mucronulate to mucronate [the mucro 0–0.1 (–0.2) mm long], laterally 3-nerved, light brown to tawny, medially (1–) 3-nerved, green. Style (0.4–) 0.6–1.0 (–1.2) mm long; stigmas 3, (1.5–) 2.0–3.0 (–4.0) mm long, persistent on the achene and prominently exserted from the scales. Stamens 3; filaments 2.8–4.0 mm long; anthers 0.6–0.8 mm long, oblong, the connective apex up to 0.1 mm long, red. Achenes 1.8–2.1 mm long, (0.4–) 0.5 (–0.7) mm wide, trigonous, narrowly oblong, apex apiculate, acute, base stipitate, 0.1–0.2 mm

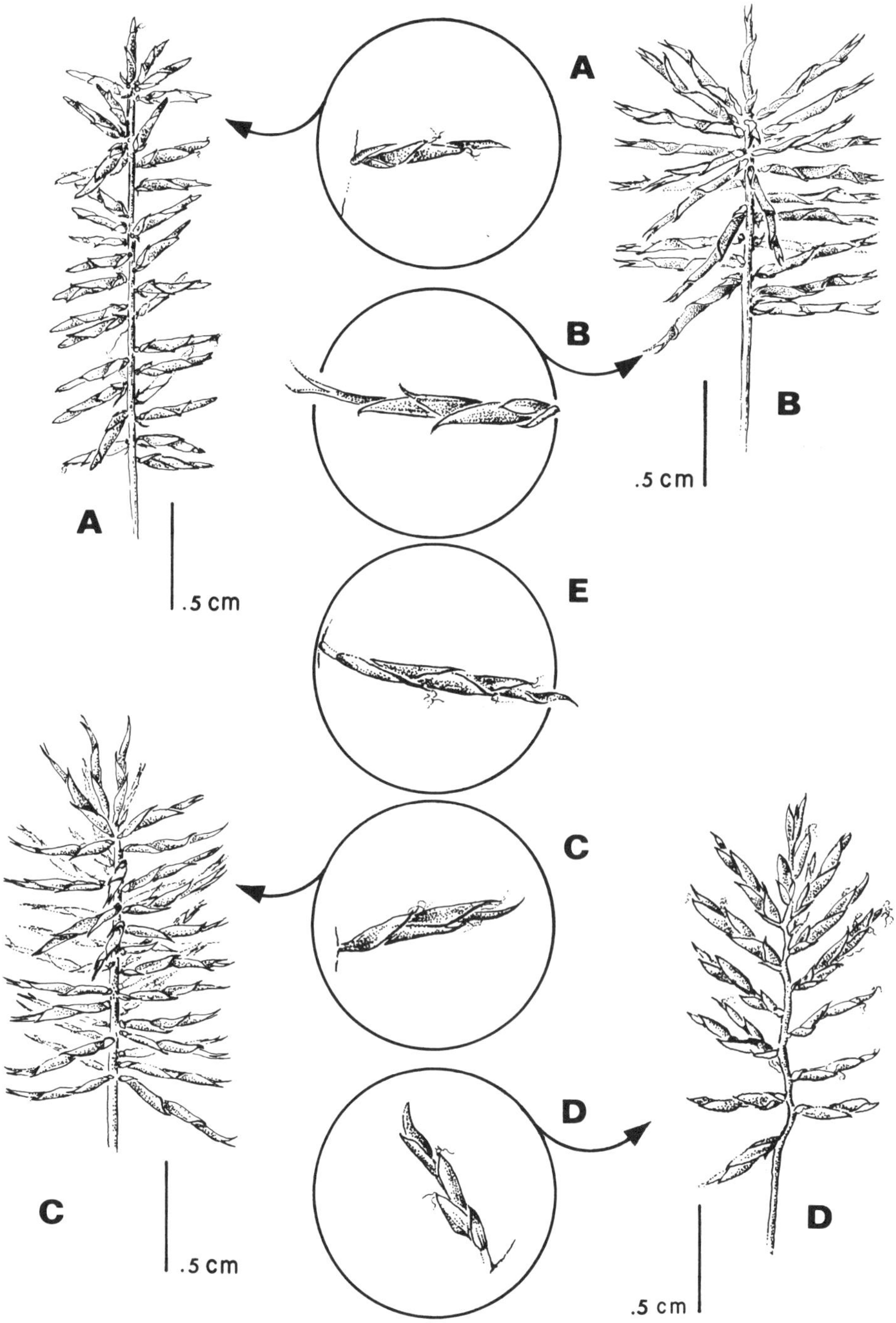

FIG. 70. Spikes and spikelets (×6) of species of *Cyperus* subg. *Cyperus*. A. *C. hermaphroditus*. B. *C. panamensis*. C. *C. thyrsiflorus*. D. *C. retroflexus*. E. *C. lentiginosus* (spikelet only). Drawn by Patricia M. Kernan.

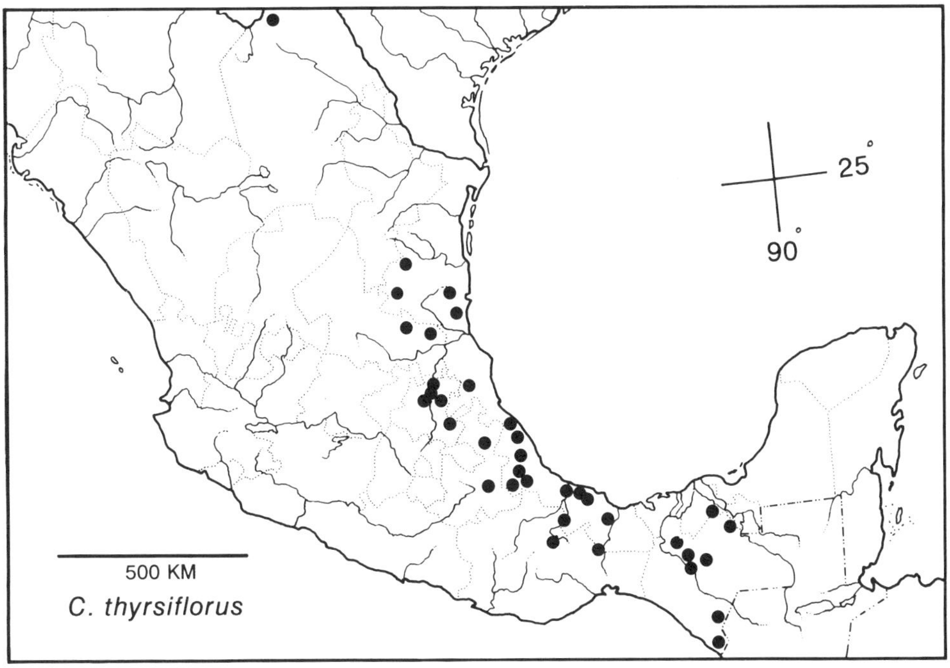

FIG. 71. Distribution of *Cyperus thyrsiflorus* in Mexico.

long, 0.1–0.2 mm wide, the adaxial face flat, slightly curved, the abaxial faces slightly convex, the surface punctate, brown to reddish brown. Fig. 70C.

Phenology. Flowering and fruiting throughout the year, but most frequently from May to September.

Distribution. In the southern U.S.A. from southeastern Georgia, southern Alabama, southern Arkansas, and Texas, south through eastern Mexico and Central America to Colombia, also in the West Indies to Trinidad. In Mexico (Fig. 71) from Tamaulipas to Chiapas along the Gulf Coast and eastern slopes of the Sierra Madre Oriental; wet to somewhat dry montane forests, thorn scrub, and roadsides; sea level to ca. 1000 m near Jalapa, Veracruz, and ca. 1500 m near San Cristóbal, Chiapas.

REPRESENTATIVE SPECIMENS. **Mexico.** CHIAPAS: 25 km N of Tapilula on Hwy 195, *Tucker 2621*, *2628* (DUKE); 1 km E of Ixtacomitán, *Tucker 2609* (DUKE).—COAHUILA: Del Carmen Mts, *Marsh 835* (TEX).—HIDALGO: between Huejutla and Molango, *Daniel 401* (MICH); Jacala, 14 Aug 1937, *Edwards s.n.* (US); between Santa Ana and Chapulhuacán, *Moore & Wood 3965* (A, BH, MICH, UC).—OAXACA: 59 km N of Matías Romero, *Francoeur 21* (MICH); 25 km N of Ixtlán de Juárez, *Tucker 2818* (DUKE).—PUEBLA: 3 km SW of Villa Juárez, *Marcks 798* (WIS); 10 km N of Huauchinango, *Moreno 112* (ARIZ, ASU, ENCB).—SAN LUIS POTOSÍ: 12 mi SW of Tamazunchale, *Duncan 2412* (UC); 7–8 mi NE of Xilitla, *King 4347* (MICH, UC).—TAMAULIPAS: Hwy 180, ca. 8 km N of Aldama, *Tucker 2372* (DUKE, ENCB, MICH); Jaumave, *Viereck 708* (US).—VERACRUZ: région d'Orizaba, *Bourgeau 2631* (C, M, NY); Volcán San Martín, *Denton 1831* (UC); Colipa, *Liebmann 14401* (C); Jalapa, *Pringle 7815* (GH, MICH, VT); above Tuxpango, *Sharp 441719* (TENN); 10 km S of Tecolapan, near Catemaco, *Tucker 2513* (DUKE, ENCB, MICH).

Kükenthal (1935–36) cited eight collections of *C. thyrsiflorus* from Mexico. Benedict Ayers (1946) did not see any of these and included the species in his treatment of *Cyperus* in Mexico only because Kükenthal had cited it; however, Ayers cited as "*C. tenuis*" six collections of *C. thyrsiflorus*. *Cyperus thyrsiflorus* does appear to be closely related to *C. tenuis*; the two can be distinguished as discussed under *C. tenuis* (no. 82).

84. Cyperus hermaphroditus (Jacquin) Standley, Contr. U.S. Natl. Herb. 18: 88. 1916. *Carex hermaphrodita* Jacquin, Collectanea 4: 174. 1791. *Mariscus jacquinii* H. B. K., Nov. gen. sp. 1 [quarto]: 216. 1816, nom. superfl. *Mariscus hermaphroditus* (Jacquin) C. B. Clarke in Urban, Symb. Antill. 2(1): 165. 1900.—TYPE: VENEZUELA. Based on a cultivated plant derived from a wild plant, which "in udis crescit ad Caracas," (holotype: not located). [The application of the name follows Kükenthal (1935–36) and is also based Jacquin's plate, Icon. pl. rar. 3(11), t. 615. 1792].

Kyllinga incompleta Jacquin, Collectanea 4: 101. 1791. *Cyperus incompletus* (Jacquin) Link, Hort. bot. berol. 1: 319. 1827. *Mariscus patulus* Schrader ex Nees in Martius, Fl. bras. 2(1): 47. 1842, nom. superfl.—TYPE: Based on a cultivated plant, derived from a wild plant collected near Caracas (holotype: not located). [The application of the name is based Jacquin's plate, Icon. pl. rar. 2, t. 300. 1792].

Cyperus alpinus Liebmann, Mexic. halvgr. 27. 1850.—TYPE: MEXICO. Puebla: Chinautla, *Liebmann s.n.* (holotype: C!; isotype: HAL!).

Mariscus longiradiatus Liebmann, Mexic. halvgr. 44. 1850. *Cyperus hermaphroditus* var. *longiradiatus* (Liebmann) Kükenthal, Pflanzenreich 4(20): 489. 1936.—TYPE: MEXICO. Veracruz: Mirador, *Liebmann 14379* (lectotype, designated by Tucker, 1986: C!; isolectotype: C!).

Mariscus ambiguus Liebmann, Mexic. halvgr. 46. 1850. *Cyperus subambiguus* Kükenthal, Pflanzenreich 4(20): 527. 1936, non *Cyperus ambiguus* Liebmann, 1850.—TYPE: MEXICO. Oaxaca: Hda. de Castresana, Jun 1842, *Liebmann 14376* (holotype: C!).

Cyperus longespicatus Boeckeler, Allg. Bot. Zeitschr. Syst. 2: 18. 1896. *Cyperus hermaphroditus* var. *longespicatus* (Boeckeler) Kükenthal, Pflanzenreich 4(20): 489. 1936.—TYPE: BRAZIL. *Herb. Mendonça 910* (holotype: not located).

Cyperus randuzii Boeckeler, Allg. Bot. Zeitschr. Syst. 2: 17. 1896.—TYPE: COSTA RICA. Guanacaste: Nicoya, Jan 1900, *Tonduz 13743* (holotype: B, not located; isotype: GH!). [Boeckeler named this species for the collector Tonduz but misread his name on the label as "Randuz." Kükenthal (1936) considered this an orthographic error and changed the epithet to *tonduzii*.]

Cyperus randuzii var. *tenuis* Boeckeler, Allg. Bot. Zeitschr. Syst. 2: 18. 1896. *Cyperus tonduzii* var. *tenuis* (Boeckeler) Kükenthal, Pflanzenreich 4(20): 490. 1936.—TYPE: COSTA RICA. San José, *Tonduz s.n.* (holotype: B, not located).

Mariscus jacquinii var. *angustior* C. B. Clarke in Urban, Symb. Antill. 2 (1): 46. 1900. *Cyperus hermaphroditus* var. *angustior* (C. B. Clarke) Kükenthal, Pflanzenreich 4(20): 489. 1936.—SYNTYPES: ST. VINCENT, *Smith 243* (K); MARTINIQUE, *Belanger 792* (K).

Perennial, (10–) 30–80 (–140) cm tall. Rhizomes 1–3 cm long, 4–10 mm in diameter, indurate, horizontal, bearing culms 3–10 mm apart. Culms (0.4–) 1.0–

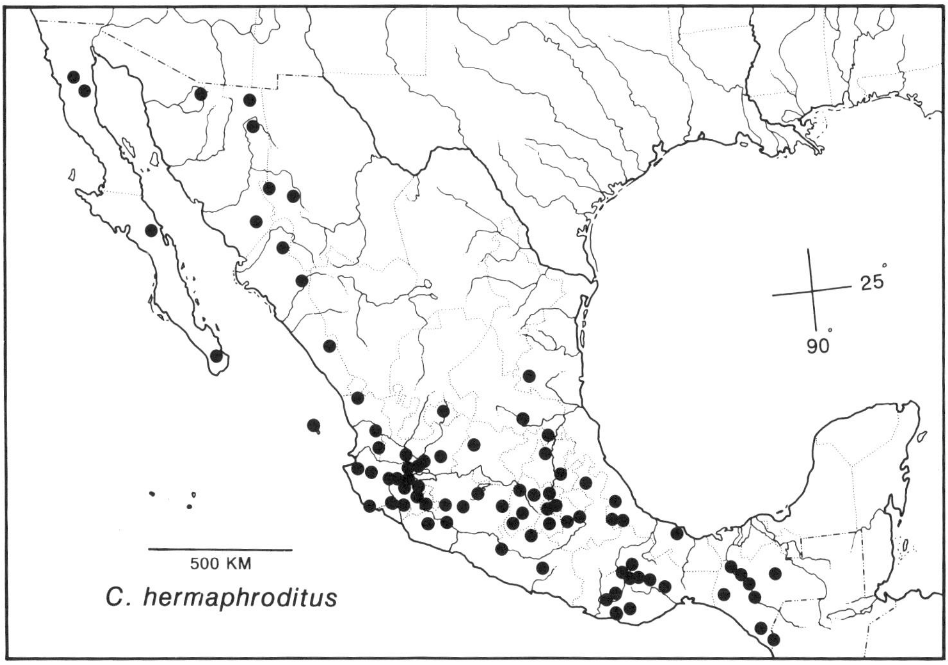

FIG. 72. Distribution of *Cyperus hermaphroditus* in Mexico.

3.0 (–5.0) mm in diameter, trigonous, smooth. Leaves 3–10, (5–) 15–70 (–90) cm long, (2–) 4–10 (–13) mm wide, v-shaped or flanged v-shaped, the margins and keel scabrellate. Inflorescence bracts (2–) 4–8, (2–) 8–40 (–65) cm long, (1–) 2–8 mm wide, v-shaped or inversely w-shaped, the margins and keel scabrellate, ascendent at 30–45°. Rays (3–) 5–10 (–17), (1–) 5–12 (–24) cm long; secondary rays absent. Spikes (10–) 25–45 (–62) mm long, 7–22 mm wide, loosely to densely ovoid or oblong-cylindric; rachis (5–) 15–40 (–55) mm long. Spikelets (10–) 50–150, (3–) 4–8 (–18) mm long, (0.8–) 1–1.2 (–1.6) mm wide, linear to narrowly ellipsoid, quadrate (about as wide as thick); rachilla deciduous, 0.5–0.7 mm wide, ca. 0.1 mm thick, flexuous, light green, successive scales 1.0–1.5 mm apart, wings 0.4–0.5 mm wide, hyaline. Scales persistent, (1–) 3–8, (2.0–) 2.6–3.2 (–4.0) mm long, 1.2–2.0 mm wide, ovate to elliptic, apex obtuse, laterally weakly to strongly 3 (–4) -nerved, pale greenish white to golden-brown, deep brown or reddish brown, medially 1–3-nerved, green (or light brown). Stamens 3; filaments 2.0–3.5 mm long; anthers (0.4–) 0.6–1.0 mm long, oblong-ellipsoid, the connective apex not prolonged. Style 0.4–1.6 mm long; stigmas 3, 1.0–1.5 mm long. Achenes 1.4–1.8 (–2.0) mm long, 0.6–0.8 (–1.0) mm wide, trigonous, ellipsoid, apex acute to obtuse, slightly apiculate, base sessile or nearly so, the faces flat or slightly concave, the surface puncticulate, brown or reddish brown. Fig. 70A.

Phenology. Flowering and fruiting throughout the year.

Distribution. Texas, U.S.A., south through Central and South America to northern Argentina, also in the Lesser Antilles. In Mexico (Fig. 72) in the mountains of Baja California, throughout the Sierra Madre and the Trans-Mexican Volcanic Region, along the Pacific Coast, and in the uplands of Chiapas but absent from the eastern slopes of the Sierra Madre Oriental, the Gulf coastal lowlands, the Isthmus of Tehuantepec, and the Yucatan Peninsula; most often in disturbed areas in moist forests, but also in croplands, roadsides, ditches, arroyos, lava flows, and montane meadows; near sea level to ca. 3000 m.

REPRESENTATIVE SPECIMENS. **Mexico.** AGUASCALIENTES: Presa Rodríguez, *Gallegos 29* (IBUG).—BAJA CALIFORNIA: SE side of La Laguna, *Chambers 901* (UC); Sierra de Guadalupe, *Moran 18805* (SD).—CHIAPAS: Tenejapa, *Breedlove 10780* (DS); Mt. Ovando, *Matuda 2567* (GH, MICH, NY); 20 km S of Tapilula, *Tucker 2633* (DUKE, ENCB).—CHIHUAHUA: Sierra Charuco, Arroyo Hondo, *Gentry 1807* (F, UC, US).—COLIMA: 9–10 km E of Minatitlán, *McVaugh 26243* (MICH).—DISTRITO FEDERAL: Tlalpan, *Fisher 73* (DS, M, US); deleg. Contreras, Rancho La Cañada, *Ventura 3561* (ARIZ, ENCB, SD, WIS).—GUANAJUATO: Collines de Guanajuato, *Dugès 34* in 1893 (GH).—GUERRERO: Dtto. Galeana, Sto. Domingo, *Hinton 14652* (GH, MICH, UC); Omiltemi, *Rzedowski 15930* (DS, ENCB, MICH).—HIDALGO: Dtto. Jacala, Puerto de la Zorra, *Moore 1699* (BH, UC).—JALISCO: Mpio. Zapopan, Las Agujas, 28 Oct 1979, *Humberto s.n.* (IBUG); Est. Biológica Chamela, *Lott 1508* (DUKE, MEXU); NE slopes of Nevado de Colima, *McVaugh 13507* (ENCB, MICH, US).—MÉXICO: 5 km S of Tequisquiac, *Cruz Cisneros 1043* (ARIZ, DS, ENCB); base of Iztaccihuatl, *Herrera 38* (ARIZ, ENCB, IBUG); Cerro Sacramonte, near Amecameca, *Salinas 56* (ARIZ, DS, ENCB, MICH).—MICHOACÁN: vic. Morelia, *Arsène 2791* (NY, US); 8–10 km SW of Jiquilpan, *Feddema 248* (DS, MICH); 18 mi S of Pátzcuaro, *King & Soderstrom 5211* (MICH, NY, UC, US).—MORELOS: Autopista Cuernavaca-México, *Puga 1795* (IBUG).—NAYARIT: mountains 10 mi SE of Ahuacatlán, *Feddema 273* (DS, MICH); Isla María Madre, *Ferris 5664* (DS).—OAXACA: 8 mi SW of Sola de Vega, *Davidse 9677* (MO); San Pedro Nolasco, *Galeotti 5874* (F, GH, MICH, NY, UC, US); 8 km N of San José del Pacífico, *Tucker 2867* (DUKE, ENCB); 18 km W of Puerto Escondido, *Tucker 2902* (DUKE).—PUEBLA: vic. Puebla, *Arsène 174* (GH, NY, US); 2 km E of Villa Juárez, *González Quintero* (DS, F, MICH).—SAN LUIS POTOSÍ: 12 mi SW of Tamazunchale, *Duncan 2412* (MICH); Salto de Agua, *Purpus 1647* (E, F, GH, NY, UC, US).—SINALOA: Sierra Surutato, 3 mi N of Los Ornos, *Breedlove & Thorne 18292* (CAS).—SONORA: region of Río Bavispe, *Phillips 581* (MICH).—TAMAULIPAS: 6 km NW of Gómez Farias, *Gilbert 47* (TEX).—VERACRUZ: Orizaba, *Bourgeau 2591* (GH, US); above Acultzingo, near El Puerto, *Sharp 44755*, *44756* (TENN); Mpio. Altotonga, Santa Cruz, *Ventura 167* (ARIZ, ASU, ENCB).

Cyperus hermaphroditus is the most variable Mexican species of subg. *Cyperus*, as suggested by the lengthy synonymy. There is considerable variation in the length of the rays, the dimensions of the spikes, and the density of spikelets on the rachis. About 200 collections, from throughout the range of the species, were studied to evaluate the merit of the segregate taxa previously recognized. Four of the five varieties recognized by Kükenthal (1935–36) were cited by him from Mexico: var. *hermaphroditus*, var. *angustior*, var. *longespicatus,* and var. *longiradiatus*. The fifth, var. *condensatus*, was based on a single Bolivian collection. Neither O'Neill (1940) nor Benedict Ayers (1946) recognized any of these varieties.

Kükenthal cited three collections of var. *angustior* from Mexico, describing these plants as having "folia bracteaque angustiora." To me these collections differ in no appreciable way from typical representatives of *C. hermaphroditus*.

Kükenthal characterized var. *longespicatus* as having longer, narrower spikes (approximately 4 cm long) than var. *hermaphroditus* and based var. *longiradiatus* on a single collection (*Liebmann 14379*) with rays "up to 22 cm long." I observed that the variation in length of spikes and longest rays is continuous and does not support recognition of these varieties.

Kükenthal treated plants with densely cylindric, 1-flowered spikes as *C. tribrachiatus*, but a previous study showed them to be indistinguishable from *C. hermaphroditus* (Tucker 1983).

Cyperus hermaphroditus is most similar morphologically to *C. thyrsiflorus* (no. 83) and often confused with it. *Cyperus thyrsiflorus* is readily distinguished by its narrower stipitate achenes (approximately 0.5 mm wide, vs. 0.6–0.8 mm wide in *C. hermaphroditus*), and long, conspicuous stigmas, 1.5–4 mm long (vs. 1–1.5 mm long). In addition, *C. hermaphroditus* is usually 30–80 cm tall, whereas *C. thyrsiflorus* is 20–40 cm tall. The two species are mostly allopatric; *C. thyrsiflorus* occurs in the southern U.S.A., the Gulf lowlands, and on eastern slope of the Sierra Madre Occidental. Both species occur in nearly all of Mexico, except the Chihuahuan Desert, but are found together only in the uplands of Chiapas.

85. Cyperus panamensis (C. B. Clarke) Britton ex P. C. Standley, J. Wash. Acad. Sci. 15: 457. 1925. *Mariscus panamensis* C. B. Clarke, Bull. Misc. Inform. Kew, add. ser. 8: 15. 1908.—Type: Panama. Empire Station, Oct 1861, *Hayes 427* (lectotype, designated by Tucker, 1983: K!; isolectotype: BM!).

Caespitose perennial, (4–) 10–50 (–70) cm tall. Rhizomes (not often collected) 1–2 cm long, 3–7 mm wide, indurate, oblique. Culms 1.0–2.0 (–2.8) mm in diameter, trigonous, smooth. Leaves (1–) 3–4, (7–) 20–40 (–50) cm long, 4–7 mm wide, inversely w-shaped or v-shaped, the margins and nerves on both faces scabrellate. Inflorescence bracts (2–) 4–5 (–6), (2–) 8–30 (–40) cm long, 1–2 mm wide, inversely w-shaped, the margins, keel and nerves scabrellate on both faces, ascendent at 45–75°. Rays (4–) 6–8, 0.2–11 cm long; in some small plants; secondary rays absent; rays sometimes absent, the inflorescence of 1 (–3) sessile spikes. Spikes (10–) 14–45 mm long, 14–23 mm wide, loosely to densely broadly ellipsoid to hemispherical, occasionally in Central American collections with 1–2 smaller secondary sessile spikes at the base of the spike; rachis 8–40 mm long. Spikelets (3–) 12–40 (–100), 7–11 (–17) mm long, 1.0–1.2 mm wide, oblong to linear, cylindric; rachilla deciduous, 0.4–0.6 mm wide, 0.1–0.2 mm thick, flexuous, greenish yellow, often reddish punctate, successive scales 4.5–5.6 mm apart, wings 0.4–0.7 mm wide, hyaline. Scales persistent, 3–4 (–8), (3.5–) 4.0–4.7 mm long (the cusp an additional 0.4–0.6 mm long in the proximal scales, 0.7–1.4 mm long in the distal ones), (1.5–) 1.8–2.0 mm wide, ovate, apex rounded, cuspidate, laterally weakly to moderately 3–5-nerved nearly to the margin, greenish to golden-brown, medially 1–3-nerved, green. Stamens 3; filaments 3–4 mm long; anthers 0.4–0.7 mm long, ellipsoid, the connective apex not prolonged. Style 0.5–0.9 mm long; stigmas 3, 0.9–2.1 mm long. Achenes 2.0–2.8 mm long, 0.8–0.9 mm wide, trigonous, ellipsoid-oblong to narrowly oblong, apex apiculate, base stipitate, (0.1–) 0.2 mm long, the faces flat to slightly concave, the surface punctate-papillose, light brown to reddish brown. Fig. 70B.

Phenology. Flowering and fruiting in Mexico in July, August, and September, in Central America throughout the year but primarily from July to January.

Distribution. Along or near the Pacific Coast from western Mexico (Fig. 73) to Peru; occasional, croplands, brushy thickets, roadsides, gravel along rivers and similar disturbed habitats; sea level to ca. 400 m (a single collection from 1800 m in northeastern Oaxaca).

Additional Specimens Examined. **México**. Chiapas: rd from Escuintla to El Triunfo, *Tucker 2236* (DUKE), *2693*, *Wilbur 35895*, *35901* (DUKE, ENCB); Huixtla, *Wilbur 35872* (DUKE, ENCB); on

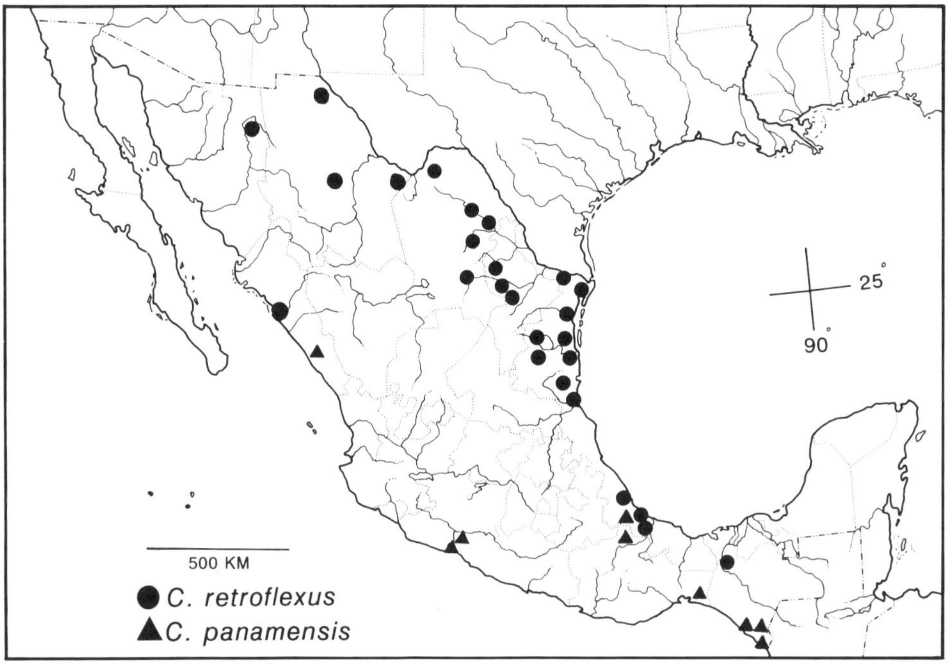

FIG. 73. Distribution of *Cyperus panamensis* and *C. retroflexus* in Mexico.

Hwy 211, ca. 25 km N of Huixtla, *Tucker 2687*, *Wilbur 35846* (DUKE, ENCB, MICH); Hwy 200 ca. 30 km N of Escuintla, *Tucker 2701* (DUKE, ENCB).—GUERRERO: Dtto. Montes de Oca, Vallecitos, *Hinton 11401* (MICH); on Hwy 200, ca. 25 km SE of Petacalco, *Tucker 2946* (DUKE).—OAXACA: on Hwy 200, ca. 8.5 km W of Niltepec, *Tucker 2757* (CAS, DUKE, ENCB, F, GH, MICH); Mpio. Santiago Laxopa, Guiloxi, *Maldonado V. & Martin 274* (MO).—SINALOA: vic. of Labradas, *Ferris & Mexia 5071* (F).—VERACRUZ: Cotaxtla Expt. Sta., 23 Aug 1962, *Janzen s.n.* (UC), *Rodríguez & Ventura 1425* (CHAPA).

Cyperus panamensis is easily distinguished by the combination of cuspidate scales and cylindric spikelets from any other species of *Cyperus* occurring in Mexico or Central America. Benedict Ayers (1946) did not include this species in his treatment.

86. Cyperus lentiginosus Millspaugh & Chase, Publ. Field Columbian Mus., Bot. Ser. 3: 74. 1903. *Cyperus tenuis* var. *lentiginosus* (Millspaugh & Chase) Kükenthal, Pflanzenreich 4(20): 418. 1936.—TYPE: MEXICO. Yucatán: Chichankanab, *Gaumer 1290* (holotype: F!).

Cyperus strigosus var. *gracilis* Britton, Bull. Torrey Bot. Club 13: 212. 1886.—TYPE: U.S.A. Texas: valley of the lower Rio Grande, *Buckley s.n.* (lectotype, here designated: NY!).

Caespitose perennial, (1–) 20–40 (–75) cm tall. Rhizomes 5–15 mm long, 1.5–2.4 mm in diameter, indurate, oblique. Culms 1.0–1.5 (–2.5) mm in diameter, trigonous, smooth. Leaves 3–7, 10–40 (–70) cm long, (1–) 3–6 (–10) mm wide, inversely w-shaped, the margins and veins below scabrellate. Inflorescence bracts

(3–) 5–7 (–10), 10–30 (–50) cm long, (1–) 3–5 (–11) mm wide, inversely w-shaped, the margins and keel scabrellate, ascendent at 30° (–45°). Rays (2–) 3–6 (–12), (1–) 3–10 (–17) cm long; secondary rays absent. Spikes (15–) 20–30 (–50) mm long, 20–30 (–45) mm wide, broadly and loosely ovoid to oblong-ellipsoid; rachis 10–25 (–50) mm long. Spikelets (10–) 25–50 (–100), (8–) 12–15 (–24) mm long, (0.8–) 1.1–1.4 (–1.8) mm wide, linear-lanceolate, subterete (but appearing compressed because of the excurved apices of the scales); rachilla deciduous, ca. 0.3 mm wide, 0.1 mm thick, flexuous, green, successive scales (2.2–) 2.6–3.4 mm apart, wings ca. 0.3 mm wide, hyaline. Scales persistent, (4–) 6–8 (–14), (3.2–) 4.0–4.5 mm long, 1.2–1.6 mm wide, oblong-lanceolate, the mucronulate to mucronate apex 0.2–0.5 mm long: the proximal scales mucronulate and the distal scales with mucros 0.4–0.6 mm long, laterally 3–4-nerved, off-white to stramineous, densely red-speckled, medially green, 3 (–5) -nerved. Stamens 3; filaments 3.5–4.4 mm long; anthers 0.5–0.8 (–1.4) mm long, oblong to linear-oblong, the connective apex less than 0.1 mm long, reddish. Style 0.8–1.0 mm long; stigmas 3, 1.0–2.0 mm long. Achenes 1.7–2.0 mm long, 0.6–0.7 mm wide, trigonous, oblong-ellipsoid, apex rounded, apiculate owing to the dark purple style base, base 0.1 (–0.2) mm long, ca. 0.2 mm wide, subsessile to stipitate, the adaxial face convex, the abaxial faces flat, the surface papillose, brown. Fig. 70E.

Phenology. Flowering and fruiting from April to December.

Distribution. Along the Caribbean Coast from southern Texas, U.S.A., to Belize and Guatemala, also in southern Florida, U.S.A., and Cuba; edges of forests, thickets, roadsides, and open land; sea level to ca. 500 m.

REPRESENTATIVE SPECIMENS. **Mexico.** TAMAULIPAS: 15 mi W of González toward Mante, *Johnston 4930B* (TEX); 9 mi W of Victoria, *Kenoyer & Crum 3353* (A); vicinity [of] Victoria, *Palmer 413* in 1907 (F, GH, NY, US).—QUINTANA ROO: Tancah, *Swallen 2806* (MICH); Cozumel Islands, San Miguel, *Swallen 2859* (MICH); orilla de la Laguna de Coba, *Téllez et al. 3762* (BM, MEXU).—SAN LUIS POTOSÍ: Mpio. Tampamolon, Tamarindio, *Alcorn 2026* (CHAPA, TEX, WIS).—YUCATÁN: Chichankanab, *Gaumer 1282, 1286* (F, GH); San Felipe, *Gaumer 1402* (F, GH, NY); Chicxulub, *Gaumer 23416* (BM, C, F, NY).

Cyperus lentiginosus is recognized by the mucronate, excurved apices of the distal scales. Only two other species of subg. *Cyperus* have such conspicuous apices, *C. panamensis* and *C. confertus* Sw. The latter has ovoid achenes and is endemic to the West Indies. The former, although very similar to *C. lentiginosus*, has closely appressed scales with longer mucros (0.5–1 mm long) and occurs along the Pacific Coast of Mexico and Central America.

Kükenthal (1935–36) treated *C. lentiginosus* as a variety of *C. tenuis*; however, the two species differ in several features. In addition to the longer scales with cuspidate apices, *C. lentiginosus* has inversely w-shaped leaves and bracts and open spikes; in *C. tenuis* the leaves and bracts are v-shaped and the spikes are dense.

87. Cyperus retroflexus Buckley, Proc. Acad. Nat. Sci. Philadelphia 14: 9. 1862. *Cyperus uniflorus* var. *retroflexus* (Buckley) Kükenthal, Pflanzenreich 4(20): 521. 1936.—TYPE: U.S.A. "Northern Texas," *Buckley s.n.* (holotype: PH!).

Cyperus uniflorus Torrey & Hooker, Ann. Lyceum Nat. Hist. New York 3: 431. 1836, non *Cyperus uniflorus* Thunberg, 1825. *Mariscus uniflorus* (Torrey & Hooker) Steudel, Syn. pl. glum. 2: 64. 1854.—TYPE: U.S.A. Texas, *Drummond 287* (holotype: NY!).

Cyperus uniflorus var. *floribundus* Kükenthal, Pflanzenreich 4(20): 521. 1936.—TYPE: MEXICO. Tamaulipas: vic. of Victoria, May–Jun 1907, *Palmer 397* in 1907 (lectotype, here designated: B!; isolectotypes: C! E! F! NY!).

Cyperus uniflorus var. *pseudothyrsiflorus* Kükenthal, Pflanzenreich 4(20): 521. 1936.—TYPE: MEXICO. Nuevo León: Sierra Madre near Monterrey, 30 Jun 1888, *Pringle 1966* (holotype: B!; isotypes: BR! UC!).

Caespitose perennial, (5–) 15–40 (–80) cm tall. Rhizomes 0.5–2.5 cm long, 3–8 mm in diameter, indurate, horizontal. Culms 1.0–2.5 mm in diameter, trigonous, smooth, tuberous-thickened basally. Leaves 3–10, (5–) 10–30 (–60) cm long, (0.5–) 1.0–3.0 (–4.0) mm wide, flat to v-shaped, the margins and keel scabrellate. Inflorescence bracts 3–6, (1–) 3–15 (–30) cm long, 0.5–4.5 mm wide, v-shaped, the margins and keel scabrellate, horizontal to ascendent at 30°. Rays 3–6 (–9), (5–) 10–32 cm long; secondary rays absent. Spikes 8–36 mm long, 8–20 mm wide, loosely to densely ovoid to cylindric; rachis 5–20 mm long. Spikelets (5–) 20–35 (–65), (2.2–) 4–10 (–18) mm long, 0.7–1.1 mm wide, linear to oblong, subterete; rachilla deciduous, ca. 0.6 mm wide, 0.1–0.2 mm thick, flexuous, greenish to stramineous, successive scales 3.5–4.5 (–5.4) mm apart, wings 0.5–0.7 mm wide, hyaline. Scales persistent, 1–3 (–5), 2.6–4.0 mm long, 1.6–2.0 mm wide, narrowly ovate to oblong-elliptic, apex rounded to subacute, mucronate or conspicuously cuspidate, the cusp 0.7–1.1 mm long in distal scales, laterally 3–6-nerved, pale greenish white, stramineous with reddish dots, or deep red, medially greenish or light brown, smooth or scabrellate distally; the sterile terminal scale of a spikelet only ca. 0.4–0.6 mm wide, involute, frequently uncinate. Stamens 3; filaments 2.4–4.0 mm long; anthers (0.3–) 0.4–0.7 mm long, narrowly ellipsoid, the connective apex not prolonged. Style 0.4–0.5 mm long; stigmas 3, 0.5–0.9 mm long. Achenes (1.8–) 2.0–2.4 mm long, 0.6–0.8 (–0.9) mm wide, trigonous, ellipsoid, apex slightly apiculate, base sessile or slightly stipitate, the faces slightly concave, the surface puncticulate, light brown to reddish brown. Fig. 70D.

Phenology. Flowering and fruiting in July, August, and September, a few collections from the remaining months.

Distribution. Mississippi and Arkansas to Arizona, U.S.A., south to eastern Veracruz, Tabasco, Nuevo León, Chihuahua, and Sinaloa, Mexico (Fig. 73); frequent, fallow fields, brushy slopes and pastures, pine, oak and cherry forests, semidesert scrub and dry roadsides; sea level to 2000 (–2300) m.

REPRESENTATIVE SPECIMENS. **Mexico.** CHIHUAHUA: 6 mi S of Samalayuca, *Henrickson 7499b* (LL); sandy plains near Chihuahua, *Pringle 807* (CAS, DS, F, M, MICH, UC, VT); near Paso del Norte, *Pringle 808* (BM, CAS, E, F, MICH, UC, VT).—COAHUILA: Cañón del Alamo, Sierra Maderas del Carmen, *Henrickson 15047* (LL); Mpio. Castaños, *Muller 3109* (F, LL, TENN, UC).—NUEVO LEÓN: near Monterrey, *Pringle 1966* (BR, E, UC); between China and Reynosa, *Smith 5453* (TEX).—SINALOA: Altata, *Gentry 5453* (ARIZ).—SONORA: Río de Bavispe, Cañón del Carricito, *White 3098* (ARIZ).—TABASCO: carreterra Chiltepec a Aquiles Serdán, *Magaña et al. 59* (CHAPA).—TAMAULIPAS: Papalote de la Mirandena, *Crutchfield & Johnston 5565* (MICH, TEX); E of San José, *LeSueur 21* (F, TEX); vicinity of Victoria, *Palmer 397* (C, F, NY); Hwy 101, 40 km S of San Fernando, *Tucker 2358* (CAS, DUKE, ENCB, MICH); 25 km N of Tampico, *Tucker 2384* (DUKE, ENCB).—VERACRUZ: 1 km E of Alvarado, *Marcks 862* (MICH, WIS); SE of Veracruz, 3 km SE of Buenavista, *Tucker 2500* (DUKE, MICH).

Cyperus retroflexus (Fig. 70D) is distinguished from others with deciduous spikelets by the involute, sterile terminal scale of the spikelet. In plants with cuspidate scales, the terminal scale is usually somewhat recurved, giving the appearance of a "hook" at the end of the spikelet.

Kükenthal recognized four varieties in *C. uniflorus* [=*C. retroflexus*]. Horvat (1941), who studied about 50 collections of *C. retroflexus* from the U.S.A. and Mexico, did not recognize Kükenthal's varieties; I agree with Horvat. For example, Kükenthal defined var. *floribundus* as having lax spikes, spikelets up to 13 mm long, and scales with a long, excurved mucro; he distinguished var. *uniflorus* as having subdense spikes, spikelets 6–8 mm long, and subobtuse scales. Study of 55 collections from Texas and Mexico revealed no discontinuity in these characters between var. *uniflorus* and var. *floribundus*. Spikelet length ranges from 2.2–18.0 mm, without obvious discontinuities, and the length of the mucronate to cuspidate scale apices show a range of 0.1–1.1 mm. The plants included by Kükenthal in var. *floribundus* merely represent a portion of the range of morphological variation present in *C. retroflexus*.

Kükenthal (1935–36), Benedict Ayers (1946), and Correll and Johnston (1970) applied to this species the illegitimate name *Cyperus uniflorus* Torrey & Hooker, even though Kükenthal listed the name *C. uniflorus* Thunberg in his list of "Species incertae sedis vel inextricabiles." Thunberg's name was validly published; a diagnosis was provided for each new species on the last page of the *Florula Javanica*. I also annotated specimens as "*C. uniflorus*" prior to 1985.

88. Cyperus retrorsus Chapman, Bot. Gaz. 3: 17. 1878. *Mariscus retrorsus* (Chapman) C. B. Clarke, Bull. Misc. Inform. Kew, add. ser. 8: 15. 1908.—TYPE: U.S.A. Florida: Caximbas Bay, Robert's Key, *Chapman s.n.* (holotype: NY!).

Mariscus cylindricus Elliott, Sketch Bot. S. C. Ga. 1: 74. 1816. *Cyperus ovularis* var. *cylindricus* (Elliott) Torrey, Ann. Lyceum Nat. Hist. New York 3: 279. 1836. *Cyperus cylindricus* (Elliott) Britton, Bull. Torrey Bot. Club 6: 339. 1879, non *Cyperus cylindricus* Boeckeler, 1859. *Cyperus retrorsus* var. *cylindricus* (Elliott) Fernald & Griscom, Rhodora 37: 152. 1935.—TYPE: U.S.A. South Carolina or Georgia, *Elliott s.n.* (holotype: CHARL).

Caespitose perennial, 20–50 (–85) cm tall. Rhizomes (0.5–) 1–2 cm long, 2–4 mm in diameter, indurate, cormose, ± horizontal. Culms 1–2 (–4) mm in diameter, trigonous, smooth. Leaves 3–6 (–10), 10–40 (–55) cm long, 1.5–4 (–5) mm wide, v-shaped to inversely w-shaped, the margins and keel scabrellate. Inflorescence bracts (3–) 4–6 (–10), (4–) 14–30 (–55) cm long, 1–5 mm wide, flat to broadly v-shaped, the margins and keel scabrellate, ascendent at 30–45°. Rays 4–8 (–11), 0.5–3.5 (–8.5) cm long; secondary rays absent. Spikes 8–16 mm long, 4–12 (–16) mm wide, densely oblong-ovoid; rachis 6–15 mm long. Spikelets 40–120, 3–7 mm long, 0.8–1.6 mm wide, oblong-lanceolate, subterete; rachilla deciduous, 0.2–0.4 mm wide, 0.1–0.2 mm thick, flexuous, greenish white, successive scales 1.0–1.2 mm apart, wings 0.2–0.3 mm wide, hyaline. Scales persistent, 2–5, 1.8–2.5 mm long, 1.0–1.5 mm wide, oblong to oblong-lanceolate, apex obtuse, laterally 3–4-nerved, greenish to yellowish white, medially 1 (–3) -nerved, greenish. Stamens 3;

filaments 1.6–2.4 mm long; anthers 0.3–0.5 mm long, oblong, the connective apex not prolonged. Style 0.5–0.8 mm long; stigmas 3, 0.6–0.9 mm long. Achenes 1.2–1.7 mm long, 0.5–0.6 mm wide, trigonous, slightly dorsiventrally flattened, oblong-ellipsoid, apex subacute, base cuneate, the faces flat to slightly convex, the surface papillose, brown.

Phenology. In Mexico probably flowering and fruiting from June through August.

Distribution. New York, Kentucky, and Missouri, south to Florida and Texas, U.S.A.; in Mexico known from a single collection from northeastern Tamaulipas in a depression in a cultivated field at 50–100 m.

ADDITIONAL SPECIMEN EXAMINED. **Mexico.** TAMAULIPAS: on Hwy 101, ca. 60 km S of Matamoros, 14 Jul 1984, *Tucker 2342* (DUKE).

Cyperus retrorsus can be distinguished from other Mexican species of *Cyperus* by its inflorescence composed of dense oblong-ovoid spikes terminating the primary rays.

V. Cyperus subgenus **Diclidium** (Nees) C. B. Clarke, J. Linn. Soc., Bot. 21: 34, 191. 1884. *Diclidium* Schrader ex Nees in Martius, Fl. bras. 2(1): 51. 1842.—LECTOTYPE, designated by Goetghebeur, 1989: *Diclidium odoratum* Nees [=*Cyperus odoratus* L.].

Cyperus subgenus *Torulinium* (Desvaux) Kükenthal, Pflanzenreich 4(20): 614. 1936. *Torulinium* Desvaux in W. Hamilton, Prodr. pl. Ind. occ. 15. 1825. *Mariscus* subg. *Torulinium* (Desvaux) C. B. Clarke in Hooker f., Fl. Brit. India 6: 624. 1893.—LECTOTYPE, designated by Kern, 1974: *Torulinium ferax* (L. C. Richard) W. Hamilton [=*Cyperus odoratus* L.].

Culms and leaves with kranz (chlorocyperoid) anatomy. Spikelets borne in loose to dense ovoid to cylindric spikes. Rachilla articulate at the base of each scale, the mature spikelets breaking up into 1-fruited segments each consisting of an internode of the rachilla, an attached scale, and an achene clasped by the rachilla wings. Scales persistent. Carpels 3, stigmas 3. Achenes trigonous, slightly dorsiventrally flattened.

Subgenus *Diclidium* comprises 8 species in pantropical and warm-temperate regions. Four species occur in the New World, but only the most widespread species of this subgenus, *C. odoratus*, occurs in Mexico. The other New World species are *C. hayesii* of the Pacific Coast of Central America (Tucker 1983), *C. filiformis* Sw. of the West Indies (Koyama 1978), and *C. correllii* (T. Koyama) G. C. Tucker of the Bahama Islands (Tucker 1984a).

89. Cyperus odoratus L., Sp. pl. 1: 46. 1753. *Diclidium odoratum* Schrader ex Nees in Martius, Fl. bras. 2(1): 54. 1842. *Torulinium odoratum* (L.) Hooper, Kew Bull. 26: 579. 1972.—LECTOTYPE: Sloane, Voy. Jam. Nat. Hist. 1: 116. t. 74, fig. 1, 1707 (see Dandy, 1944, and Tucker, 1984a).

Cyperus ferax L. C. Richard, Actes Soc. Hist. Nat. Paris 1: 106. 1792. *Mariscus ferax* (L. C. Richard) C. B. Clarke in Hooker f., Fl. Brit. India 6: 624. 1894. *Torulinium ferax* (L. C. Richard) C. B. Clarke in Urban, Symb. Antill. 2: 165. 1900.—TYPE: FRENCH GUIANA. Cayenne, *Leblond s.n.* (holotype: P, IDC microfiche!; isotype?: G!).

Cyperus speciosus Vahl, Enum. pl. 2: 364. 1805. *Cyperus ferax* subsp. *speciosus* (Vahl) Kükenthal, Pflanzenreich 4(20): 619. 1936.—TYPE: U.S.A. Virginia, *Vaillant s.n.* (holotype: P, see Fernald & Griscom, 1935).

Mariscus huarmensis H. B. K., Nov. gen. sp. 1 [quarto]: 214. 1816. *Cyperus huarmensis* (H. B. K.) M. C. Johnston, Southw. Naturalist 11: 124. 1966.—TYPE: PERU. Huarmey et Lima, *Humboldt & Bonpland s.n.* (holotype: P-HBK, IDC microfiche!).

Cyperus haenkei J. Presl & K. Presl in K. Presl, Reliq. Haenk. 1: 172. 1828.—TYPE: MEXICO. *Haenke s.n.* (not located at PR or PRC).

Diclidium aciculare Schrader ex Nees in Martius, Fl. bras. 2(1): 55. 1842. *Cyperus acicularis* (Nees) Steudel, Syn. pl. glum. 2: 45. 1854. *Cyperus ferax* var. *acicularis* (Nees) Kükenthal, Pflanzenreich 4(20): 619. 1936.—TYPE: BRAZIL. Bahia, *Martius s.n.* (holotype: M).

Cyperus macrocephalus Liebmann, Mexic. halvgr. 33. 1850. *Torulinium macrocephalum* (Liebmann) T. Koyama, Phytologia 29: 74. 1974.—TYPE: MEXICO. Veracruz: Río Nautla at Pital, *Liebmann 14367* (holotype: C!, photos: F! MICH!; isotypes: GH! HAL! K! LL! MICH!).

Cyperus cubanus Liebmann, Mexic. halvgr. 34. 1850.—TYPE: CUBA. Near Havana, Apr 1845, *Liebmann 14355* (holotype: C!).

Cyperus rufinus Liebmann, Mexic. halvgr. 34. 1850.—TYPE: MEXICO. Puebla: Tehuacán, San Lorenzo, Dec 1842, *Liebmann 14373* (holotype: C!).

Cyperus sanctae-crucis Liebmann, Mexic. halvgr. 35. 1850.—TYPE: VIRGIN ISLANDS. St. Croix, Lagunen ved Frederiksstad, *Oersted 14374* (holotype: C!).

Cyperus granadinus Liebmann, Mexic. halvgr. 36. 1850.—TYPE: NICARAGUA. Granada, Jun [1847], *Oersted 14363* (holotype: C!).

Cyperus oerstedii Liebmann, Mexic. halvgr. 36. 1850.—TYPE: NICARAGUA. Segovia, Jan 1848, *Oersted 14372* (holotype: C!).

Cyperus fossarum Liebmann, Mexic. halvgr. 37. 1850.—TYPE: MEXICO. Oaxaca: near Tehuantepec by Hacienda de Sta. Crux, "in fossis," Dec. 1842, *Liebmann 14359* (holotype: C!; isotypes: GH! K!).

Cyperus fragilis Liebmann, Mexic. halvgr. 38. 1850.—TYPE: MEXICO. Veracruz: coast region by Paso de Doña Juana, Feb 1841, *Liebmann 14360* (holotype: C!; isotype: K!).

Cyperus engelmannii Steudel, Syn. pl. glum. 2: 47. 1854. *Cyperus ferax* subsp. *engelmannii* (Steudel) Kükenthal, Pflanzenreich 4(20): 620. 1936.—TYPE: U.S.A. Illinois: Cahokia, Sep 1845, *Engelmann s.n.* (holotype: P?; isotype: GH!).

Cyperus californicus S. Watson, Bot. Calif. 2: 216. 1880.—TYPE: U.S.A. California, *Fitch s.n.* (holotype: GH!).

Cyperus oxycarioides Britton, Bull. Torrey Bot. Club 11: 86. 1884.—TYPE: U.S.A. Texas: valley of the lower Río Grande, 1879–83, *Buckley s.n.* (holotype: NY!).

Cyperus eggersii Boeckeler, Beitr. Cyper. 1: 53. 1888. *Torulinium eggersii* (Boeckeler) C. B. Clarke in Urban, Symb. Antill. 2: 56. 1900. *Torulinium macrocephalum* var. *eggersii* (Boeckeler) C. D. Adams, Ann. Missouri Bot. Garden 78: 254. 1991.—TYPE: DOMINICAN REPUBLIC. Santo Domingo, near Batey on Río Yasica, 23 Jun 1887, *Eggers 2627* (holotype: B!).

Caespitose perennial or annual, (4–) 10–50 (–130) cm tall. Rhizomes absent or occasionally present, 0.5–2.0 cm long, 0.5–1.5 cm in diameter, indurate, horizontal to slightly oblique. Culms (0.5–) 1–4 mm in diameter, trigonous, smooth. Leaves (2–) 3–6 (–10), 10–65 cm long, 4–12 mm wide, flanged v-shaped or inversely w-shaped, the margins and keel scabrellate to spinulose-scabrellate. Inflorescence bracts (4–) 5–8 (–10), (3–) 10–25 (–55) cm long, 1–14 mm wide, the margins and keel spinulose-scabrellate, horizontal to ascendent at 30–60°. Rays (2–) 6–9 (–12), 2–8 (–13) cm long; or rays infrequently absent, the inflorescence a single dense capitate cluster of closely imbricate spikes; secondary rays absent. Spikes 1–5 (–12), 10–20 (–40) mm long, (8–) 10–30 (–35) mm wide, loosely to densely cylindric, ovoid or subpyramidal. Spikelets (10–) 20–60 (–100), (5–) 8–15 (–38) mm long, 0.8–1.3 (–1.9) mm wide, linear-oblong to narrowly linear, cylindric to subcylindric; rachilla articulate at the bases of the scales at maturity, 0.3–0.4 mm wide, 0.1–0.4 mm thick, straight to slightly flexuous, becoming corky or spongy-thickened with achene maturation, stramineous, successive scales 1.5–2.2 (–3.0) mm apart, wings 0.2–0.5 mm wide, membranous to corky, stramineous to bronze or carmine, clasping the mature achene. Scales persistent, (4–) 8–12 (–30), (2.0–) 2.2–2.8 (–3.2) mm long, (1.2–) 1.4–1.6 (–1.8) mm wide, ovate to elliptic, semicircular in cross section, apex obtuse, laterally faintly to strongly 1 (–3) -nerved, reddish to stramineous to shiny brown or beige, medially faintly to rather conspicuously 2–5-nerved, green. Stamens 3; filaments 2.2–3.4 mm long; anthers (0.2–) 0.3–0.4 (–0.7) mm long, ellipsoid to oblong, the connective apex minute, less than 0.1 mm long, reddish. Style 0.4–0.7 (–1.0) mm long; stigmas 3, (1.0–) 1.5–3.0 mm long. Achenes (1.0–) 1.2–1.5 (–1.9) mm long, 0.5–0.6 (–0.75) mm wide, trigonous (dorsiventrally flattened or equilateral in cross section), narrowly ellipsoid to oblong or rarely obovoid-oblong, apex acute to barely obtuse, base stipitate, 0.1–0.2 mm long, the faces flat to convex, the surface finely papillose, brown, reddish brown, or black. Figs. 1F, 74.

Phenology. Flowering and fruiting throughout the year.

Distribution. Pantropical and warm-temperate regions; in the New World from Maine, Minnesota, and California, U.S.A., and southern Ontario, Canada, south through Mexico to northern Argentina, also in the West Indies. Common in the lowlands of Mexico (Fig. 75); clearings, roadside ditches, beaches, riverbanks, sand bars, croplands, and various disturbed situations; sea level to 500 (–2000) m.

Representative Specimens. **Mexico.** Baja California: Sierra Juárez, Cañón de Guadalupe, *Moran 26212* (SD); S edge of Ensenada, *Moran 29321* (SD).—Baja California Sur: Cape region, ca. 1 mi E of San Bartolo, *Thomas 7697* (ARIZ, DS, GH, MO, NY, RSA).—Campeche: Tuxpeña, *Lundell 1042* (ARIZ, DS, F, GH, MO, TEX, UC).—Chiapas: Tapachula, *Breedlove 24273* (DS); Mpio. Cintalapa, 5–6 km NW of Las Cruces, *Breedlove 28405* (DS, NY).—Chihuahua: Presa de Chihuahua, *LeSueur 1095*, (F, MO, TEX, US); Colina Lake, W of Camargo, *Knobloch 666* (MSC).—Coahuila: Quatro Ciénegas Basin, N of Puerto Salado, *Arnold 114* (ASU); Del Carmen Mts, *Marsh 809* (F, GH, TEX); Múzquiz swamp, *Marsh 933* (F, GH, TEX).—Colima: Colima, *Orcutt 4583* (DS, GH, MO).—Guerrero: Dtto. Adama, Temisco, *Mexia 8836* (CAS, GH, K, NY, UC); ca. 140 km E of Acapulco, *Tucker 2927* (DUKE, ENCB).—Hidalgo: Dtto. Molango, by Lake Atexca, *Moore 3469* (BH, UC).—Jalisco: 1 mi W of Ayo el Chico, *McVaugh 17220* (MICH, NY, TEX); 1 km al S de San Martín de Bolaños, *Rzedowski 26122* (DS, ENCB, MSC, WIS).—México: Dtto. Temascaltepec, Ixtapán, *Hinton 8883* (DS, F, GH, K, NY, POM, UC).—Michoacán: Zamora, *Pringle 11205* (CU, F, GH, K, MO, NY, Z); S of Lake Cuitzeo, *Sohns 729* (UC).—Morelos: Mpio. Villa de Ayala, *Zurate F. 4-VIII* (ARIZ, ASU, ENCB, IBUG).—Nayarit: Tepic, *Jones 23455* (MO, POM, UC).—Nuevo León: Monterrey,

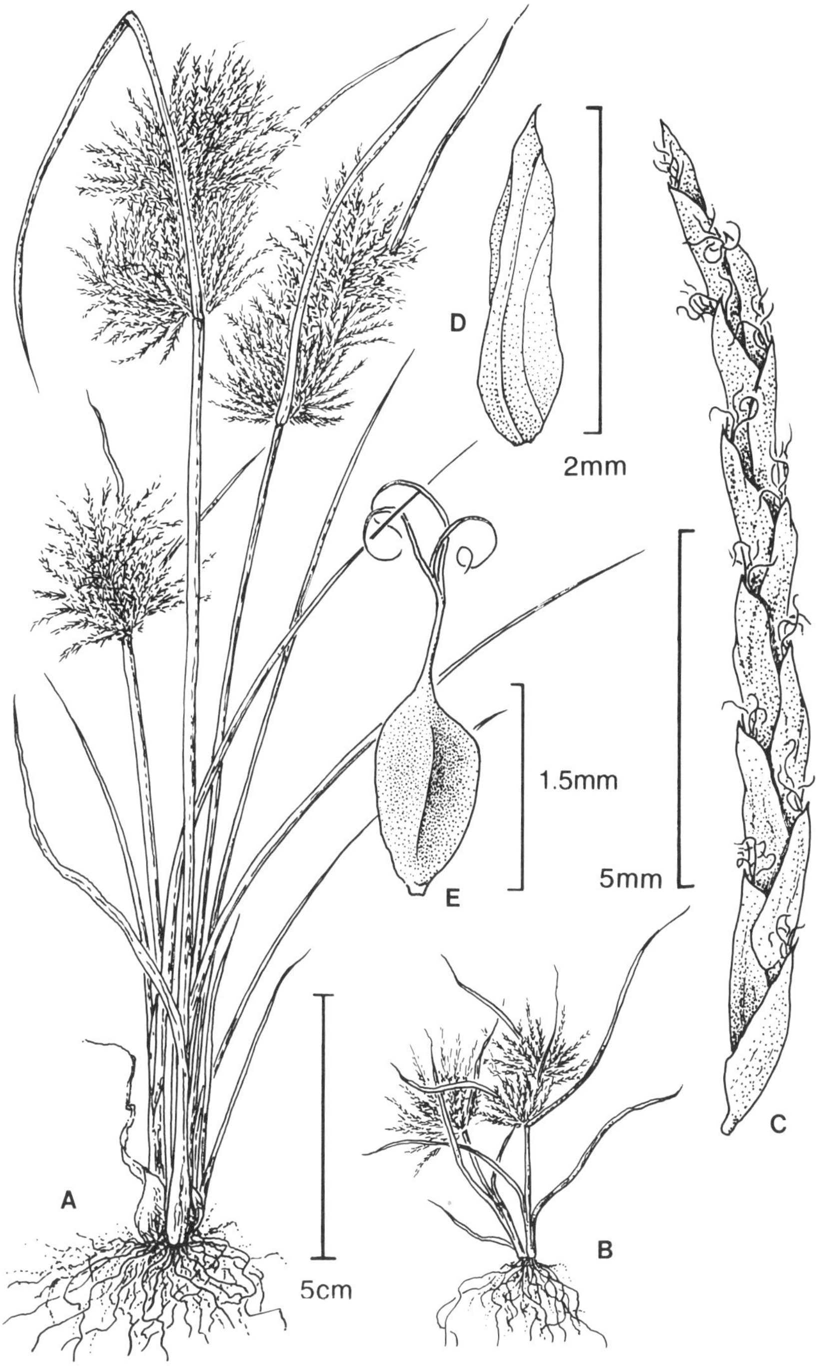

FIG. 74. *Cyperus odoratus.* A. Habit. B. Aspect of densely grouped plants. C. Spikelet. D. Scale. E. Achene. Drawn by Patricia M. Kernan for the Flora of New York State (Richard S. Mitchell, ed.).

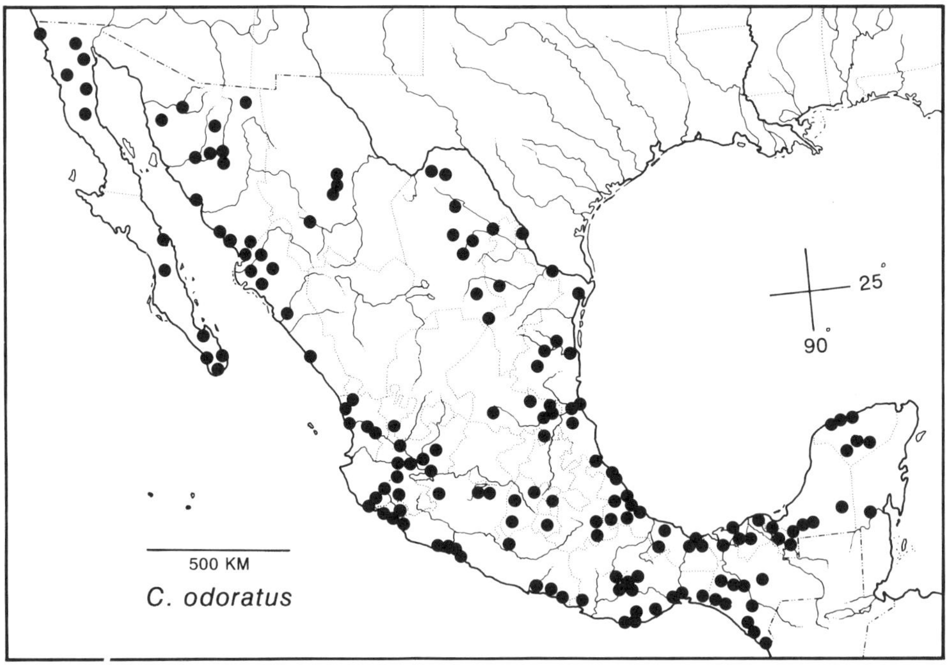

FIG. 75. Distribution of *Cyperus odoratus* in Mexico.

Kenoyer 320 (F).—OAXACA: valley of Etla, *Conzatti 93* (GH); rd from Mitla to Ayutla, 35 km from Mitla, *Tucker 2193* (DUKE); 30 km E of Puerto Escondido, *Tucker 2898* (DUKE).—PUEBLA: Tehuacán area, above Calipan, *Smith 3745* (F, NY, US).—QUINTANA ROO: 40 km by rd SW of Chetumal, *Chater et al. 72* (BM).—SAN LUIS POTOSÍ: Mpio. El Banito, *Chase 7551* (ARIZ, F, GH, MO); 21.5 mi E of Tamuín, *Tucker 2003* (DUKE, ENCB).—SINALOA: Sierra Surutato, Mpio. Sinaloa y Vela, Cañón de Tarahumare, *Breedlove & Thorne 17991* (CAS, RSA); Imala, *Gentry 5093* (ARIZ, DS, GH, MO, NY); Mazatlán, *Ortega 5765* (CAS, DS, GH, K, MO, POM, Z).—SONORA: 5 mi N of Estación Don, *Breedlove & Thorne 18641* (CAS, RSA); San Pedro Bay, *Johnston 4305* (C, CAS).—TABASCO: 35 km W of Paraíso, Laguna Redonda, *Barlow 21/6* (GH, WIS); Laguna de Curahuero, *Rovirosa 441* (NY, US).—TAMAULIPAS: 4 mi S of Jaumave, *Stanford et al. 2357* (DS, RSA, US); 60 km S of Matamoros on Hwy 101, *Tucker 2336* (CAS, DUKE, ENCB, F, MICH).—VERACRUZ: Orizaba, *Bourgeau 3145* (G, K); Playa Vicente, *Martínez Calderón 2061* (F, NY); between Pánuco and Tampico, *Tucker 2409* (DUKE, ENCB); Antigua, *Purpus 6252* (BM, GH, MO, UC).—YUCATÁN: Izamal, *Gaumer 1035* (F, GH, LCU, NY); Chichén Itzá, *Millspaugh 1667* (F).

As circumscribed here, *C. odoratus* is easily identified by its cylindric to subcylindric spikelets in which the corky rachilla of the mature spikelet disarticulates at the base of each scale. Thus, the mature spikelet breaks up into segments, each consisting of a scale and an internode of the rachilla clasping the achene with its corky wings.

Cyperus odoratus is an exceedingly variable pantropical and warm-temperate species. Numerous segregates have been named, four of which have been accepted by various authors who dealt with Mexican sedges: *C. acicularis*, *C. eggersii*, *C. engelmannii*, and *C. macrocephalus*. Kükenthal (1935–36) recognized *C. eggersii* and *C. macrocephalus* as species, but reduced *C. acicularis* and *C. engelmannii* to a variety and subspecies, respectively, of *C. odoratus*. Correll and Johnston (1970) recognized *C. macrocephalus* as a species. Koyama (1978) considered *C. acicularis* a synonym of *C. odoratus*.

In his treatment of the sedges of the Yucatan Peninsula, O'Neill (1940) recognized five species in subg. *Diclidium*. Benedict Ayers (1946) followed O'Neill's taxonomic concepts in his treatment of *Cyperus* in Mexico. O'Neill distinguished these five species as follows:

A. Achene linear, 1.5–2.0 mm long, 0.3 mm wide; scales usually distant, the apex of one not reaching the base of the next on the same side of the rachilla; wings of the rachilla linear, 1.2 mm long, 0.2 mm wide, reaching the top of the achene.
  B. Leaves and bracts membranous, 1–4 (–6) mm wide; culms 1–2 mm in diameter at the base. *C. engelmannii*.
  B. Leaves and bracts coriaceous, 8–16 mm wide; culms 5–10 mm in diameter at the base. *C. acicularis*.
A. Achene oblong to obovoid; scales imbricated on the same side of the rachilla; wings of rachilla elliptic, frequently not extending beyond the middle of the achene, 1.0 mm long, 0.4–0.5 mm wide.
  C. Inflorescence congested into a single globose compact head, 2–3 cm in diameter; the spikes sessile; achene ellipsoid; culms sharply trigonous; leaves 4–6 mm wide. *C. macrocephalus*.
  C. Inflorescence with some rays; some of the spikes penduncled; achenes obovoid-oblong to linear-oblong.
    D. Spikelets suberect, penicillate, very densely fascicled at the ends of the rays. *C. eggersii*.
    D. Spikelets divaricate, subdense, scattered along the rachis. *C. odoratus*.

According to O'Neill's key, *C. engelmannii* and *C. acicularis* both have narrower achenes than the other three species; however, I found no plants with achenes less than 0.5 mm wide, and none of the specimens studied would key out to *C. acicularis* or *C. engelmannii* by O'Neill's criteria. Mature achenes were ellipsoid to oblong, not linear, bringing into question O'Neill's distinction between these species. According to O'Neill's key, these two species also differ from the others in having scales distant on the rachilla, i.e., the apex of a scale not reaching the base of the scale above it. I noted earlier (Tucker 1983) that in southern Central America, *C. acicularis* could not be distinguished from *C. odoratus* on the basis of the overlap of the scales; this is also true for the Mexican and northern Central American material.

A comparison of specimens annotated by O'Neill as *C. engelmannii* and *C. odoratus* revealed a continuous range in the amount of overlap or separation of successive scales on the same side of the rachilla. In some plants determined as *C. engelmannii*, successive scales are separated by as much as 0.6–1.0 mm; in others, the separation is less, 0.2–0.6 mm; in still other plants the scales overlap by approximately 0.2 mm and yet are separated by as great a distance on different spikelets at a comparable state of maturity on the same plant.

Benedict Ayers (1946, p. 55) stated: "In most of the United States, *C. engelmannii* is a very distinct entity. In Yucatan it is less distinct and in Brazil the line between *C. odoratus* and *C. engelmannii* is still more difficult to draw."

Fernald (1950) recognized *C. odoratus* and *C. engelmannii*, although the number of collections of these species from North America at the Gray Herbarium make it evident that he did not see a large sample. Voss (1972) noted that, in Michigan, *C. engelmannii* was "almost suggestive" of an abnormal form of *C. odoratus*. The large number of specimens from Mexico and all of Central America examined in the present study as well as previously (Tucker 1983) has not revealed a single consistent character separating *C. engelmannii* from *C. odoratus*. Further study of ca. 600 collections from Canada and the United States (Tucker 1987) confirmed that *C. engelmannii* and *C. odoratus* intergrade extensively. The infraspecific variation of *C. odoratus* in the U.S.A. is like that observed in Mexican collections.

Benedict Ayers (1946, p. 55) noted that *C. macrocephalus* ". . . is quite distinct along the Río Grande, but further south forms intermediate between it and *C. eggersii* occur." Correll and Johnston (1970) stated that *C. macrocephalus* appeared to intergrade through the "form" *C. eggersii* into *C. odoratus*. A critical analysis of specimens annotated by O'Neill and many more recent collections revealed some quantitative differences between these three taxa. No further differences were found, although numerous other characters, particularly those in O'Neill's key, were checked. On the average, *C. eggersii* and *C. macrocephalus* have shorter scales, shorter anthers, and shorter achenes than *C. odoratus*. However, there is some overlap in the ranges of each character, and none exhibits a discontinuous distribution between any pair of species. Although no single difference consistently separates either *C. eggersii* or *C. macrocephalus* from *C. odoratus*, there is a partial correlation of denser inflorescences with shorter scales, anthers, and achenes. This might be indicative of genetic linkage of these characters or a similar response of these parts of the plants to certain environmental conditions.

The absence of morphological discontinuity between *C. odoratus* and *C. eggersii* and *C. macrocephalus* in length of scales, anthers, or achenes and the observed intergradation of inflorescence forms among these three taxa provide evidence that *C. macrocephalus* and *C. eggersii* are conspecific with *C. odoratus*. Several other New World species of *Cyperus* also have variants with dense rayless inflorescences, e.g., *C. luzulae* and *C. manimae*, an observation made by Correll and Johnston (1970) as well.

Kükenthal (1935–36) treated *Cyperus hayesii* (C. B. Clarke) Standl. as a variety of *C. odoratus*. A previous study (Tucker 1983) showed *C. hayesii* to be a well-marked species of the Pacific Coast of Central America, which is closely related to *C. correllii* (T. Koyama) Tucker of the Bahama Islands (Tucker 1987).

Kükenthal (1935–36) included *Cyperus purpureo-vaginatus* Boeck. as a synonym of *C. odoratus*. Examination of the type from Warming's herbarium (C) shows that the plant is really *C. distans*. The continuous rachilla, the secondary rays, and deciduous scales clearly indicate that it is a member of the subg. *Cyperus*.

Species of Cyperus Possibly Occurring in Mexico

Cyperus eragrostis Lamarck, Tabl. encycl. 1: 146. 1791. Temperate South America, and California, U.S.A., to southwestern British Columbia, Canada. Common in southern California—several collections are from San Diego County (Tucker 1992) not far from the Mexican border; also naturalized in southeastern Texas (Tucker 1987). *Cyperus eragrostis* is closely related to *C. ochraceus* and *C. entrerianus*, from which it can be distinguished by its markedly stipitate achenes and excurved scale apices. For a description and illustration see Denton (1978).

Cyperus hayesii (C. B. Clarke) Standley, J. Wash. Acad. Sci. 15: 457. 1925. Guatemala to Colombia along the Pacific Coast; possibly present along the Pacific Coast of Chiapas. *Cyperus hayesii* is closely related to *C. odoratus* but has a rhizomatous habit, broader spikelets, and achenes with a distinctive notch on the outer surface. For a description see Tucker (1983).

Cyperus setigerus Torrey & Hooker, Ann. Lyceum Nat. Hist. New York 3: 434. 1836. Great Plains from Missouri and New Mexico to southern Texas, U.S.A.; known from the Rio Grande Plains (Correll & Johnston 1970) and possibly present in northern Tamaulipas or northern Sonora. *Cyperus setigerus* is similar to *C. parishii* Britton ex Parish of southern California and Arizona and also to *C. esculentus*, from which it differs in its caespitose habit and lack of stolons. For a description see Fernald (1950) or Correll and Johnston (1970).

Cyperus strigosus L., Sp. pl. 47. 1753. A common species widely distributed in the U.S.A. and southern Canada. I have seen several collections from San Diego County, California (Tucker 1992), and I believe the species might occur in northern Baja California. *Cyperus strigosus* would key to *C. hermaphroditus* in the present treatment, from which it could be distinguished by its bulbous culm bases, shorter size (generally less than 50 cm tall), and stramineous scales. For a description see Fernald (1950).

Excluded and Doubtful Names

Cyperus aristatus Rottbøll, Descr. pl. icon. rar. 23. 1773. *Dichostylis aristata* (Rottbøll) Palla, Bot. Jahrb. Syst. 10: 296. 1888.—*Cyperus aristatus* is a superfluous name for *Scirpus intricatus* L., which Rottbøll cited in synonymy in the protologue. Rottbøll's specimen is *C. squarrosus* [Cape of Good Hope and Tranquebar, *König s.n.* (C!)].

Cyperus acutiusculus Lagasca, Gen. sp. nov. 2. 1816.—Type: "H[abitat] in N[ueva] H[ispania]. Semina misit *D. Sessé*." (not located at MA).—The description is not sufficient to assign this name with certainty to a species of *Cyperus*. It apparently applies to a species with six reflexed rays, two inflorescence bracts ca. three times as long as the rays, headlike spikes, and 12-flowered linear spikelets. The name was not noted by Kükenthal (1935–36) and Benedict Ayers (1946).

Cyperus croceus Vahl, Enum. Pl. 2: 357. 1805.—Kükenthal (1935–36) reported *C. croceus* from Yucatán on the basis of a collection he did not see: Mangrove-Sumpf bei Sisal, *Schott* [*s.n.*]. O'Neill (1940) excluded *C. croceus* from the treatment of the Cyperaceae of the Maya Areas, because he saw no specimens of it.

Cyperus exaltatus Retzius, Obs. bot. 5:11. 1789.—*Cyperus exaltatus* was reported from Nayarit (Bot. Beechey's Voy. 312. 1840), but Kükenthal (1935–36), in the absence of specimens from Mexico, considered its occurrence there doubtful. Among specimens received on loan from Kew is one of *C. imbricatus* labelled

"Mexico, Beechey." *Cyperus imbricatus* is morphologically similar to *C. exaltatus,* and misidentification was evidently the source of the report of *C. exaltatus* from Mexico.

Cyperus filicinus Vahl, Enum. Pl. 2: 332. 1805.—A species of brackish coastal soils of the eastern U.S.A., it was doubtfully reported from Mexico by Kükenthal (1935–36) from a specimen he did not see, *F. Mueller 2155*. A duplicate at NY is *C. polystachyos*.

Cyperus orbicephalus (Beetle) T. Koyama & McVaugh, Bull. Torrey Bot. Club 90: 231. 1963. *Scirpus orbicephalus* Beetle, Madroño 7: 169. 1944. = *Karinia mexicana* (Britton) Reznicek & McVaugh in McVaugh, Fl. novo-galiciana 13: 387. 1993.

Cyperus retrorsus var. *deeringianus* (Britton & Small) Fernald & Griscom, Rhodora 37: 152. 1935.—Kükenthal (1935–36) reported this variety from Colima, based on *Palmer 2* [in 1897]. The duplicate of this number at US is *C. hermaphroditus*. There is no extant specimen of this number at B, and I am at a loss to explain Kükenthal's apparent confusion of these rather different species. It seems unlikely for a taxon of the southeastern United States to occur as a disjunct in western Mexico.

Cyperus schaffneri Boeckeler, Flora 61: 30. 1878.—TYPE: "in valle Mexico," *Schaffner s.n.* (not located; not at B).—Boeckeler thought his new species "e vicinia *C. caracasani* [=*C. tenuis*] et *C. camphorati*." I suspect the type was a specimen of *C. thyrsiflorus* or *C. hermaphroditus*.

Cyperus sphacelatus Rottbøll, Descr. pl. icon. rar. 26. 1773.—This species occurs in Central America along the Caribbean Coast of Nicaragua, Costa Rica, and Panama. A specimen (*Abbon s.n.*, E) purportedly from Tampico, Mexico, has a label from a documented set of counterfeit Mexican collections (Standley 1927; Bonner 1972).

ACKNOWLEDGMENTS

This paper is based largely on a dissertation submitted in June, 1985, to the Graduate School of Duke University in partial fulfillment of the requirements of the degree of Doctor of Philosophy in the Department of Botany. I thank Robert L. Wilbur, my major professor, for his careful criticism, support, and interest through all aspects of this study. I appreciate his company in the field, as well as that of Owen M. Schwartz, during our 1984 trip to Mexico. I also thank the members of my advisory committee, W. L. Culberson, D. E. Stone, L. E. Anderson, and M. D. Rausher, for reading the manuscript and providing many helpful comments, and T. W. Johnson, Jane Philpott, R. A. White, and R. B. Searles for their interest, advice, and support during my graduate education. Thanks are offered to W. Dwight Billings for providing considerable ecological information, particularly with regard to C_3 and C_4 plants.

The manuscript was reviewed by Paul Goetghebeur, A. A. Reznicek, and Rogers McVaugh, whose thoughtful criticisms and attention to content and consistency are greatly appreciated. Christiane Anderson improved the final manuscript with her editorial precision and skill.

Rogers McVaugh has several times shared his extensive knowledge of Mexico and helped with my studies of Liebmann's Mexican collections of *Cyperus*. Jerzy Rzedowski helped with field work in

Mexico. Luz M. Villarreal de Puga not only kindly allowed me to borrow nearly 600 sheets from the herbarium of the Instituto de Botánica de Guadalajara, but had them packed on a moment's notice.

Vicky Martin Tucker, my wife, generously typed most of this dissertation. In addition she compiled the elevational and phenological data for the species descriptions from my note cards. Her constant support and interest in my studies are a great kindness. I especially want to thank my family for their encouragement. Particularly my mother and my grandparents have been supportive during my seemingly interminable college career. Mrs. Sherri H. Herndon, Associate Curator of the Duke Vascular Herbarium lent a kindly ear on many occasions and assisted in securing loans. Many colleagues in the Duke Department of Botany have made many individual contributions: Melvin D. Turner, John M. MacDougal, and W. John Kress. Karin Douthit, Patricia M. Kernan, Virginia H. Cantarella, Mary Barnas Pomeroy, and Charlotte M. Taylor prepared the illustrations.

During this study, and my preceding study of the Central American species of *Cyperus*, some 15,000 herbarium specimens have been borrowed from, or examined at, the herbaria listed below. Thanks are extended to the staff members of these institutions, without whose kind cooperation this study would not have been possible: A, AC, ARIZ, ANSM, ASU, B, BKL, BH, BM, BR, C, CAS, CCNL, CHAPA, CLEMS, CONN, CR, CU, DAV, DS, DUKE, E, ECON, ENCB, F, FLAS, G, GH, HAL, IBUG, K, LCU, LL, M, MASS, MEXU, MICH, MO, MSC, MT, MTMG, NHA, NY, NYS, PENN, PH, PMA, POM, PR, PRC, RM, RSA, S, SD, SMU, TENN, TEX, TRT, TRTE, UC, UCR, US, UTEP, VT, WIS, WRSL, WVA, YU, and Z.

Support from the National Science Foundation (Grant no. NSF-BSR-83-03070) made possible field work in Mexico during the summers of 1983 and 1984. I thank Norton G. Miller, Clifford A. Siegfried, and the New York State Biological Survey for facilities and support for the rewriting and many of the illustrations for the present monograph. This paper is published as contribution number 571 of the New York State Museum.

LITERATURE CITED

Baijnath, H. 1975. A study of *Cyperus alternifolius* L. *sens. lat.* (Cyperaceae). Kew Bull. 30: 521–526.

Barros, M. 1938. Ciperaceas argentinas. III. Anales Mus. Argent. Ci. Nat. "Bernardino Rivadavia" 39: 253–381.

Benedict Ayers, Bro. 1946. The genus *Cyperus* in Mexico. Catholic Univ. Amer, Biol. Stud. 1: 1–104.

Bonner, C. E. B. 1972. Un cas de "faux" en botanique. Ann. Mus. Genève 130: 18, 19.

Britton, N. L., and A. Brown. 1901. *Manual of the flora of the northern states and Canada.* New York: Holt & Co.

Bryson, C. T. 1984. Weed alert: smallflower umbrella sedge (*Cyperus difformis* L.). So. Weed Sci. Soc Newsl. 7(1): 6.

Chapman, A. W. 1860. *Flora of the southern U.S.A.* New York: Ivison, Phinney & Co.

Clarke, C. B. 1884. On the Indian species of *Cyperus*; with remarks on some that specially illustrate the subdivisions of the genus. Jour. Linn. Soc., Bot. 21: 1–202.

———. 1908. New genera and species of Cyperaceae. Bull. Misc. Inform. Kew, add. ser. 8: 1–196. 1908.

Corcoran, M. L. 1942. A revision of the subgenus *Pycreus* in North and South America. Catholic Univ. Amer., Biol. Ser. 37: 1–68.

Correll, D., and M. C. Johnston. 1970. *Manual of the vascular flora of Texas.* Renner: Texas Research Foundation.

Dandy, J. 1944. Cyperaceae. In *Catalogue of the vascular plants of S. Tomé* by A. W. Exell, 354–365. London: British Museum.

Denton, M. F. 1978. A taxonomic treatment of the Luzulae group of *Cyperus*. Contr. Univ. Michigan Herb. 11: 197–271.

Elliott, S. 1816–1821. *A sketch of the botany of South-Carolina and Georgia*, vol. 1. Charleston, South Carolina: J. R. Schenck.

Fernald, M. L. 1939. Last survivors in the flora of Tidewater Virginia. Rhodora 41: 529–559.

———. 1950. *Gray's manual of botany*, ed. 8. New York: American Book Co.

Fernald, M. L., and L. Griscom. 1935. Three days of botanizing in southeastern Virginia. Rhodora 37: 150–157; 186–195.

Gleason, H. A., and A. Cronquist. 1963. *Manual of vascular plants.* New York: Van Nostrand.

Goetghebeur, P. 1989. Studies in the Cyperaceae 9. Problems in the lectotypification and infrageneric taxonomy of *Cyperus* L. Bull. Soc. Roy. Bot. Belgique 122: 103–114.

Harborne, J. B., C. A. Williams, and K. L. Wilson. 1982. Flavonoids in leaves and inflorescences of Australian *Cyperus* species. Phytochemistry 21: 2491–2507.

Hooper, S., and J. Raynal. 1969. New species and names in African *Pycreus* P.-Beauv. (Cyperaceae). Kew Bull. 23: 313–314.

Horak, M. J., and J. S. Holt. 1986. Isozyme variability and breeding systems in populations of yellow nutsedge (*Cyperus esculentus*). Weed Sci. 34: 538–543.

Horowitz, M. 1972. Growth, tuber formation, and spread of *Cyperus rotundus* L. from single tubers. Weed Res. 12: 348–363.

Horvat, M. L. 1941. A revision of the subgenus *Mariscus* found in the U.S.A. Catholic Univ. Amer., Biol. Ser. 33: 1–147.

Johnston, I. M. 1931. The flora of the Revillagigedo Islands. Proc. Calif. Acad. Sci., ser. 4, 20: 9–104.

Junghuhn, F. W. 1831. *Cyperus*. In "Plantarum Mexicanarum a cel. viris Schiede et Deppe collectarum" by D. Schlechtendal and A. Chamisso, 23–26. Linnaea 6: 22–64.

Kearney, T. H., and R. H. Peebles. 1951. *Arizona Flora*. Berkeley and Los Angeles: University of California Press.

Kern, J. H. 1974. Cyperaceae. In *Flora Malesiana*, ed. C. J. G. van Steenis, ser. 1. 7(3): 435–753. Groningen: Wolters-Noordhoff.

Koyama, T. 1958. Taxonomic study of the genus *Scirpus* Linné. J. Fac. Sci. Univ. Tokyo, Sect. 3, Bot. 7: 271–366.

———. 1978. Cyperaceae. In *Flora of the Lesser Antilles*, ed. R. A. Howard, 3: 220–320. Jamaica Plain, Massachusetts: Arnold Arboretum of Harvard University.

Kükenthal, G. 1935–36. *Cyperus* (Cyperaceae: Scirpoideae). In *Das Pflanzenreich*, ed. L. Diels, 4 (20): 1–621. Leipzig: Wilhelm Engelmann.

Kunth, C. S. 1837. *Enumeratio plantarum*. Vol. 2. *Cyperographia synoptica*. Stuttgart and Tübingen: J. Cotta, Stuttgart and Tübingen.

Lerman, J. C., and J. Raynal. 1972. La teneur en isotope stable du carbone chez les Cyperacées: sa valeur taxonomique. Compt. Rend. Hebd. Séances Acad. Sci. 275-D: 1391–1395.

Liebmann, F. M. 1850. Mexicos halvgraes bearbeidede efter Forgaengernes of egne Materialier med Tillaeg af de i Nicaragua og Costa Rica af Mag. A. S. Örsted samlede samt nogle faa ubeskrevne vestindiske Former. Kongel. Danske Vidensk.- Selsk. Skr. 2: 189–277.

Lipscomb, B. L. 1980. *Cyperus difformis* L. (Cyperaceae) in North America. Sida 8: 320–327.

Lye, K. A. 1981. Two new subgenera of *Cyperus*. Nordic Jour. Bot. 1: 57–61.

———. 1992. The history of the genus *Mariscus* (Cyperaceae). Lidia 3: 37–72.

Marcks, B. G. 1972. Preliminary reports on the flora of Wisconsin, no. 66. Cyperaceae II—Sedge Family II. The genus *Cyperus*—The umbrella sedges. Proc. Wisconsin Acad. Sci. Arts Lett. 62: 261–284.

Martin, W. C., and C. R. Hutchins. 1980. *A flora of New Mexico*. Braunschweig: J. Cramer.

Matuda, E. 1958. Las ciperáceas del Valle de México y sus alrededores. Anales Inst. Biol. Univ. Nac. México 29: 107–163.

McGivney, M. V. 1938. A revision of the subgenus *Eucyperus* found in the U.S.A. Catholic Univ. Amer., Bio. Ser. 26: 1–74.

McIntire, S. 1985. Seed reserves in temperate Australian rice fields following pasture rotation and continuous cropping. Jour. Appl. Ecol. 22: 875–884.

McLaughlin, A. D. 1944. The genus *Cyperus* in the West Indies. Catholic Univ. Amer., Biol. Stud. 5: 1–108.

Merrill, E. D., and S. Y. Hu. 1950. Work and publications of Henry Muhlenberg, with special attention to unrecorded or incorrectly recorded binomials. Bartonia 25: 1–66.

Metcalfe, C. R. 1977. *Anatomy of Monocotyledons*. Vol. 5. *Cyperaceae*. London: Oxford Univ. Press.

Mulligan, G. A., and B. E. Junkins. 1976. The biology of Canadian weeds. 17. *Cyperus esculentus* L. Canad. J. Pl. Sci. 56: 339–350.

O'Neill, H. T. 1940. *The sedges of the Yucatan Peninsula*. Publ. Carnegie Inst. Washington 19.

———. 1942. The status and distribution of some Cyperaceae in North America. Rhodora 44: 43–64.

O'Neill, H. T., and Bro. Benedict Ayers. 1944. Some New World *Cyperi*. Leafl. W. Bot. 4: 33–38.

Oteng-Yeboah, A. A. 1976. Taxonomic studies in Cyperaceae–Cyperoideae. Notes Roy. Bot. Gard. Edinburgh 33: 311–316.

Pedersen, T. M. 1968. Cyperaceae. In *Flora de la Provincia de Buenos Aires*, ed. A. L. Cabrera, 4(1): 315–421. Buenos Aires: I. N. T. A.

Presl, J. S., and K. B. Presl. 1828. Cyperaceae. In *Reliquiae haenkeanae*, ed. K. B. Presl, 1(3): 165–206. Prague: J. G. Calve.

Radford, A. E., H. E. Ahles, and C. R. Bell. 1967. *Manual of the vascular flora of the Carolinas*. Chapel Hill, North Carolina: University of North Carolina Press.

Raynal, J. 1967. Notes cypérologiques: V. Sur un groupe de *Cyperus* montagnards afro-américains. Adansonia, sér. 2, 6: 385–392.

———. 1973. Contribution à la classification de la sous-famille des Cyperoideae. Adansonia, sér. 2, 13: 145–171.

———. 1978. Notes cypérologiques: 33. Mélanges nomenclaturaux (2). Adansonia, sér. 2, 17: 273–280.

Rikli, M. 1895. Beiträge zur vergleichende Anatomie der Cyperaceen mit besonderer Berücksichtigung der inneren Parenchymschiede. Jahrb. Wiss. Bot. 27: 485–580.

Robinson, B. L., and M. L. Fernald. 1908. *Gray's manual of botany*, ed. 7. New York: American Book Co.

Scanlon, G. M. 1942. A study of the genus *Cyperus* in the Hawaiian Islands. Catholic Univ. Amer., Bio. Ser. 41: 1–62.

Seymour, F. C. 1982. The flora of New England, ed. 2. Phytologia Memoirs 5: 1–612.

Shreve, F., and I. L. Wiggins. 1964. *Vegetation and flora of the Sonoran Desert*. Stanford: Stanford University Press.

Simpson, D. A. 1990. A revision of *Cyperus* sect. *Leucocephali*. Kew Bull. 45: 485–501.

Stafleu, F. A., and R. S. Cowan. 1976. Taxonomic Literature, ed. 2, vol. 1: A–G. Regnum Veg. 94: 1–1136.

———. 1979. Taxonomic Literature, ed. 2, vol. 2: H–Le. Regnum Veg. 98: 1–991.

———. 1981. Taxonomic Literature, ed. 2, vol. 3: Lh–O. Regnum Veg. 105: 1–980.

———. 1983. Taxonomic Literature, ed. 2, vol. 4: P–Sak. Regnum Veg. 110: 1–1214.

———. 1985. Taxonomic Literature, ed. 2, vol. 5: Sal–Ste. Regnum Veg. 112: 1–1066.

———. 1986. Taxonomic Literature, ed. 2, vol. 6: Sti–Vuy. Regnum Veg. 115: 1–926.

———. 1988. Taxonomic Literature, ed. 2, vol. 7: W–Z. Regnum Veg. 116: 1–653.

Standley, P. C. 1927. A counterfeit collection of Mexican plants. J. Bot. 65: 118–121.

———. 1931. The Cyperaceae of Central America. Publ. Field Columbian Mus., Bot. Ser. 8: 239–292.

Stoller, E. W. 1973. Effect of minimum soil temperature on the differential distribution of *Cyperus rotundus* and *Cyperus esculentus* in the U.S.A. Weed Res. 13: 209–217.

Strausbaugh, P. D., and E. L. Core. 1970. *Flora of West Virginia*, ed. 2, part 1. Grantsville, West Virginia: Seneca Books.

Stuckey, R. L. 1970. Type specimens of flowering plants described from Long's Expedition to the source of the St. Peter's River in 1823. Michigan Bot. 9: 41–47.

Svenson, H. K. 1943. Cyperaceae. In "Flora of Panama," ed. R. E. Woodson, Jr., and R. W. Schery. Ann. Missouri Bot. Gard. 30: 281–324.

Thomas, W. W. 1984. The systematics of *Rhynchospora* section *Dichromena*. Mem. New York Bot. Gard. 37: 1–116. 1984.

Tieszen, L. L., M. M. Senyimba, S. K. Imbamba, and J. H. Troughton. 1979. The distribution of C_3 and C_4 grasses and carbon isotope discrimination along an altitudinal and moisture gradient in Kenya. Oecologia 37: 337–350.

Torrey, J. 1836. Monograph of the North American Cyperaceae. Ann. Lyceum Nat. Hist. New York 5: 181–448.

Trimen, H. 1888. Hermann's Ceylon Herbarium and Linnaeus's 'Flora Zeylanica.' J. Linn. Soc., Bot. 24: 129–155.

Tucker, G. C. 1983. The taxonomy of *Cyperus* (Cyperaceae) in Costa Rica and Panama. Syst. Bot. Monogr. 2: 1–85.

———. 1984a. Taxonomic notes on two common neotropical species of *Cyperus*. Sida 10: 298–307.

———. 1984b. A revision of *Kyllinga* Rottb. (Cyperaceae) in Mexico and Central America. Rhodora 86: 507–538.

———. 1985a. *Cyperus flavicomus*, the correct name for *Cyperus albomarginatus*. Rhodora 87: 539–541.

———. 1985b. The correct name for *Cyperus cayennensis* (*C. flavus*), Cyperaceae. Southw. Naturalist 30: 607–608.

———. 1986. The species of *Cyperus* described by Liebmann in "Mexicos halvgraes." Syst. Bot. 11: 14–19.

———. 1987. The genera of Cyperaceae in the Southeastern U.S.A. J. Arnold Arb. 68: 361–445.

———. 1992. *Cyperus*. In *The Jepson Manual*, ed. J. C. Hickman, T. J. Rosatti, and D. H. Wilken, 1138–1140. Berkeley: University of California Press.

Tucker, G. C., and R. McVaugh. 1993. *Cyperus*. In *Flora novo-galiciana*, by R. McVaugh, 3: 270–344. Ann Arbor, Michigan: The University of Michigan Herbarium.

Van der Veken, P. 1965. Contribution à l'embryographie systématique des Cyperaceae-Cyperoideae. Bull. Jard. Bot. État 35: 285–354.

Vorster, P. J. 1978. Revision of the taxonomy of *Mariscus* Benth. and related genera in southern Africa. Dissertation (D. Sc.), University of Pretoria.

Voss, E. G. 1972. *Michigan flora*. Vol. 1. *Monocotyledons*. Bloomfield Hills, Michigan: Cranbrook Institute.

Wiggins, I. L. 1980. *Flora of Baja California*. Stanford: Stanford University Press.

Wilson, K. L. 1991. *Cyperus haspan*—Spelling of the epithet. Cyperaceae Newsletter 9: 8.

NUMERICAL LIST OF THE SPECIES

1. C. involucratus
2. C. canus
3. C. dioicus
4. C. acuminatus
5. C. reflexus
6. C. surinamensis
7. C. virens
8. C. ochraceus
9. C. entrerianus
10. C. luzulae
11. C. megalanthus
12. C. laxus
13. C. matudae
14. C. breedlovei
15. C. simplex
16. C. lundellii
17. C. haspan
18. C. difformis
19. C. humilis
20. C. seslerioides
21. C. arsenei
22. C. perennis
23. C. uncinulatus
24. C. microbrunneus
25. C. nayaritensis
26. C. tenerrimus
27. C. michoacanensis
28. C. amabilis
29. C. cuspidatus
30. C. flavicomus
31. C. fugax
32. C. polystachyos
33. C. unioloides
34. C. aschenbornianus
35. C. niger
36. C. bipartitus
37. C. lanceolatus
38. C. flavescens var. piceus
39. C. laevigatus
40. C. squarrosus
41. C. iria
42. C. prolixus
43. C. distans
44. C. articulatus
45. C. esculentus
46. C. rotundus
47. C. compressus
48. C. wilburii
49. C. erythrorhizos
50. C. digitatus
51. C. semiochraceus
52. C. imbricatus
53. C. giganteus
54. C. ischnos
55. C. sanguineoater
56. C. dipsaceus
57. C. schweinitzii
58. C. sphaerolepis
59. C. fendlerianus
60. C. ciliatus
61. C. hypopitys
62. C. spectabilis
63. C. calderoniae
64. C. pennellii

65a. C. manimae var. manimae

65b. C. manimae var. asperrimus

65c. C. manimae var. divergens

66. C. oxylepis
67. C. elegans
68. C. planifolius
69. C. sordidus
70. C. pycnostachyus
71. C. duripes
72. C. dentoniae
73. C. tempeae
74. C. svensonii

75. C. regiomontanus
76. C. aggregatus
77. C. pallidicolor
78. C. swartzii
79. C. mutisii
80. C. ligularis
81. C. camphoratus
82. C. tenuis
83. C. thyrsiflorus
84. C. hermaphroditus
85. C. panamensis
86. C. lentiginosus
87. C. retroflexus
88. C. retrorsus
89. C. odoratus

INDEX TO NUMBERED COLLECTIONS

The numbers in parentheses refer to the corresponding species in the text and in the Numerical List of Species presented above.

Abbon, Fr. 6412 (1).
Abbon & Arsène 6122 (8).
Abbott, R. Q. 150 (85); 284 (20).
Abrams, L. 12827 (47); 12876 (28).
Adame G., G. 21 (8); 22 (89).
Agniel 10516 (45).
Aguirre, M. 21 (45).
Agundis, O. 1045 (89); 5030 (45).
Ahshapanek, D. 123 (2).
Alatorre F., E. R. 9 (77).
Alcorn, J. B. 2026 (87).
Alexander, E. J., & E. Hernández 1965 (42).
Alfaro Razura, O. J. 14 (45).
Alfonso, H. S. 2 (1).
Allison, F. 100 (20).
Amante, A. 1A (18).
Anderson, W. R., & C. E. 3594 (62); 5122 (20); 5145 (79); 5204 (20).
Anderson, W. R., & C. Laskowski 3594A (58); 3594B (62); 4396 (38).
Arguelles, E. 368 (7); 1652 (20).
Arias R., L. M., et al. 624 (40).
Arnold 114 (87).
Arredondo G., A. 21 (51); 28 (51).
Arreola N., H., & L. Guzmán H. 462 (44).
Arreola & Soltero 188 (20).
Arsène, G. 38 (45); 40 (45); 63 (45); 161 (45); 171 (65b); 174 (85); 242 (65b); 272 (70); 1009 (79); 1024 (58); 1320 (35); 1335 (85); 1380 (40); 1430 (62); 1726 (35); 1735 (38); 2087 (38); 2239 (51); 2321 (40); 2400 (7); 2401 (30); 2485 (45); 2638 (65c); 2692 (65c); 2692a (65b); 2791 (85); 2860a (85); 2923 (54); 2925 (55); 2939 (51); 2941 (85); 2989 (62); 2994 (85); 3002 (45); 3050 (45); 3082 (51); 3091 (65b); 3092 (18); 3093 (18); 3117 (9); 3169 (62); 3171 (45); 3173 (18); 3174 (30); 3314 (85); 3501 (35); 5122 (62); 5138 (45); 5220 (40); 5231 (45); 5239 (62); 5275 (35) (38); 5291 (85); 5316 (55); 5317 (65c); 5318 (54); 5319 (62); 5459 (62); 5460 (62); 5471 (35, 36, 51); 5555 (55); 5579 (65b); 5580 (54); 5582 (65b); 5589 (54); 5640 (65c); 5643 (30); 5651 (65c); 5751 (65c); 5830 (35); 5846 (55); 5848 (65b) (65c); 5914 (62); 5915 (45); 5918 (62); 6028 (38); 6066 (38); 6238 (89); 6931 (85); 6945 (45); 7311 (30); 7411 (21); 8393 (85); 9860 (62); 9887 (65b); 9888 (65b); 9889 (65c); 9890 (79); 9896 (51); 9898 (62); 9899 (62); 9901 (62); 10145 (79); 11009 (47).
Arsène & Nicolas 5861 (65c).
Arturo, M. 42 (89).
Aschenborn, A. 118 (70); 121 (34); 333 (20).
Atilano D., I. 19 (85).
Atmar, G. L. 75 (89); 76 (8).
Aurelio A., I. A. 12 (85).
Baldwin, J. T. 14401 (44).
Balls, E. K. 4678 (84); 4870 (20); B5070 (65c); 5079 (20); B5384 (65a); 5457 (45); 5499 (35).
Bancalari, M. 9 (45).
Banda, R., et al. 62 (20); 127 (20).
Banda S., R. 1426 (35) (38).
Barba G., A. 18 (8).
Barkley, F. A. 5 (44); 12-B (44); 2453 (35); 2668 (39); 2678 (39); 2683 (35); 7324 (20); 7437 (35); 7597 (35); 7601 (44); 7608 (50); 14092 (72); 14167 (44); 14599 (35); 16M645 (67); 17M336 (35); 17M621 (46); 17M738 (28); 17M752 (6); 44M846 (35).
Barkley & Rowell 7458 (65c).
Barkley et al. 34 (80); 2668 (39); 2678 (39).
Barlow, F. D. 19/13 (44); 21/6 (89); 23/1 (80); 23/9 (32); 24/2 (12); 28/9 (89); 28/9A (89); 28/12C (53); 28/24 (44); 28/24B (44); 28/26 (53); 30/135 (10); 30/139 (17); 62c/6 (44); 62c/10 (80); 459 (80).
Barragan Contreras, I. F. 21 (85).
Barragán Hdez., A. 46 (85).
Barrera, A. 279 (80).
Bartlett, H. H. 10981 (45); 10279 (85); 10874 (85).
Basile, Fr. 192 (62); 194 (35); 195 (89); 197 (45).

Batiz G., J. P. 1A (52).
Beaman, J. H. 2097 (20); 4228 (20); 4301 (20); 4355 (64); 4538 (20).
Beaman & C. Alvarez 6307 (83).
Beetle, A. A. M-6905 (72); M-6921 (40); M-6980 (72); M-7067 (65b); M-7665 (77); M-7720 (38); M-7731 (65c); M-7732 (65b); 7759 (35); 7779 (35).
Bell, C. R., & J. A. Duke 16588 (45).
Benhumea S., V. 310 (1).
Benítez, A. 50 (45).
Bennett 832 (59).
Bennett et al. 790 (20).
Benz, B., & M. Burd 423 (37).
Beristain, F. 7771 (12); 7772 (10); 773 (66); 7822 (12); 7947 (10); 7977 (81).
Berlandier, J. L. 212 (89); 288 (39); 525 (84); 753 (33); 860 (58); 862 (45); 2306 (50); 3220 (2); 3222 (44); 3223 (50).
Bernely, R. 11 (28).
Boege, W. 1102 (10); 1919 (45); 2423 (6); 2998 (46).
Botteri, J. 169 (28); 171 (45); 174 (35); 177 (8); 178 (6); 180 (89); 181 (83); 184 (84); 185 (79); 525 (83); 536 (83) (84); 736 (28); 737 (28); 741 (33) (35); 742 (33); 744 (89); 747 (84); 748 (84); 749 (84); 750 (65b); 751 (83); 752 (79); 754 (84); 761 (65c); 1000 (2).
Bourgeau, M. 209 (70); 212 (20); 431 (35) (62); 526 (45); 529 (58); 661 (79); 2165 (2); 2291 (6); 2386 (45); 2391 (84); 2586 (83); 2591 (84); 2631 (83); 2640 (83); 2731 (33, 34); 2733 (37); 2734 (65c); 2979 (62) (79); 3145 (89).
Bravo González, P. A. 39 (84).
Brandegee, T. S. 599 (89); 602 (6).
Breedlove, D. E. 6073C (74); 6073D (74); 6782 (7); 6907 (84); 7107 (33); 7427 (79) (84); 10358 (10); 10515 (65b); 10544 (33); 10622 (46); 10622 (46); 10623 (45); 10780 (84); 10802 (84); 10863 (85); 10867 (7); 10869 (35); 11076 (84); 11263 (84); 12396 (35); 12479 (84); 12709 (84); 13624 (10); 14627 (35); 14640 (44); 14677 (8); 14678 (7); 14700 (35); 14701 (33); 14799 (37); 14836 (37); 14841 (10); 15061 (35); 15064 (33); 15275 (35); 15307 (84); 16374 (59); 16784 (38); 16837 (35); 17833 (31); 17881 (28); 17954 (6); 17956 (8); 17991 (89); 18092 (31); 18294 (38); 18422 (35); 18640 (9); 18641 (89); 20519 (28); 20616 (12); 20780 (89); 20886 (6); 20887 (45); 20892 (47); 20951 (89); 21035 (10); 21091 (6); 21239 (17); 21473 (35); 21534 (17); 21623 (37); 22204 (17); 22856 (46); 24273 (89); 25199 (10); 25201 (37); 25560 (45); 25643 (26); 25647 (47); 26115 (84); 26365 (45); 26403 (37); 26422 (12); 26442 (15); 26620 (10); 26634 (37); 26627 (6); 26903 (89); 26788 (84); 26991 (79); 27039 (79); 27296 (6); 27299 (31); 27883 (10; CHAPA, 20); 27974 (79); 28059 (8); 28343 (84); 28401 (37); 28405 (89); 28603 (14); 28664 (80); 30487 (6); 30494 (35); 33275 (10); 33314 (44); 33792 (2); 33901 (44); 34166 (19); 34314 (6); 34318 (19); 36622 (37); 36781 (47); 36789 (10); 36790 (30); 36791 (80); 36800 (26); 36896 (28); 37025 (17); 37119 (7); 37132 (35); 37284 (20); 37303 (77); 37352 (89); 37353 (8); 37384 (89); 37518 (38); 37519 (17); 37532 (81); 37542 (6); 37553 (26); 37577 (47); 37608 (12); 37791 (37); 37810 (44); 37964 (62); 38256 (17); 38356 (17); 38495 (89); 38504 (38); 38868 (89); 39613 (45); 40529 (28); 40557 (40); 41050 (35); 41164 (50); 41243 (35); 41715 (35); 41716 (35); 43342 (22); 43795 (26); 44307 (80); 46650 (17); 47091 (19); 51154 (46); 51443 (79); 53064 (84); 53198 (74); 53842 (35); 54052 (28); 57710 (10); 57712 (19); 58025 (19); 58468 (89).
Breedlove & D. I. Axelrod 43327 (84).
Breedlove & Davidse 54180 (80).
Breedlove & P. Raven 8139 (7); 8180 (33); 13601 (79).
Breedlove & Smith 32092 (16).
Breedlove & R. Thorne 17833 (31); 17991 (89); 18091 (26); 18092 (31); 18123 (79); 18124 (79); 18255 (79); 18292 (84); 18293 (54); 18294 (38); 18422 (35); 18516 (79); 18641 (89); 18654 (51); 20703 (37); 20803 (66); 20828 (66); 20832 (66); 20856 (80); 21242 (33); 21466 (7); 21257 (36); 21283 (35); 21335 (35).
Briseño Centeno, A. 7 (56).
Brizuela, F. 4 (80).
Brown, B. 81-23 (6).
Burgess, T. L. 6076 (89).
Burton, T., & R. Zarkin 6 (6).
Bye, R. A. 2436 (58); 2437 (58); 2438 (20); 2718 (59); 5886 (35); 5934 (61); 6756 (45); 7710 (58); 7711 (59); 7863 (20); 8756 (20); 8773 (58); 8896 (20).
Byrne, K. 61 (46).
Cabral, I. 425 (8).
Calzada, J. I. 353 (44); 566 (F: 82; NY: 83); 1476 (44); 1915 (79); 2242 (10); 2247 (89); 2589 (84); 7708 (84); 7755 (84); 7774 (44); 7793 (89); 7855 (35).
Calzada & Márquez 4573 (8).
Carlquist, S. 342 (71).
Carlson, M. C. 4115 (39).
Carrillo, C. 1 (38).

Gibson, A. C. 2237 (20).
Gibson, A. C., & L. C. 2045 (67); 2113 (46); 2158 (44); 2181 (54).
Gilbert, L. E. 47 (84).
Gilly, C. L. 44 (10).
Gilly & E. Hernández X. 3 (12); 9 (10).
Gimate L., J. 639 (8); 943 (12).
Gómez, F. F. 50 (6).
Gómez G., A. 258 (73).
Gómez M., F. H. 50 (6) (8).
Gómez Murga, S. 14 (65c).
Gómez-Pompa, A. 4573 (46); 4599 (80); 4607 (82).
Gómez Vicuña, R. 10 (35); 20 (89).
González, C. R. 1 (65c).
González, J. J. 5 (6).
González, L. M. 54 (85); 349 (41); 351 (67); 659 (55); 995 (6); 1103 (35); 1559 (80); 1560 (6); 2068 (38).
González & L. M. V. Puga 625 (30); 627 (45); 628 (6); 777 (47); 901 (80); 1263 (26); 2189 (40); 7526 (79); 7535 (26); 7545 (46); 7559 (47).
González, M. 168 (44).
González, S. 29 (35); 977 (84); 1007 (39); 1009 (39); 1954 (77); 2088 (20); 2139 (20); 2146 (20); 2480 (59); 2481 (58).
González & S. Acevedo 2465 (45); 2550 (35); 2720 (45); 2725 (39).
González & R. Fernández 2174 (76 ?); 2198 (77).
González & M. González 1791 (20).
González & Y. Herrera 1327 (20); 1522 (77).
González & J. Rzedowski 1844 (85); 2244 (35).
González & J. Tena 2530 (38); 2799 (89).
González y Miranda, S. 1 (32); 9 (80).
González Jiho, J. G. 45 (6).
González Q. [Quintero], L. 201 (84); 3157 (77).
González T., R. 260 (20); 277 (85); 309 (84); 455 (84); 586 (12).
González V., & J. Pérez de la Rosa 2566 (20); 2570 (77).
Gould, F. W. 8729 (20); 8910 (35); 12245 (75).
Gray, C. E., & C. M. Rowell 2911 (26); 2963 (47).
Greenman, J. M. 447 (1).
Gregg 539 (89).
Gregory, D. P., & G. Eiten 58 (20); 88 (54); 226 (20); 227 (84); 315 (80); 327 (47).
Guerrero C., B. 691 (75).
Guerrero, F. 8000 (20).
Guevara F., F. 1467 (89).
Guízar N., E. 513 (85); 1849 (77).
Gustavo, B. 13 (8).
Gutiérrez, A. 6 (65c); 11 (28); 242 (8); 301 (8).
Gutiérrez C., H. 15 (20).
Gutiérrez H., M. A. 13 (7).
Gutiérrez M., J. B. 5 (9).
Gutiérrez R., A. 7 (84).
Gutiérrez R., L. 26 (80); 333 (83).
Guzmán, A. A. 7 (46).
Guzmán, A. A., et al. 107 (35).
Guzmán H., L. 6649 (45).
Guzmán Mejía, R. 345 (47); 346 (31); 505 (20); 583 (44); 949 (20); 959 (45); 1030 (65b); 1450 (6); 1477 (6) 1478 (38);; 6811 (69) 8529 (67).
Guzmán Mejía, R., et al. 1433 (30); 6340 (83).
Hallberg, B. 917 (84); 1048 (54).
Hammerly, B. 250 (22).
Hansen, B., & J. 1485 (84); 1613 (89); 1629 (10); 1691 (84).
Hansen & M. Nee 1629 (10); 1691 (83); 7653 (44).
Harker & Mellowes 54 (72).
Harriman, N., & R. Jansen 14341 (89).
Harris, J. T. 92 (84).
Harshberger, J. W. 93 (33).
Hartman, C. V. 37 (58); 829 (45).
Hartweg, C. 240 (39); 254 (45); 256 (20); 257 (20).
Hawkes, J. G., et al. 1249 (20) (40); 1269 (45).
Henrickson, J. 6845 (45); 11406 (58); 11653a (40); 13439 (62); 13523 (39); 14117 (39); 14286 (89); 14986 (83); 15030 (65c); 15067 (40); 15167 (58).
Hernández, C. 44M (35).
Hernández, C., & A. Barkeley 14590A (8).
Hernández, L. 9 (45).
Hernández, R. 1001 (35); 1239 (44).
Hernández, R., & R. Trigos 1078 (11).
Hernández, S. C. 175 (35); 201 (84); 210 (20); 383 (77).
Hernández A., J. J. 14 (89).
Hernández X., E. 191 (12); 362 (58); 367 (40); 368 (20); 384 (45); 398 (77); 400 (77).
Herrera G., A. L. 38 (84).
Hidalgo y Anda, M. S. 106 (77).
Hildebrand, H. 101 (44).
Hill, S. R. 1662 (15); 1725 (10).
Hinton, G. B. 961 (84); 1060 (26); 1067 (6); 1153 (45); 1229 (20); 1359 (35) (38); 1367 (55); 1782 (79); 1826 (54); 2197 (89); 2272 (89); 2340 (79); 3732 (6); 4672 (29); 4685 (30); 5454 (46); 5536 (19); 5815 (46); 5818 (52); 5831 (44); 5991 (47); 6022-b (6); 6038 (6); 6168 (20); 6173 (39); 6454 (26); 6475 (29); 6475 (40); 6699 (31); 8883 (89); 8884 (6); 8885 (2); 9111 (26); 9244 (65c); 9280 (72); 9302 (37); 9305 (28); 9323 (62); 9432 (38); 9460 (30); 9465 (54) (55); 9496 (43); 9689 (6); 10792 (12); 10884 (26); 11401 (85); 11747 (2); 12110 (26); 13112 (65b); 14342 (26); 14645 (12); 14652 (84); 15049 (84); 15143 (35); 15832 (2).

Holstein, G., & W. S. Armbruster. 20358 (12).
Holway, E. W. D. 533 (45).
Howe, D. 4041 (89).
Howell, J. T. 8415 (69); 10304 (26); 10363 (31); 10477 (69); 10528 (69); 10560 (69).
Huerta, A. 107 (87).
Huerta, R. 5 (6).
Huerta A., J. 34 (84).
Hunnewell, F. W. 18264 (86).
Iltis, H. H. 256 (20); 770 (20); 934 (20); 27216 (89); 27252 (12); 27270 (89); 27287 (89); 29163 (1); 29179 (45).
Iltis, Benz & Burd 28764 (35).
Iltis & T. S. Cochrane 238 (89).
Iltis & Doebley 33 (9); 36 (35) (38); 37 (62); 38 (9); 41 (45); 44 (40); 318a (79); 388 (35).
Iltis & Guzmán M. 3208 (20); 3258 (30); 3258a (38); 3269 (9).
Iltis & Lasseigne 152 (58); 628 (65c); 629 (30); 631 (65c); 635 (28); 636 (40).
Iltis et al. 246 (62); 401 (34); 428 (84); 511 (79); 575 (44); 576 (62); 584 (51); 595 (79); 650 (44); 770 (20); 1090 (9); 1098 (35); 1113 (44); 1115 (39); 1116 (39); 1279 (55).
Iñíguez, J. C. 11 (9).
Jara, J. 1 (45).
Johansen, D. A. 567 (3).
Johnston, I. M. 3437 (39); 4124 (3); 4145 (3); 4305 (87); 7488 (62); 8072A (58); 8074 (58); 8105A (45); 8123 (40); 8127 (58); 8130 (58); 8130A (62); 8138 (45); 8561 (62); 8705 (58); 8868 (45).
Johnston, I. M., & C. H. Muller 455 (62); 540 (58).
Johnston, M. C. 670 (53); 2683 (20); 4930B (86); 6067 (67); 8138 (45); 8547 (20); 8893A (58); 9328 (58); 11232 (20); 11978A (65b); 11985F (40); 12010A (45).
Johnston & J. Crutchfield 5771A (40).
Johnston et al. 8214C (40); 11991A (40).
Jones, G. N. 22472 (44); 22475 (67); 22478 (45); 22575a (40).
Jones, M. E. 664 (89); 666 (89); 22745 (69); 22832 (47); 23455 (89); 23456 (89); 24063 (22); 24607 (3); 27598 (37); 27601 (79); 27601a (6); 27602 (84); 27603 (28).
Jorge, B. G. 20 (6).
Juan José, J. G. 18 (6).
Judziewicz, E. J. 3319 (20); 3506–1/2 (6).
Judziewicz & Cochrane 4864 (77).
Judziewicz et al. 4884 (38).
Jurgensen, C. 263 (20); 633 (10); 657 (15).
Keil, D. 13338 (35); 13373 (59).
Keil et al. 16750 (22).
Kenoyer, L. A. A573 (6); 320 (89); 323 (46); 576 (89); 1013 (44); 1603 (65b); 1663 (54); 1664 (65b); 1665 (20); 1790 (62); 1791 (62); 1792 (20); 1917 (45); 2164 (65c); 2360 (9); 2432 (45); 2433 (77).
Kenoyer & H. A. Crum 2738 (62); 2960 (62); 3301 (37); 3353 (86); 3555 (50); 3666 (8); 3674 (86); 3788 (83); 3871 (8); 3872 (50).
Kerber, E. 57 (79) (83); 91 (37).
Kessler, J. 4888 (8); 4897 (89).
Kibbey, W. B. 6054 (45); 7053 (40).
King, R. M. 16 (41); 226 (46); 306 (46); 398 (45); 443 (80); 470 (26); 560 (26); 677 (41); 828 (54); 854 (12); 855 (10); 871 (46); 917 (48); 929 (28); 1101 (53); 1133 (44); 1244 (26); 1271 (26); 1332 (44); 1358 (67); 1379 (89); 1418 (46) 1572 (45); 1724 (26); 1768 (30); 1776 (28); 1823 (26); 1836 (28); 1875 (81); 1949 (12); 2097 (84); 2101 (35); 2132 (82); 2151 (84); 2694 (80); 2988 (84) (83); 3001 (74); 3116 (26); 4347 (83); 4382 (11).
King & Soderstrom 4586 (38); 4644 (40); 4711 (84); 4714 (38); 4874 (6); 4879 (42); 4913 (84); 5001 (84); 5211 (84).
Kitchen, S., & Smith 22 (20); 4762 (20).
Knobloch, I. W. 146 (20); 393 (89); 471 (84); 666 (89); 1092 (20); 1111 (35); 1150 (20); 1222 (35); 1250 (77); 5243 (77); 5254 (20); 5409 (35); 5427 (58); 5474 (61); 5474B (61); 5529 (54); 5926 (58).
Koch, S. D. 747 (8); 24936 (8); 25053 (10); 25116 (20); 25213 (20); 25407 (6); 25408 (19); 25602 (20); 25631 (40); 27451 (20); 27506 (28); 27507 (47); 27632 (39); 27760 (40); 75383 (64).
Kral, R. 25101 (62); 25237 (34); 25313 (66); 25332 (54); 25631 (40); 27593 (34); 27632 (39); 27760 (40).
Kuntze, O. 23662 (65a).
Lacas, M. M. 68 (89).
Lachica, F., & Sánchez Fal–2135 (44).
Lane, M. 2605 (62); 2620 (20); 2661 (65a); 2675 (9); 2674 (44); 2675 (9); 2674 (44); 2675 (9); 2726 (20); 2787 (65b).
Langlassé, E. 135 (52); 177 (44); 193 (47); 193a (6); 203 (52); 203-b (89); 223 (26); 293 (47); 298 (75); 299 (6; 37); 302 (12); 303 (31); 919 (2); 929 (2).
Langman, I. K. 3161 (6); 3200 (6).
Lathrop, E. 5838 (83); 5881 (84); 5895 (37); 5910 (75); 5952 (45); 5963 (33).
Laughlin, R. M. 219 (2); 648 (7); 685 (84); 1228 (84); 1445 (84); 2207 (84).
Leavenworth, W. C. 591 (84); 823 (84).
Leavenworth & Hoogstraal 1062 (77); 1066 (84); 1671 (54).
Ledesma, M. A. 9 (55).
Le Doux 2000 (20).
Lehto, E. 21826 (20); 21844 (5); 24751 (26); 24826 (40).

León Casillas, J. 25 (89).
LeSueur, H. 14 (12); 22 (6); 27 (40); 35 (80); 39 (16); 40 (44); 224 (20); 226 (45); 227 (62); 228 (59); 229 (58); 1093 (59); 1095 (89); 1097 (58) (59); 1104 (35); 1105 (79); 1106 (62); 1107 (45) (61) (62); 1108 (35); 1109 (59); 1110 (59); 1287 (56); 1720 (65c); 1722 (45); 1724 (57); 2000 (45); 2001 (45).
Liebmann, F. M. 663 (77); 14350 (84); 14351 (83); 14352 (65c); 14353 (82); 14354 (81); 14356 (83); 14357 (56); 14358 (40); 14359 (89); 14360 (89); 14361 (45); 14364 (28); 14365 (28); 14367 (89); 14369 (50); 14370 (76); 14371 (76); 14373 (89); 14376 (83); 14377 (79); 14378 (55); 14379 (83); 14380 (83); 14382 (37); 14396 (920); 14397 (20); 14398 (15); 14401–14404 (83); 14406–14408 (37); 14410 (38); 14414 (65b); 14416 (65b); 14418 (83); 14420–14422 (2); 14428–14430 (17); 14431–14433 (54); 14434 (11); 14438–14441 (35); 14442 (32); 14443–14448 (19); 14451 (7); 14452–14458 (6); 14459–14461 (65a); 14462 (67); 14466–14468 (76); 14469 (79); 14470 (79); 14471 (42); 14472 (42).
Linden, J.-J. 70 (53); 79 (7); 280 (54).
Lloyd, C. E. 470 (79).
López López, L., & G. J. Martin 172 (84); 204 (84).
López Luna, R., & G. J. Martin 327 (84).
López, M. 124 (20).
López, S. 77 (20); 98 (77).
Lot H., A. 745 (44); 1039 (1); 1266 (44); 1333 (80); 1945 (44); 2268 (44); 2490 (44); 2622 (53).
Lot, A., & Novelo 863 (44).
Lott, E. J. 511 (80); 569 (31); 628 (26); 1250 (80); 1275 (26); 1508 (84); 2199 (80).
Lundell, C. L. 803 (8); 1042 (89); 1169 (50); 1191 (19); 1606 (11); 4702 (46); 5094 (62); 7277 (78); 7836 (45); 7837 (79).
Lyonnet, E. 35 (45); 37 (84); 74 (20); 280 (62); 288 (65b); 719 (79); 721 (65a); 1455 (35); 1533 (45); 1534 (79); 1744 (84); 1819 (84); 2803 (1).
Lyonnet & J. Elcoro 1744 (83); 1745 (79).
MacDaniels, L. H. 120 (54); 146 (47); 147 (28); 225 (26); 284 (65b); 355 (65c); 422 (8); 470 (40); 507 (65a); 628 (7); 693 (65b).
MacDougall, T. L–13 (39).
Macias, R. C. 12 (51).
Magaña, M. A., & C. Cowan 2112 (35); 2128 (32).
Magaña, M. A., et al. 59 (88); 102 (17).
Makrinius, E. 715 (89).
Maldonado, A. 1 (6).
Maldonado, E. A. 24 (6).
Maldonado V., N., & G. J. Martin 274 (85); 328 (55).
Manning, W. 531007 (35); 531147 (44); 531148 (65b).
Marcks, B. G., & C. Marcks 798 (83); 817 (84); 830 (8); 831 (89); 839 (89); 840 (89); 860 (80); 861 (80); 868 (50); 875 (6); 876 (32); 881 (80); 882 (46); 887 (12); 940 (83); 1035 (55); 1126 (77); 1127 (84); 1140 (35); 1157 (26); 1171 (54); 1174 (31); 1175 (80); 1176 (44); 1185 (35); 1195 (35); 1211 (35); 1249 (59); 1262 (58); 1262b (58); 1263a (65b); 1265 (55).
Marco, G. V. 8 (84).
Marín, M. 161 (84).
Mario, P. R. 40 (4).
Márquez R., W. 833 (7); 902 (7).
Marsh, E. G., Jr. 401 (8); 511 (62); 748 (45); 801 (58); 809 (89); 835 (83); 889 (8); 933 (89); 934 (89); 1649 (89).
Martín, A. T. 19 (83).
Martin, G. J. 230 (26).
Martínez, F. 105 (46); 169 (84).
Martínez, I. S. 1 (84).
Martínez, S. 1312 (45).
Martínez Calderón, G. 66 (79); 100 (10); 138 (45); 856 (12); 1048 (46); 1477 (46); 1735 (12); 1964 (12); 1983 (31); 2053 (17); 2061 (89).
Mason, H. L. 1651 (69); 1678 (80).
Mason & G. D. Hanna 14594 (69); 14608 (80); 14623 (80).
Matamoros, N. H. 244 (35).
Matuda, E. 298 (6); 301 (12); 302 (10); 304 (80); 305 (15); 306 (12); 307 (83); 1737 (37); 1753 (13); 2424 (84); 2518 (15); 2553 (13); 2567 (84); 3350 (53); 3451 (89); 3483 (44); 3574 (19); 3629 (10); 3744 (33); 3851 (50); 3868 (44); 3869 (89); 4412 (2); 4557 (35); 4649 (20); 4480 (54); 4486 (37); 4683 (84); 4823 (45); 16573 (26); 16574 (12); 17706 (1); 17256 (50); 17271 (6); 17593 (44); 19075 (77); 19261 (65c); 19262 (65c); 21164 (39); 21234 (65c); 21329 (77); 21235 (45); 21320 (77); 21366 (65b); 21368 (20); 21378 (65a/65c, intermediate); 21390 (77); 21402 (65b); 21809 (84); 23537 (45); 25609 (77); 25654 (42); 25657 (45); 26033 (2); 26198 (79); 26202 (79); 26242 (84); 26285 (84); 26656 (77); 26673 (77); 26813 (84); 27183 (89); 27218 (79); 27364 (38); 27365 (38); 27366 (84); 27530 (89); 27538 (79); 28344 (84); 28867 (84); 31183 (79); 31212 (26).
Maxwell 159 (46).
Maysilles, J. 7164 (20); 7232-A (20); 7439 (58); 7506 (20); 7699 (58); 7734 (62); 7921 (59); 8098 (77); 8107 (20); 8185 (58); 8356 (61); 8408 (61); 8470-A (20).
McGill, L. 9269 (40); 9771 (58).

McGill, L., & D. Keil 7889 (45); 7890 (40); 8270 (45); 8312B (58).
McGill et al. 9269 (40); P12314 (39).
McPherson, G. 1130 (20).
McVaugh, R. 13022 (65c); 13073 (84); 13080 (35); 13108 (38); 13507 (84); 13643 (65c); 13647 (38); 15676 (44); 15963 (54); 15965 (26); 16290 (20); 16561 (62); 16767 (63); 16872 (20); 16991 (20); 17061 (45); 17090 (20); 17091 (65a); 17213 (44); 17215 (51); 17220 (89); 17221 (8); 17268 (38); 17271 (30); 17284 (65b); 17663 (62); 17737 (77); 17842 (55); 18375 (77); 18417 (20); 18575 (30); 18577 (45); 18580 (51); 18585 (35); 18578 (44); 18598 (40); 18640 (28); 18654 (62); 18658 (54); 18707 (31); 18708 (38); 18709 (41); 18830 (79); 18853 (54); 19073 (54); 19129 (47); 19297 (31); 19327 (17); 19389 (28); 19404 (51); 19431 (38); 19465 (39); 19495 (35); 19496 (35); 19519 (39); 19593 (38); 19655 (38); 19661 (65a); 19662 (62); 19663 (55); 19700 (79); 19732 (47); 19742 (53); 19895 (79); 20005 (38); 20007 (30); 20144 (79); 20231 (38); 20365 (54); 20690 (35); 20741 (51); 20743 (80); 21133 (12); 21162 (83); 22030 (84); 22085 (51); 22334 (77); 22345 (84); 22508 (17); 22551 (2); 23806 (84); 24304 (20); 24331 (38); 24358 (7); 24562 (34); 25009 (69); 25375 (12); 26243 (84); 26075 (9).
McVaugh & W. Koelz 503 (84); 819 (84); 987 (38); 1105 (51); 1113 (52); 1136 (38).
Mears, E. 524e (8); 454e (9).
Mearns, E. A. 2282 (30).
Medina C., A. 1461 (63).
Meisner 1979 (89); 1984 (79).
Mell, C. D. 2115 (48).
Menéndez, F. 243 (89); 245 (17); 256 (50); 434 (19).
Menéndez L., F. 520 (8).
Merton, E. C. 2050 (45).
Mexia, Y. 82 (80); 86 (44); 128 (72); 144 (72); 149 (26); 1282 (2); 2663 (20); 8818 (79); 8832 (79); 8836 (87); 8915a (52); 8917 (46); 8924 (6); 9128 (82) (89); 9129 (10); 9201 (15); 9228A (12).
Mick, G., & K. Roe 67 (20); 148 (84); 249 (7); 252 (35); 297 (79); 345 (79).
Millspaugh, C. F. 1519 (8); 1667 (89); 1671 (89); 1686 (67); 1697 (68); 1773 (89).
Miranda, F. 1889 (74); 7464 (8); 8537 (53);
Mita Stein, M. 112 (5).
Mockford, J., & C. Rowell 2158 (26); 2772 (6).
Mohr, C. 9 (28); 17 (47); 29 (83).
Mohr & J. Botteri 12 (42).
Moises, R. A. 10 (44).
Monjo Soso, S. 2 (6).
Montes, M. N. 464 (84).
Montoya M., M. B. 29 (83).
Moore, H. E., Jr. 74 (35); 1302 (84); 1348 (20); 1356 (62); 1370 (39); 1699 (65b) (84); 1941 (37); 2016 (42); 2025 (83); 2117 (50); 2895 (2); 3321 (62); 3330 (35); 3366 (35); 3468 (7); 3469 (89); 3473 (35); 4927 (62); 6386 (1).
Moore & C. E. Wood, Jr. 3965 (83); 4115 (20); 4236 (5).
Moran, R. 3630 (3); 3674 (40); 5720 (69); 5789 (71); 5904 (69); 5906 (80); 5936 (69); 7038 (22); 7317 (3); 7466 (67); 9524 (3); 11514 (3); 12535 (3); 13601 (40); 14151 (59); 14273 (59); 15452 (59); 15504 (59); 15707 (89); 15720 (39); 16145 (40); 18620 (40); 18805 (84); 19226 (40); 19799 (39); 21141 (59); 21545 (89); 24471 (40); 24577 (59); 25651 (84); 26136 (40); 26212 (89); 27675 (40); 28027 (1); 29168 (40); 29192 (40); 29236 (40); 29251 (89); 29313 (45); 29321 (89); 29332 (39); 29383 (45); 29410 (39).
Moran & J. Reveal 19614 (39); 19799 (39); 20128 (67); 26136 (40).
Moran & R. Thorne 14273 (59).
Moreno G., S. 112 (83); 256 (77) (84).
Moreno P., N. 54 (8); 64 (45) (46).
Morley, T. 706 (75); 711 (45); 711A (45).
Morton, C. V., & E. Makrinius 2475 (12).
Mudie, P. J. 711 (39); 910 (40); 988a (39).
Mueller, C. H., & M. T. Mueller 28 (77); 696 (83); 2836 (77) MO: (62); 3134 (62); 3366 (62) (65b); 3367 (62); 3462 (65b) (65c), intermediate; 3497 (40).
Müller, F. 1979 (89); 1983 (83); 1984 (79); 1986 (28); 1987 (83); 1988 (65c); 1989 (62); 1994 (54); 3067 (79); 3068 (65b); 3134 (62); 3343 (58); 3361 (58); 3362 (20); 3463 (58); 3497 (40); 3518 (58); 3519 (38).
Nabhan, G. 148 (39); 235 (39); 362 (45).
Nabhan, G., & R. Felger 375 (89).
Nabhan et al. 148 (39); 235 (39).
Narváez, M., & A. E. Salagar 218 (2).
Nee, M., & B. Hansen 18344 (19); 18353 (50); 18594 (84).
Nee et al. 22973 (1).
Nelson, E. W. 522 (44); 616 (84); 712 (84); 1005 (65a); 1414 (62); 1458 (62); 2894 (54); 4259 (84); 6188 (59); 6242 (45); 6417 (39).
Nesom, G. 5136 (36).
Nevling, L. E. Jr., & A. Gómez-Pompa 79 (6); 105 (44); 111 (89); 338 (20); 379 (46); 457 (46); 503 (44); 655 (10); 2461 (44); 2501 (84); 2544 (12).
Nicolas, Fr. 250 (62); 288 (65b); 5218 (58); 5229 (62); 5248 (62); 5811 (45); 5858 (45); 5861 (62).

Nieves 77 (35).
Nieves & Vázquez 67 (9).
Norris, D. H., & D. J. Taranto 13321 (80); 13682 (26); 13696 (72); 13764 (47); 13766 (76); 13823 (79); 13854 (6); 13943 (28); 13953 (26); 14169 (20); 14360 (54); 14369 (20); 14533 (54); 14737 (79); 14982 (21); 15154 (62).
Novelo, A. 255 (44); 373 (44).
Núñez Casillas, F. 8 (6).
Ochoa B., R. I. 395 (62).
Orcutt, C. R. 3243 (11); 3247 (19); 3256 (89); 3257 (6); 3338 (46); 3519 (65b); 3541 (65a); 3688 (62); 3710 (65b); 3711 (62); 4082 (38); 4203 (47); 4445 (44); 4543 (28); 4544 (72); 4583 (89); 4593 (72); 4602 (72); 4665 (89); 5001 (47); 5003 (31); 5035 (47); 5036 (26); 5388 (62).
Ordovica, M. E. 37 (65b).
Ornelas Uribe, R. 19 (20); 20 (84); 31 (20); 50 (6).
Orosco, A. D. L. 111 (6).
Orozco, A. 278 (44); 2247 (44).
Orozco-Segovia, A. 403 (44).
Ortega, J. G. 4189 (72); 4821 (84); 5103 (2); 5255 (42); 5541 (89); 5763 (47); 5765 (89); 5935 (6); 6560 (67); 7050 (44).
Ortega-Ortiz, R. 396 (89).
Ortenburger, A. I., et al. 16M292 (62); 16M761 (77).
Ortiz J., J. M. 70 (35).
Ortiz & Gowan 1866 (1).
Oscar, B. 19 (28).
Ownbey, G. 2005a (20).
Pacheco M., F. 64 (45); 155 (51).
Padilla, J. J. 9 (6).
Palacio F., H. 1E (47).
Palmer, E. [arranged by year] 1880: 1330 (40). 1885: 49 (72); 49 (72); 49-3/4 (20); 63 (31); 496 (40). 1886: 14 (7); 82 (20); 191 (6); 194 (45); 213 (42); 249 (44); 253 (54); 290 (20); 396 (31); 433 (28) (31). 1887: 36 (89); 39 (39); 193 (40); 203 (45); 417 (40); 509 (39); 636 (44); 637 (40); 639 (89); 640 (32) (89). 1890: 701 (72); 707 (72); 747 (89); 863 (90); 1079 (31); 1095 (69). 1891: 1081 (47); 1304 (2); 1381 (89); 1382 (67); 1383 (80); 1540 (9); 1759 (31). 1894: 118 (17); 291 (6); 295 (80); 597 (80). 1895: 117 (32); 448 (47). 1896: 178 (39); 256 (45); 376 (39); 377 (62); 544 (58) (62); 722 (77); 732 (35); 733 (38). 1897: 2 (84); 3 (31); 4 (47); 5 (26); 7 (37); 147 (88); 232 (45); 247 (40). 1904: 168 (20). 1905: 651 (9). 1906: 334 (45); 333 (36); 335 (62); 351 (58); 560 (55). 1907: 25 (19); 413 (86) (89). 1910: 73 (89); 150 (46); 176 (6).
Parker, H. M. 351 (44); 411 (46).
Parry, C., & E. Palmer 433 (28); 906 (62); 907 (58); 908 (45); 909 (77); 910 (20); 911 (20); 911-1/2 (20).
Pennell, F. W. 17692 (20); 18201 (65c); 18370 (65a); 18372 (20); 18470 (64); 18472 (64); 18508 (65b); 18531 (20); 18855 (58); 18869 (65c); 19564 (79); 19741 (26); 19884 (30); 19885 (54); 19887 (30); 19888 (6); 19889 (65b); 19890 (79); 19967 (54); 20067 (20); 20260 (40).
Pennell, M., et al. 224 (26); 230 (45); 397 (55).
Pennington, C. W. 9 (65c); 107 (45); 212 (20).
Pérez, A. M. 43 (84).
Pérez, H. 44 (89).
Pérez A., J. 5 (52).
Pérez de la Rosa, J. 746 (20).
Perino, C., & J. 3377 (20).
Philbrick, R. N. 762 (80).
Phillips, A. 75-163 (89).
Phillips, E. A. 581 (84); 676 (61).
Pineda R., A. 780 (77).
Pinkava, D., et al. P4299 (89); P12848 (80).
Porter 309 (22).
Pringle, C. G. 60 (35); 124 (35); 310 (58); 311 (62); 512 (79); 554 (20); 588 (89); 806 (20); 807 (87); 808 (87); 809 (35); 810 (30); 811 (28); 1008 (79); 1032 (89); 1167 (61); 1391 (59); 1392 (65b); 1393 (20); 1394 (65b); 1395 (58); 1396 (58); 1397 (56); 1682 (65c); 1683 (56); 1786 (6); 2053 (65c); 2055 (77); 2563 (34); 2626 (17); 3131 (44); 3159 (62); 3160 (65a); 3161 (65b, 65c); 3162 (77); 3163 (65b); 3260 (42); 3261 (18); 3262 (38); 3283 (17); 3380 (28); 3414 (62); 3415 (58); 3416 (89); 3418 (84); 3419 (84); 3420 (59) (62); 3421 (5); 3422 (79); 3423 (55); 3716 (11); 3724 (8); 3770 (33); 3789 (35); 3806 (9); 3843 (34); 3844 (65c) (55); 3848 (35); 3895 (51); 3917 (19); 4170 (39); 4269 (27); 4320 (39); 4446 (38); 4475 (84); 4476 (79); 4617 (47); 5892 (84); 5946 (67); 6044 (2); 6313 (70); 6314 (7); 6671 (35); 6755 (39); 7118 (45); 7140 (39); 7182 (21); 7814 (83); 7815 (83); 8921 (62); 9332 (58); 9460 (89); 9500 (70); 10032 (89); 11204 (34); 11205 (89); 11206 (20); 11299 (38); 11722 (28); 11723 (17); 11724 (5); 13231 (55); 13232 (51); 13233 (62); 13234 (62); 13235 (78) (80); 13236 (65b); 13237 (65b); 13238 (7); 15633 (28).
Puga, L. M. V. [Villarreal de] 160 (79); 164 (84); 177 (45); 178 (6); 194 (89); 245 (79); 251 (28); 510 (55); 570 (84); 597 (20); 601 (84); 633 (79); 651 (79); 652 (62); 695 (77); 703 (62); 894 (20); 924 (69); 991 (80); 1291 (35); 1351 (60); 1429 (28); 1464 (20); 1465 (84); 1466 (7); 1509 (55); 1511 (79); 1512

17475 (28); 17607 (35); 17715 (62); 17790 (80); 17873 (12); 18375 (77); 19495 (35); 19973 (50); 19990 (80); 19992 (89); 20361 (10); 20382 (20); 20394 (35); 20703 (54); 20806 (54); 20888 (40); 20987 (84); 21032 (84); 21050 (35); 21153 (47); 21168 (28) (31); 21191 (15); 21429 (39); 21434 (51); 22509 (26); 22710 (54); 22726 (20, 26); 22922 (35); 24257 (62); 24270 (45); 24501 (46); 24507 (9); 24933 (45); 24934 (47); 24953 (89); 24964 (44); 24967 (53); 25077 (62); 25855 (77); 26054 (45); 26122 (89); 26166 (20); 26395 (47); 27459 (45); 27510 (77); 27589 (44); 29002 (35); 30122 (17); 30478 (5); 32409 (39); 34265 (35); 35034 (35); 35260 (77); 35290 (35); 35347 (35); 35347a (35); 35367 (45); 35370 (35); 35701 (35); 36301 (5); 36347 (58); 37554 (77); 43684 (63).

Rzedowski, J., & McVaugh 728 (6); 828 (38); 972 (38); 1084 (38); 1323 (26); 1325 (29); 1415 (69).

Salcedo V., F. J. 10 (52).

Salinas, G. 56 (84).

Sánchez, C. 45 (10).

Sánchez, F. 3432 (89).

Sánchez Mtz., F. 3435 (32).

Sánchez S., M. E. 497 (20).

Sanders, A. C. 1054 (31) (89); 4539 (6); 8658 (69); 9170 (44); 9218 (40); 9219 (66); 9241 (51); 9251 (67); 9324 (72); 9331 (31); 9345 (26); 9351 (9); 9353 (41); 9354 (76); 9378 (72).

Sanders, et al. 1850 (26); 1866 (47); 4365 (35); 4539 (6); 8037 (26); 8081 (77) (84); 8119 (20); 8180 (28); 8491 (28); 8664 (47); 10401 (1).

Santana Michel, F. J. 180 (6); 181 (46); 182 (29); 183 (89); 441 (41); 494 (89); 1361 (28); 1392 (38); 1405 (41); 1570 (20); 1594 (5).

Santana Michel et al. 2761 (26); 2754 (54); 2805 A (72).

Santana R., A. 23 (65c).

Santos, J. V. 3072 (83); 3221 (77); 3288 (12); 3291 (83); 3505 (79); 3557 (28).

Sauer, J., & D. Gade 3086 (44); 3189 (45).

Schaffner, J. G. 2 (70); 4 (33); 7 (62); 9 (5) (65b); 10 (58); 12 (39); 198 (38); 208 (45); 210 (40); 213 (58); 509 (20); 560 (62); 561 (58); 563 (40); 564 (5); 565 (77); 1032 (20).

Schiede, C. W. J. 66 (62); 67 (35); 71 (35); 72 (44); 82 (20); 803 (37); 845 (44); 846 (35); 848 (32); 876 (54); 877 (65b).

Schiede & Deppe 852 (77) (79).

Schmitz 16 (51); 371 (34); 373 (20) 808 (39).

Schneider, R. A. 975 (58); 987 (62); 1003 (59).

Schoenwetter, J. JSOX-39 (65b) (77).

Schnooberger, I. 8703 (20).

Schott, A. 77 (40); 412 (68); 563 (2); 565 (86); 566 (89); 696 (67); 910 (40).

Schumann, W. 1675 (45).

Schutt & Drawe 7 (89); 24 (89).

Seaton, H. E. 6 (79); 52 (46); 53 (83); 64 (6); 188 (45); 283 (65c).

Seler, E., & C. Seler, 781 (89).

Sessé, Mociño et al. 5297 (54); 5298 (89); 5302 (45); 5305 (35).

Shannon, K. 89 (80).

Sharp, A. J. 4453 (35); 44267 (77); 44268 (84); 44275 (21); 44332 (45); 44420 (45); 44421 (65c); 44422 (65c); 44423 (77); 44425 (84); 44426 (20); 44541 (65b); 44564 (38); 44755 (84); 44756 (84); 44783 (8); 44791 (65b); 44792 (77); 44980 (84); 441211 (84); 441379 (84); 441583 (84); 441719 (83); 441351-b (8); 45458 (33); 45899 (77); 45934 (84); 52063 (84).

Sharp & C. L. Gilly 153 (77); 154 (84); 156 (65b); 157 (65b); 160 (62); 161 (62); 162 (62); 163 (163).

Sharp & E. Hernández X. 50306 (83).

Sharp et al. 50306 (86); 52278 (83).

Shreve, F. 6686 (40); 7162 (40); 7828 (47); 8670 (62); 8671 (20); 8901 (45); 9055 (20); 9157 (65c); 9237 (62).

Smith, C. E., Jr., et al. 3702 (89); 3745 (89); 3836 (77); 4131 (67); 4133 (89); 4134 (39).

Smith, C. L. 963 (84); 1058 (80); 1059 (6) (10); 1909 (79).

Smith, J. G. 165 (28); 228 (38).

Smith, L. C. 555 (20).

Smith, W. T. 55 (20).

Soderstrom, T. R. 565 (30); 850 (40).

Sohns, E. R. 72a (89); 623 (45); 638 (45); 729 (89); 932 (47); 1146 (58); 1173 (40); 1440 (83); 1622 (37); 1656 (44).

Solano, S. C., & M. A. Vara 88 (9).

Solano, C. 398 (89).

Solheim, S., & V. Powers 819 (20); 901 (32).

Solis Juárez, J. 6 (65c).

Sorensen, P., et al. 7824 (84).

Soriano F., H. J. 51 (62).

Soto González 12 (6).

Sousa, M. 2610 (83).

Souza N., N. 62 (46); 96 (46); 256 (8); 320 (8); 373 (44); 438 (44).

Souza N., N., & G. Arce 478 (10).

Stanford, L. R., et al. 523 (62); 2357 (89); 2448 (62); 2526 (59).

Steere, W. C. 1971 (40); 1972 (80).

Stevens, W. D., & M. Fairhurst 1800 (54); 1801 (62); 1802 (62) (79); 1989 (79); 2037 (41).

Stewart, R. M. 783 (40); 1216 (62); 2176 (45); 2372 (45); 2494 (62).
Stewart & I. M. Johnston 2035 (45); 2060 (40); 2076 (62); 2115 (62).
Straw, R. M., & M. Forman 1972 (59); 2032 (65b); 2074 (84).
Suárez, G. 2 (83).
Swallen, J. R. 2374 (67); 2587 (89); 2588 (8); 2599(76); 2653 (86); 2662 (40); 2675 (86); 2798 (68); 2806 (86); 2814 (86); 2859 (86); 2939 (40); 2973 (68).
Tapia, L., & F. Vázquez 84 (83).
Taylor, C. M. 1837 (8); 1838 (89); 1980 (44); 1987 (45); 2097 (9); 2114 (44); 2117 (80); 2134 (9); 2299 (7); 2381 (7); 2511 (62); 2517 (67); 2589 (54); 2598 (83); 2608 (84); 2611 (84); 2613 (83); 2621 (84); 2633 (7); 2636 (65b); 2642 (84); 2643 (84); 2651 (12); 2675 (72); 2677 (45); 2681A (26); 2682 (26); 2698 (45); 3386 (7).
Taylor, J., & C. Taylor 10 (19); 6058 (44); 7367 (2).
Téllez V., O., et al. 3762 (86).
Templeton, B. C. 8736 (20); 9470 (53).
Thomas, J. H. 7696 (3); 7697 (89); 7829 (3).
Thorne, R. 57354 (39); 57745 (39); 57854 (39).
Thorne, R., & Lathrop 41365 (7); 41240 (84); 41718 (7); 46718 (7).
Thurber, G. 932 (79); 1082 (30).
Ton, A. S. 426 (84); 488 (84); 849 (84); 850 (7); 934 (84); 936 (7); 945 (35); 949 (84); 1091 (65a); 1092 (35); 1093 (33); 1151 (65c); 1161 (37); 1176 (37); 1265 (7); 1295 (7); 1296 (84); 1363 (84); 1569 (42); 1675 (84); 1679 (75); 1682 (37); 2293 (37); 2375 (84); 2379 (7); 2519 (7); 2851 (84); 2852 (79); 3476 (2).
Torres C., R. 189 (89).
Toscan L., M. G. 33 (52).
Townsend, C., & C. Barber 148 (20); 149 (59).
Troublefield, S. P., & C. M. Rowell 2815 (37); 2816B (37); 2846 (44).
Tucker, G. C. 1926 (40); 1928 (46); 1929 (4); 1930 (87); 1932 (50); 1934 (50); 1935 (67); 1936 (8); 1937 (44); 1942A (46); 1943 (19); 1948 (89); 1950 (89); 1953 (11); 1968 (89); 1971 (35); 1973 (8); 1974 (67); 1977 (44); 1978 (678); 1979 (19); 1983 (44); 1991 (67); 1992 (8); 1993 (8); 1995 (8); 2000 (8); 2001 (19); 2002 (50); 2003 (89); 2004 (67); 2005 (44); 2006 (39); 2026 (46); 2028 (9); 2029 (19); 2030 (50); 2031 (9); 2035 (9); 2039 (67); 2040 (9); 2048 (19); 2049 (6); 2054 (7); 2055 (67); 2062 (11); 2063 (89); 2064 (80); 2067 (53); 2069 (32); 2070 (37); 2071 (9); 2072 (80); 2075 (6); 2078 (83); 2079 (65b); 2080 (76); 2081 (76); 2089 (79); 2090 (83); 2092 (79); 2098 (79); 2105 (83); 2106 (46); 2108 (44); 2110 (44); 2112 (80); 2113 (50); 2116 (8); 2119 (44); 2137 (45); 2139 (35); 2140 (8); 2141 (6); 2142 (45); 2143 (83); 2144 (76); 2145 (83); 2148 (9); 2157 (35); 2158 (84); 2161 (35); 2171 (7); 2187 (20); 2193 (89); 2194 (35); 2198 (42); 2199 (79); 2202 (60); 2202A (60); 2205 (83); 2206 (65b); 2208 (75); 2209 (84); 2210 (79); 2212 (37); 2214 (35); 2217 (79); 2218 (37); 2220 (10); 2223 (54); 2224 (65b); 2226 (33); 2227 (10); 2229 (40); 2230 (89); 2230A (85); 2232 (6); 2233 (47); 2234 (47); 2236 (28); 2237 (28); 2238 (29); 2239 (28); 2240 (31); 2241 (47); 2242 (10); 2243 (76); 2245 (12); 2246 (29); 2247 (31); 2248 (89); 2249 (6); 2250 (26); 2251 (12); 2254 (82); 2255 (15); 2256 (29); 2257 (30); 2259 (6); 2260 (72); 2261 (45); 2262 (6); 2263 (28); 2264 (47); 2265 (31); 2266 (41); 2267 (29); 2268 (76); 2269 (37); 2270 (8); 2272 (25); 2273 (26); 2274 (45); 2276 (25); 2277 (42); 2278 (76); 2279 (6); 2331 (50); 2334 (50); 2335 (8); 2336 (89); 2340 (45); 2341 (6); 2342 (87); 2343 (89); 2344 (4); 2345 (67); 2346 (8); 2349 (8); 2356 (8); 2358 (87); 2359 (50); 2360 (8); 2361 (89); 2362 (8); 2365 (8); 2366 (67); 2368 (67); 2369 (8); 2372 (83); 2376 (45); 2377 (87); 2378 (40); 2379 (6); 2380 (47); 2381 (47); 2382 (87); 2383 (46); 2384 (87); 2386 (46); 2387 (76); 2389 (45); 2390 (44); 2401 (72); 2403 (87); 2404 (45); 2405 (46); 2407 (73); 2408 (39); 2409 (89); 2410 (89); 2411 (89); 2413 (44); 2414 (8); 2416 (44); 2417 (89); 2418 (8); 2419 (50); 2420 (45); 2421 (46); 2422 (8); 2423 (9); 2424 (8); 2427 (8); 2428 (50); 2429 (46); 2430 (9); 2431 (73); 2432 (8); 2434 (46); 2436 (8); 2446 (11); 2447 (50); 2448 (46); 2449 (45); 2450 (8); 2451 (9); 2452 (9); 2453 (19); 2454 (8); 2455 (46); 2456 (8); 2457 (46); 2459 (45); 2460 (46); 2462 (41); 2463 (8); 2464 (9); 2466 (46); 2467 (8); 2468 (9); 2470 (6); 2471 (17); 2472 (9); 2473 (53); 2475 (47); 2476 (9); 2478 (80); 2479 (19); 2480 (32); 2481 (46); 2483 (87); 2484 (41); 2485 (47); 2486 (9); 2487 (76); 2488 (45); 2489 (41); 2490 (76); 2491 (60); 2493 (76); 2494 (41); 2495 (81); 2497 (6); 2498 (50); 2499 (44); 2500 (87); 2501 (76); 2502 (46); 2503 (10); 2504 (32); 2506 (44); 2507 (53); 2509 (50); 2510 (6); 2512 (41); 2513 (83); 2514 (79); 2516 (45); 2517 (46); 2518 (47); 2519 (82); 2520 (41); 2521 (76); 2522 (6); 2523 (47); 2524 (32); 2527 (32); 2528 (10); 2529 (37); 2530 (17); 2532 (12); 2533 (32); 2534 (10); 2535 (6); 2536 (32);

(45); 4225 (79); 4268 (44); 4337 (46); 8674 (20).
Vera Santos, J. 3193 (54); 3233 (46).
Verhoek-Williams, S. 601 (20).
Vidaña R., J. M. 100 (84).
Viereck, H. W. 708 (83).
Vilas, F. 69 (83).
Villa, R., et al. 2171 (55); 2172 (20).
Villarreal, J. A., et al. 1784 (35).
Villareal, L. M., & S. Carvajal 10708 (65c).
Villegas D., M. 552 (84); 572 (45).
Virlet 1178 (65b).
Von Rozynski, H. 126-a (89).
Wallace 253 (44).
Wallace et al. 274 (72).
Walther, J. R. 68 (37).
Ward, D. B. 7804 (37); 7839 (44); 7871 (6); 7877 (46); 7900 (83); 7914 (12).
Waterfall, U. T., & Wallis 12601 (20); 13525 (20); 13561 (20); 13591 (55); 13646 (65b); 13655 (20); 13656 (65b); 13901 (18); 14627 (44); 15464 (20); 15466 (20); 15510 (20).
Weaver, J. N. 778 (64); 849 (45); 854 (70); 867 (39).
Weaver, R. J., Jr. 2164 (55).
Weber, W. A., & L. A. Charette 11783 (77); 11841 (77); 11922 (84).
Webster, G. L., & G. J. Breckon 15854 (77).
Webster et al. 17M500 (79).
Wendt, T. 461 (58); 510 (58); 1786 (40); 1843 (40).
Wendt & Adamcewicz 415 (40); 510 (17).
Wendt et al. 1786 (40).
Weintraub, F. C., & J. Roller 42 (20); 61 (20); 82 (20); 115 (20); 143 (77).
West, R. C. G–2 (79); P–21 (79); S–17 (45).
White, S. S. 780 (89); 944 (58); 1109 (77); 1378 (89); 2226 (45); 2267 (35); 2267-A (39); 2332 (20); 2607 (45); 2610 (40); 2639 (45); 2640 (65b); 3098 (84); 3223 (79); 3414 (59); 3639 (84); 4728 (77); 4782 (58); 4790 (40); 4835 (56).
White, S. S., & R. M. Chatters 67 (35).
Whitehead, J. 796 (40).
Whiting, A. F. 540 (62); 634 (35); 762 (58); 818 (20); 892 (62).
Wiggins, I. L. 5659-a (3); 5687B (80); 5991 (89); 6058 (89); 6060 (40); 7029 (84); 7143 (79); 7223 (89); 7271 (40); 7271-b (22); 7321 (67); 7373 (72); 9041 (35); 14678 (40); 14711 (6); 15158 (40); 15348 (40); 15433 (40); 15447 (67); 15492 (67); 15949 (39); 16613 (59); 18062 (46); 20881 (39).
Wiggins & D. Demaree 4705 (39); 5003 (59).
Wiggins & Ernst 554 (40).
Wiggins & Gillespie 4175A (39).
Wiggins & R. Rollins 117 (89); 119 (89); 182 (40); 241 (45); 275 (30); 358 (72).
Wilbur, C. R., & R. L. Wilbur 1513 (26); 1909 (27); 1939 (62); 1951 (27); 2187 (79); 2211 (35); 2261 (29).
Wilbur, R. L. 35129 (8); 35139 (4); 35141 (6); 35142 (8); 35166 (50); 35240 (67); 35241 (8); 35260 (8); 35292 (46); 35295 (8); 35327 (41); 35329 (9); 35343 (45); 35345 (45); 35346 (80); 35347 (6); 35358 (9); 35375 (9); 35377 (9); 35389 (8); 35390 (9); 35396 (9); 35397 (8); 35404 (44); 35410 (46); 35411 (60); 35412 (76); 35413 (45); 35425 (44); 35429 (6); 35435 (12); 35438 (47); 35442 (80); 35449 (6); 35470 (10); 35471 (8); 35472 (6); 35474 (80); 35478 (46); 35479 (41); 35494 (53); 35498 (6); 35505 (32); 35506 (10); 35507 (6); 35517 (32); 35518 (89); 35519 (6); 35524 (80); 35527 (41); 35528 (6); 35529 (67); 35530 (80); 35531 (47); 35533 (10); 35534 (8); 35535 (19); 35536 (6); 35538 (45); 35541 (41); 35548 (12); 35557 (8); 35558 (10); 35559 (76); 35566 (6); 35567 (41); 35576 (12); 35580 (10); 35583 (67); 35604 (84); 35609 (84); 35616 (37); 35629 (82); 35634 (12); 35635 (10); 35665 (75); 35672 (76); 35674 (6); 35676 (62); 35682 (76); 35690 (79); 35691 (54); 35692 (65b); 35693 (76); 35721 (65b); 35725 (84); 35770 (65b); 35776 (7); 35781 (79); 35783 (89); 35784 (37); 35785 (82); 35785A (83); 35786 (6); 35791 (7); 35796 (45); 35799 (60); 35807 (6); 35824 (6); 35826 (89); 35827 (79); 35832 (26); 35833 (47); 35834 (29); 35835 (28); 35836 (82); 35837 (28); 35838 (79); 35843 (14); 35846 (85); 35847 (12); 35844 (41); 35852 (89); 35855 (89); 35860 (79); 35861 (6); 35868 (28); 35869 (47); 35871 (47); 35872 (85); 35873 (46); 35874 (10); 35875 (85); 35876 (12); 35877 (26); 35879 (47); 35880 (28); 35882 (76); 35883 (31); 35888 (85); 35893 (31); 35894 (12); 35895 (85); 35896 (89); 35897 (28); 35899 (28); 35901 (85); 35902 (89); 35903 (10); 35904 (6); 35905 (47); 35910 (47); 35913 (47); 35916 (81); 35918 (30); 35920 (6); 35926 (41); 35928 (47); 35929 (29); 35930 (31); 35931 (60); 35932 (28); 35934 (46); 35935 (45); 35936 (28); 35937 (31); 35940 (60); 35941 (72); 35945 (9); 35946 (41); 35947 (47); 35948 (31); 35953 (60); 35954 (6); 35956 (81); 35957 (47); 35958 (41); 35959 (6); 35960 (28); 35969 (29); 35970 (31); 35971 (76); 35972 (6); 35973 (47); 35975 (41); 35982 (47); 35988 (6); 35989 (41); 35994 (6); 36000 (60); 36002 (85); 36003 (41); 36007 (45); 36013 (28); 36017 (41); 36018 (31); 36021 (60); 36022 (29); 36024 (6); 36029 (31);

36030 (6); 36032 (60); 36034 (41); 36035 (47); 36046 (41); 36047 (60); 36048 (47); 36053 (6); 36056 (28); 36058 (29); 36059 (31); 36060 (60); 36061 (89); 36064 (47); 36070 (89); 36071 (46); 36074 (30); 36075 (41); 36081 (6); 36082 (45); 36101 (80); 36106 (46); 36119 (6); 36120 (41); 36126 (46); 36131 (44); 36137 (67); 36150 (47); 36152 (41); 36155 (46); 36169 (45); 36170 (45); 36174 (40); 36177 (77); 36178 (6); 36180 (62); 36248 (77); 36262 (62); 36263 (62); 36274 (77); 36295 (79); 36301 (65b); 36312 (79); 36313 (62); 36316 (46); 36323 (40); 36330 (6); 36332 (28); 36343 (40); 36400 (79); 36403 (40); 36411 (79); 36412 (84); 36443 (77); 36445 (79); 36450 (77); 36477 (47); 36479 (30); 36481 (41); 36482 (47); 36483 (89); 36484 (47); 36485 (31); 36486 (89); 36489 (26); 36490 (28); 36491 (29); 36492 (47); 36493 (31); 36502 (67); 36503 (80); 36504 (47); 36509 (6); 36512 (31); 36512B (47); 36513 (28); 36521 (41); 36528 (47); 36534 (60); 36535 (47); 36536 (31); 36539 (29); 36540 (31); 36543 (47); 36544 (81); 36545 (30); 36546 (50); 36549 (30); 36550 (51); 36551 (89); 36552 (6); 36553 (41); 36554 (8); 36556 (47); 36558 (60); 36564 (60); 36566 (47); 36567 (30); 36569 (45); 36577 (45); 36578 (89); 36579 (30); 36580 (47); 36581 (31); 36589 (47); 36590 (26); 36590B (30); 36591 (31); 36594 (30); 36595 (41); 36615 (52); 36616 (41); 36617 (50); 36618 (67); 36619 (89); 36620 (47); 36623 (41); 36626 (67); 36629 (41); 36632 (75); 36648 (6); 36652 (47); 36653 (80); 36654 (75); 36656 (46); 36669 (79); 36670 (76); 36672 (79); 36675 (26); 36676 (45); 36677 (76); 36682 (26); 36691 (47); 36692 (41); 36696 (45); 36697 (89); 36698 (47); 36699 (9); 36700 (51); 36701 (30); 36702 (41); 36708 (72); 36709 (45); 36710 (47); 36714 (41); 36727 (41); 36728 (47); 36730 (46); 36733 (45); 36734 (47); 36739 (46); 36740 (45); 36741 (41); 36742 (46); 36744 (79); 36745 (84); 36746 (76); 36745 (84); 36746 (76); 36750 (46); 36751 (51); 36752 (44); 36757 (46); 36758 (89); 36759 (41); 36760 (40); 36763 (45); 36764 (76); 36767 (45); 36768 (79); 36771 (45); 36776 (20); 36778 (45); 36781 (45); 36782 (40); 36784 (45); 36785 (45); 36786 (46).

Williams, L. 8262 (89); 8263 (6); 9194 (89).
Willis, W. D., & G. E. Crisman 238 (26).
Wilson, J. S. 12254 (19); 12257 (2).
Wislizenus, A. 142 (45).
Worth, C. R., & J. L. Morrison 8618 (31); 8810 (26).
Worthington, J. D. 7326 (58); 7338 (20); 7707 (62); 8766 (20); 8811 (20); 9431 (35).
Wright, W. G. 1332 (47); 1333 (89); 1950 (18).
Wyatt, F. W. 57 (62).
Wynd, F. L., & C. H. Muller 512 (59); 516 (40); 580 (58); 595 (62); 595B (62); 654 (58).
Zamora, F. 16M855 (65c); 16M879 (62); 16M880 (62); 16M881 (62); 16M892 (62); 16M893 (62); 16M894 (62); 16M911 (62).
Zamudia, S., & M. A. Magaña A. 102 (17).
Zarate F., R. 4-VIII (89); 98 (65a).
Zavala Casas, H. 10 (46).
Zavala Chávez, F. 856 (20).
Zizumbo V., D. Z–697 (45); Z–906 (89).
Zizumbo & Colunga 176 (44).
Zuill, H. 139 (7); 502 (84); 577 (37).

INDEX TO SCIENTIFIC NAMES

Accepted taxa are in roman type; the main entry for each is in **boldface**. Synonyms are in *italics*.